Battery Management in Electric Vehicles: Current Status and Future Trends

Battery Management in Electric Vehicles: Current Status and Future Trends

Editor

Prodip K. Das

Basel • Beijing • Wuhan • Barcelona • Belgrade • Novi Sad • Cluj • Manchester

Editor
Prodip K. Das
University of Edinburgh
Edinburgh
UK

Editorial Office
MDPI
St. Alban-Anlage 66
4052 Basel, Switzerland

This is a reprint of articles from the Special Issue published online in the open access journal *Batteries* (ISSN 2313-0105) (available at: https://www.mdpi.com/journal/batteries/special_issues/Battery_Management_Electric_Vehicles).

For citation purposes, cite each article independently as indicated on the article page online and as indicated below:

Lastname, A.A.; Lastname, B.B. Article Title. *Journal Name* **Year**, *Volume Number*, Page Range.

ISBN 978-3-7258-1345-2 (Hbk)
ISBN 978-3-7258-1346-9 (PDF)
doi.org/10.3390/books978-3-7258-1346-9

© 2024 by the authors. Articles in this book are Open Access and distributed under the Creative Commons Attribution (CC BY) license. The book as a whole is distributed by MDPI under the terms and conditions of the Creative Commons Attribution-NonCommercial-NoDerivs (CC BY-NC-ND) license.

Contents

About the Editor vii

Preface ix

Prodip K. Das
Battery Management in Electric Vehicles—Current Status and Future Trends
Reprinted from: *Batteries* 2024, *10*, 174, doi:10.3390/batteries10060174 1

Molla Shahadat Hossain Lipu, Abdullah Al Mamun, Shaheer Ansari, Md. Sazal Miah, Kamrul Hasan, Sheikh T. Meraj, et al.
Battery Management, Key Technologies, Methods, Issues, and Future Trends of Electric Vehicles: A Pathway toward Achieving Sustainable Development Goals
Reprinted from: *Batteries* 2022, *8*, 119, doi:10.3390/batteries8090119 4

Robby Dwianto Widyantara, Siti Zulaikah, Firman Bagja Juangsa, Bentang Arief Budiman and Muhammad Aziz
Review on Battery Packing Design Strategies for Superior Thermal Management in Electric Vehicles
Reprinted from: *Batteries* 2022, *8*, 287, doi:10.3390/batteries8120287 64

Thomas Imre Cyrille Buidin and Florin Mariasiu
Parametric Evaluation of Thermal Behavior for Different Li-Ion Battery Chemistries
Reprinted from: *Batteries* 2022, *8*, 291, doi:10.3390/batteries8120291 84

Ioan Szabo, Liviu I. Scurtu, Horia Raboca and Florin Mariasiu
Topographical Optimization of a Battery Module Case That Equips an Electric Vehicle
Reprinted from: *Batteries* 2023, *9*, 77, doi:10.3390/batteries9020077 100

José M. Andújar, Antonio J. Barragán, Francisco J. Vivas, Juan M. Enrique and Francisca Segura
Iterative Nonlinear Fuzzy Modeling of Lithium-Ion Batteries
Reprinted from: *Batteries* 2023, *9*, 100, doi:10.3390/batteries9020100 119

Saeid Ghazizadeh, Kafeel Ahmed, Mehdi Seyedmahmoudian, Saad Mekhilef, Jaideep Chandran and Alex Stojcevski
Critical Analysis of Simulation of Misalignment in Wireless Charging of Electric Vehicles Batteries
Reprinted from: *Batteries* 2023, *9*, 106, doi:10.3390/batteries9020106 138

Mattia Neroni, Erika M. Herrera, Angel A. Juan, Javier Panadero and Majsa Ammouriova
Battery Sharing: A Feasibility Analysis through Simulation
Reprinted from: *Batteries* 2023, *9*, 225, doi:10.3390/batteries9040225 159

Angel A. Juan, Carolina A. Marugan, Yusef Ahsini, Rafael Fornes, Javier Panadero and Xabier A. Martin
Using Reinforcement Learning to Solve a Dynamic Orienteering Problem with Random Rewards Affected by the Battery Status
Reprinted from: *Batteries* 2023, *9*, 416, doi:10.3390/batteries9080416 173

Kunal Sandip Garud, Jeong-Woo Han, Seong-Guk Hwang and Moo-Yeon Lee
Artificial Neural Network Modeling to Predict Thermal and Electrical Performances of Batteries with Direct Oil Cooling
Reprinted from: *Batteries* 2023, *9*, 559, doi:10.3390/batteries9110559 189

Zhong Ren, Changqing Du and Yifang Zhao
A Novel Method for State of Health Estimation of Lithium-Ion Batteries Based on Deep Learning Neural Network and Transfer Learning
Reprinted from: *Batteries* **2023**, *9*, 585, doi:10.3390/batteries9120585 . . . . **208**

About the Editor

Prodip K. Das

Dr. Prodip K. Das currently holds the position of Associate Professor in Hydrogen Energy Systems at the University of Edinburgh. He earned his Ph.D. in Mechanical Engineering at the University of Waterloo and has been honored with several academic awards, including the Canada Graduate Scholarships and the Dr. Chandrashekar Memorial Award in Sustainable Energy. His research focus lies within the broad area of hydrogen energy and electric vehicles, particularly emphasizing the thermofluid aspects of fuel cells, electrolyzers, Li-ion batteries, and battery recycling and reuse. Dr. Das has authored over 120 scientific articles, seven book chapters, and four books covering topics such as fuel cells, Li-ion batteries, convective heat transfer, and related mechanical engineering fields. Notably, he led an invited book entitled *"Fuel Cells for Transportation: Fundamental Principles and Applications."* Additionally, Dr. Das serves as an Associate Editor for *Frontiers in Energy Research* (Frontiers Media), *Journal of Electrochemical Energy Conversion and Storage* (ASME), and *Frontiers in Chemical Engineering* (Frontiers Media), while also holding positions on the editorial/advisory boards of four other journals (*Batteries*, *Renewable and Sustainable Energy*, *Inventions*, and *Challenges*). He has also served as a Special Issue editor for *Energies* (MDPI), *Batteries* (MDPI), *Energy and AI* (Elsevier), and *Frontiers in Energy Research* (Frontiers Media).

Preface

In this reprint, 10 works are presented on the topic of battery management in electric vehicles, spanning from fundamental studies of batteries to the application of neural network modeling and machine learning to optimize battery performance. Specifically, they cover the optimization and state-of-health estimation of batteries using various techniques to enhance efficiency and ensure proper battery management. Additionally, this reprint includes studies on the design and construction of protective cases for batteries aimed at improving their safety and reliability, as well as the parametric evaluation of thermal behavior for different lithium-ion battery chemistries. Complementing the research papers, three review articles focusing on wireless charging, battery packing design, and battery management are also featured.

As the Guest Editor, I extend my sincere gratitude to all of the contributors who generously shared their expertise and insights, as well as the reviewers whose constructive feedback strengthened the quality and rigor of this work. I am also deeply appreciative of the unwavering assistance and support provided by the editorial team of *Batteries*. Financial support from the Faraday Institution (grant numbers FIRG005 and FIRG027) through the ReLiB project is also gratefully acknowledged. It is my hope that "Battery Management in Electric Vehicles – Current Status and Future Trends" will serve as a valuable resource for researchers, policymakers, and graduate students working in the areas of lithium-ion batteries and electric vehicles.

Prodip K. Das
Editor

Editorial

Battery Management in Electric Vehicles—Current Status and Future Trends

Prodip K. Das

School of Engineering, University of Edinburgh, Edinburgh EH9 3FB, UK; prodip.das@ed.ac.uk

Citation: Das, P.K. Battery Management in Electric Vehicles—Current Status and Future Trends. *Batteries* **2024**, *10*, 174. https://doi.org/10.3390/batteries10060174

Received: 13 May 2024
Accepted: 23 May 2024
Published: 23 May 2024

Copyright: © 2024 by the author. Licensee MDPI, Basel, Switzerland. This article is an open access article distributed under the terms and conditions of the Creative Commons Attribution (CC BY) license (https:// creativecommons.org/licenses/by/ 4.0/).

Rechargeable batteries, particularly lithium-ion batteries (LiBs), have emerged as the cornerstone of modern energy storage technology, revolutionizing industries ranging from consumer electronics to transportation [1,2]. Their high energy density, long cycle life, and rapid charging capabilities make them indispensable for powering a wide array of applications, with electric vehicles (EVs) standing out as one of the most transformative domains. The rise of EVs represents a pivotal shift in the automotive industry, driven by the urgent need to mitigate climate change and reduce greenhouse gas emissions.

Conventional internal combustion engine vehicles are major contributors to air pollution and greenhouse gas emissions, exacerbating the global climate crisis. Conversely, EVs offer a cleaner and more sustainable alternative, leveraging LiBs to propel a greener transportation revolution [3]. EVs powered by LiBs offer a promising solution to decarbonize transportation, reducing reliance on fossil fuels and mitigating the adverse impacts of vehicle emissions on human health and the environment [4]. In this context, the intersection of lithium-ion batteries, electric vehicles, and greenhouse gas emissions embodies a transformative synergy with profound implications for global sustainability.

Despite significant growth in the usage of lithium-ion batteries in EVs worldwide, this expansion is not without its challenges. The continuous demand for LiBs is anticipated to precipitate global environmental and supply chain concerns regarding critical raw materials [2,5]. The critical materials essential for LiBs, including cobalt, lithium, nickel, graphite, and manganese, are finite resources primarily mined in limited regions worldwide. This reliance on scarce resources coupled with the inevitable proliferation of battery waste poses formidable challenges for the future of electric mobility. The increasing demand for LiBs highlights the urgent need for effective battery management strategies to mitigate environmental and supply chain concerns while optimizing battery performance and lifespan, and understanding their degradation [6,7]. Improved battery management not only enhances the efficiency and longevity of EV batteries, but also facilitates their safe integration into secondary applications and promotes recycling and reuse, thereby minimizing the environmental footprint of spent EV batteries [8–11].

In response to these challenges, we have undertaken an exploration of the recent advances and future trajectories of battery management in electric vehicles within this Special Issue. This issue was crafted to provide the scientific community with up-to-date insights into the latest advancements and prospects regarding various facets of lithium-ion batteries. Researchers were invited to contribute original research articles as well as review papers for inclusion in the Special Issue titled "Battery Management in Electric Vehicles: Current Status and Future Trends".

This Special Issue presents seven research papers [12–18] and three critical reviews [19–21], meticulously scrutinized through peer review processes. These publications span a spectrum of EV battery advancements, encompassing fundamental studies of batteries to the application of neural network modeling and machine learning to optimize battery performance. Ren et al. [12], for instance, proposed a deep learning neural network and fine-tuning-based transfer learning strategy to accurately estimate the state of health of batteries, thereby ensuring reliable and safe operating conditions for EVs. In another study, artificial neural network modeling was employed to forecast thermal and electrical performances using an innovative cooling method [13], suggesting that direct cooling surpasses conventional air cooling and indirect cooling methods for developing next-generation thermal management techniques for high-power density batteries [13,22]. Conversely, Juan et al. [14] employed a hybrid methodology, combining simulation and reinforcement learning, to address the orienteering problem and optimize battery management under dynamic routing conditions.

To address prolonged recharging times and the limited availability of recharging stations for electric vehicles, the practice of battery sharing or swapping was introduced [15]. This initiative has garnered support from key industrial players such as Eni in Italy and Shell in the UK. It was concluded that battery swapping holds promise in alleviating traffic congestion and mitigating environmental impact. Iterative nonlinear fuzzy modeling of lithium-ion batteries was also employed to enhance their efficiency and ensure proper management [16].

The imperative to explore potential strategies for reusing, remanufacturing, or recycling batteries at the end of their lifecycle prompted an investigation into the structural optimization of battery module cases [17]. This study presents an approach aimed at enhancing the design and construction of protective housing/cases for electric vehicles, ensuring compliance with safety and reliability standards throughout various stages, from initial design to impact. Furthermore, the parametric evaluation of thermal behavior for different lithium-ion battery chemistries was showcased [18], revealing the potential of NMC ($LiNi_xMn_yCo_{1-x-y}O_2$) chemistry in future applications for lower-cost and higher-specific-energy batteries for EVs.

Complementing the research papers, three review articles focusing on wireless charging, battery packing design, and battery management are featured in this Special Issue [19–21]. Ghazizadeh et al.'s [19] review offers a comprehensive analysis of the factors influencing the efficiency of wireless charging for EV batteries, including coil designs and compensation techniques. Likewise, the second review [20] delves into the core challenges confronting battery thermal management systems within EVs, proposing innovative design approaches for battery packing to enhance efficiency and longevity. Finally, the third review [21] serves as a comprehensive roadmap to the latest technological advancements propelling EVs into the future, providing a panoramic view of innovations in storage technology, battery management systems, and power electronics, with a particular emphasis on charging strategies, methods, algorithms, and optimizations. This review encapsulates the dynamic landscape of EV technology, offering insights into the advancements shaping the vehicles of tomorrow.

Funding: Financial support from the Faraday Institution as part of its ReLiB Project (grant numbers FIRG005 and FIRG027) is gratefully acknowledged.

Acknowledgments: As the editor of this Special Issue, I wish to extend my heartfelt gratitude to all the authors for their invaluable contributions, and to the reviewers for their diligent efforts in assessing the relevance and quality of the papers. Furthermore, I am deeply appreciative of the unwavering assistance and support provided by the editorial team of *Batteries*, including our assistant editor, Spring Deng, throughout the review process.

Conflicts of Interest: The author declares no conflicts of interest.

References

1. Li, M.; Lu, J.; Chen, Z.; Amine, K. 30 years of lithium-ion batteries. *Adv. Mater.* **2018**, *30*, 1800561. [CrossRef]
2. Harper, G.D.; Kendrick, E.; Anderson, P.A.; Mrozik, W.; Christensen, P.; Lambert, S.; Greenwood, D.; Das, P.K.; Ahmeid, M.; Milojevic, Z. Roadmap for a sustainable circular economy in lithium-ion and future battery technologies. *J. Phys. Energy* **2023**, *5*, 021501. [CrossRef]
3. Alanazi, F. Electric Vehicles: Benefits, Challenges, and Potential Solutions for Widespread Adaptation. *Appl. Sci.* **2023**, *13*, 6016. [CrossRef]
4. Sanguesa, J.A.; Torres-Sanz, V.; Garrido, P.; Martinez, F.J.; Marquez-Barja, J.M. A Review on Electric Vehicles: Technologies and Challenges. *Smart Cities* **2021**, *4*, 372–404. [CrossRef]

5. Dunn, J.; Slattery, M.; Kendall, A.; Ambrose, H.; Shen, S. Circularity of Lithium-Ion Battery Materials in Electric Vehicles. *Environ. Sci. Technol.* **2021**, *55*, 5189–5198. [CrossRef] [PubMed]
6. Akbarzadeh, M.; Kalogiannis, T.; Jaguemont, J.; Jin, L.; Behi, H.; Karimi, D.; Beheshti, H.; Van Mierlo, J.; Berecibar, M. A comparative study between air cooling and liquid cooling thermal management systems for a high-energy lithium-ion battery module. *Appl. Therm. Eng.* **2021**, *198*, 117503. [CrossRef]
7. Andriunas, I.; Milojevic, Z.; Wade, N.; Das, P.K. Impact of solid-electrolyte interphase layer thickness on lithium-ion battery cell surface temperature. *J. Power Sources* **2022**, *525*, 231126. [CrossRef]
8. Cusenza, M.A.; Guarino, F.; Longo, S.; Mistretta, M.; Cellura, M. Reuse of electric vehicle batteries in buildings: An integrated load match analysis and life cycle assessment approach. *Energy Build.* **2019**, *186*, 339–354. [CrossRef]
9. Attidekou, P.S.; Milojevic, Z.; Muhammad, M.; Ahmeid, M.; Lambert, S.; Das, P.K. Methodologies for large-size pouch lithium-ion batteries end-of-life gateway detection in the second-life application. *J. Electrochem. Soc.* **2020**, *167*, 160534. [CrossRef]
10. Christensen, P.A.; Wise, M.S.; Attidekou, P.S.; Dickmann, N.A.; Lambert, S.M.; Das, P.K. Thermal and mechanical abuse of electric vehicle pouch cell modules. *Appl. Therm. Eng.* **2021**, *189*, 116623. [CrossRef]
11. Fordham, A.; Milojevic, Z.; Giles, E.; Du, W.; Owen, R.E.; Michalik, S.; Chater, P.A.; Das, P.K.; Attidekou, P.S.; Lambert, S.M. Correlative non-destructive techniques to investigate aging and orientation effects in automotive Li-ion pouch cells. *Joule* **2023**, *7*, 2622–2652. [CrossRef]
12. Ren, Z.; Du, C.; Zhao, Y. A Novel Method for State of Health Estimation of Lithium-Ion Batteries Based on Deep Learning Neural Network and Transfer Learning. *Batteries* **2023**, *9*, 585. [CrossRef]
13. Garud, K.S.; Han, J.-W.; Hwang, S.-G.; Lee, M.-Y. Artificial Neural Network Modeling to Predict Thermal and Electrical Performances of Batteries with Direct Oil Cooling. *Batteries* **2023**, *9*, 559. [CrossRef]
14. Juan, A.A.; Marugan, C.A.; Ahsini, Y.; Fornes, R.; Panadero, J.; Martin, X.A. Using Reinforcement Learning to Solve a Dynamic Orienteering Problem with Random Rewards Affected by the Battery Status. *Batteries* **2023**, *9*, 416. [CrossRef]
15. Neroni, M.; Herrera, E.M.; Juan, A.A.; Panadero, J.; Ammouriova, M. Battery Sharing: A Feasibility Analysis through Simulation. *Batteries* **2023**, *9*, 225. [CrossRef]
16. Andújar, J.M.; Barragán, A.J.; Vivas, F.J.; Enrique, J.M.; Segura, F. Iterative Nonlinear Fuzzy Modeling of Lithium-Ion Batteries. *Batteries* **2023**, *9*, 100. [CrossRef]
17. Szabo, I.; Scurtu, L.I.; Raboca, H.; Mariasiu, F. Topographical Optimization of a Battery Module Case That Equips an Electric Vehicle. *Batteries* **2023**, *9*, 77. [CrossRef]
18. Buidin, T.I.; Mariasiu, F. Parametric Evaluation of Thermal Behavior for Different Li-Ion Battery Chemistries. *Batteries* **2022**, *8*, 291. [CrossRef]
19. Ghazizadeh, S.; Ahmed, K.; Seyedmahmoudian, M.; Mekhilef, S.; Chandran, J.; Stojcevski, A. Critical Analysis of Simulation of Misalignment in Wireless Charging of Electric Vehicles Batteries. *Batteries* **2023**, *9*, 106. [CrossRef]
20. Widyantara, R.D.; Zulaikah, S.; Juangsa, F.B.; Budiman, B.A.; Aziz, M. Review on Battery Packing Design Strategies for Superior Thermal Management in Electric Vehicles. *Batteries* **2022**, *8*, 287. [CrossRef]
21. Lipu, M.S.; Mamun, A.A.; Ansari, S.; Miah, M.S.; Hasan, K.; Meraj, S.T.; Abdolrasol, M.G.M.; Rahman, T.; Maruf, M.H.; Sarker, M.R.; et al. Battery Management, Key Technologies, Methods, Issues, and Future Trends of Electric Vehicles: A Pathway toward Achieving Sustainable Development Goals. *Batteries* **2022**, *8*, 119. [CrossRef]
22. Tan, X.; Lyu, P.; Fan, Y.; Rao, J.; Ouyang, K. Numerical investigation of the direct liquid cooling of a fast-charging lithium-ion battery pack in hydrofluoroether. *Appl. Therm. Eng.* **2021**, *196*, 117279. [CrossRef]

Disclaimer/Publisher's Note: The statements, opinions and data contained in all publications are solely those of the individual author(s) and contributor(s) and not of MDPI and/or the editor(s). MDPI and/or the editor(s) disclaim responsibility for any injury to people or property resulting from any ideas, methods, instructions or products referred to in the content.

Review

Battery Management, Key Technologies, Methods, Issues, and Future Trends of Electric Vehicles: A Pathway toward Achieving Sustainable Development Goals

Molla Shahadat Hossain Lipu [1,*], Abdullah Al Mamun [2], Shaheer Ansari [3], Md. Sazal Miah [1,4], Kamrul Hasan [5], Sheikh T. Meraj [6], Maher G. M. Abdolrasol [7], Tuhibur Rahman [1], Md. Hasan Maruf [1], Mahidur R. Sarker [8], A. Aljanad [9] and Nadia M. L. Tan [10,*]

1. Department of Electrical and Electronic Engineering, Green University of Bangladesh, Dhaka 1207, Bangladesh
2. Department of Electrical and Electronic Engineering, Northern University Bangladesh, Dhaka 1205, Bangladesh
3. Department of Electrical, Electronic and Systems Engineering, Universiti Kebangsaan Malaysia, Bangi 43600, Selangor, Malaysia
4. School of Engineering and Technology, Asian Institute of Technology, Khlong Nueng 12120, Pathum Thani, Thailand
5. School of Electrical Engineering, College of Engineering, Universiti Teknologi MARA, Shah Alam 40450, Selangor, Malaysia
6. Department of Electrical and Electronic Engineering, Universiti Teknologi PETRONAS, Seri Iskandar 32610, Perak, Malaysia
7. Electrical Department, Civil Aviation Higher Institute (CAHI), Alwatar, Tobruk 7502, Libya
8. Institute of IR 4.0, Universiti Kebangsaan Malaysia, Bangi 43600, Selangor, Malaysia
9. Department of Science, Technology, Engineering and Mathematics, American University of Afghanistan, Doha 122104, Qatar
10. Key Laboratory of More Electric Aircraft Technology of Zhejiang Province, University of Nottingham Ningbo China, Ningbo 315100, China
* Correspondence: shahadat@eee.green.edu.bd (M.S.H.L.); nadia.tan@nottingham.edu.cn (N.M.L.T.)

Abstract: Recently, electric vehicle (EV) technology has received massive attention worldwide due to its improved performance efficiency and significant contributions to addressing carbon emission problems. In line with that, EVs could play a vital role in achieving sustainable development goals (SDGs). However, EVs face some challenges such as battery health degradation, battery management complexities, power electronics integration, and appropriate charging strategies. Therefore, further investigation is essential to select appropriate battery storage and management system, technologies, algorithms, controllers, and optimization schemes. Although numerous studies have been carried out on EV technology, the state-of-the-art technology, progress, limitations, and their impacts on achieving SDGs have not yet been examined. Hence, this review paper comprehensively and critically describes the various technological advancements of EVs, focusing on key aspects such as storage technology, battery management system, power electronics technology, charging strategies, methods, algorithms, and optimizations. Moreover, numerous open issues, challenges, and concerns are discussed to identify the existing research gaps. Furthermore, this paper develops the relationship between EVs benefits and SDGs concerning social, economic, and environmental impacts. The analysis reveals that EVs have a substantial influence on various goals of sustainable development, such as affordable and clean energy, sustainable cities and communities, industry, economic growth, and climate actions. Lastly, this review delivers fruitful and effective suggestions for future enhancement of EV technology that would be beneficial to the EV engineers and industrialists to develop efficient battery storage, charging approaches, converters, controllers, and optimizations toward targeting SDGs.

Keywords: battery storage; battery management; electric vehicles; converter; controllers; optimizations

1. Introduction

Global warming is one of the most concerning issues to scientists and researchers at present, and the main reason behind this vital issue is the greater emission of carbon. Approximately 3 billion metric tons of carbon dioxide emissions will be produced by only passenger cars worldwide in 2021 [1]. According to the statistics, 41% of the carbon dioxide is emitted from the transportation sector globally [2]. In the USA, a total of 29% of the carbon emissions were produced by passenger cars in the year 2020, according to USA Environment Protection Authorities [3]. However, some issues should be investigated further, such as selecting appropriate battery energy storage, fast charging approaches, power electronic devices, conversion capability, and hybridization of algorithms or methods [4,5]. Hence, further investigation is required to improve EV technology to achieve sustainable development goals (SDGs) [6,7].

Unlike traditional vehicle technology, EVs fully depend on batteries in the case of supplying power, and that is why batteries are considered as the heart of EV technology [8]. Many battery technologies have been introduced by researchers that can easily replace the traditional methods of supplying cars, such as the lead–acid, nickel–cadmium, lithium-ion, lithium-ion polymer, and sodium–nickel chloride batteries [9]. Lead–acid battery technology was introduced at the beginning of the journey of battery technology. Although it has a short life cycle, it can provide 20–40 Wh/kg at the stage of 100% charge [10,11]. To solve the life cycle problem, inventors introduced a new technology called the nickel–cadmium battery that has a long-life cycle. However, the fast charging and deep discharging can cause damage to battery health and performance [12]. Removing all the drawbacks of the battery technology, a new technology known as the lithium-ion battery was introduced, which has greater efficiency, longer life cycle, high energy density, and performance at high temperatures. All of these characteristics make this technology most suitable for EV applications [9]. Lithium-ion technology has risen to the peak because of its unique feature such as high energy density, performance at a high temperature, fast charging, and long lifespan. Nonetheless, the performance of lithium-ion batteries varies with the combination of different materials such as cobalt, manganese, iron, nickel, aluminum, and titanate [13–15]. Furthermore, the unavailability of the materials is the drawback that makes lithium-ion technology a little bit dull [12]. Although the battery technology has advanced to significant development, each of these batteries has some downsides. Recently, fuel cell and supercapacitor-based EVs have made a significant stride toward the advancement of energy storage in the EV market.

The management system of the battery storage system plays a crucial role in the EV system [16]. For proper supervision of energy storage devices for safe and healthy operation, various techniques and control operations such as cell monitoring, voltage, and current monitoring, data acquisition, charge–discharge control, power management control, temperature control, fault diagnosis, and network and communication network should work spontaneously [17–19]. In order to perform all the operations efficiently, a set of highly efficient power electronic devices are needed. DC/DC converters play a vital role in EV technology. The most widely used DC/DC converters are isolated and non-isolated. A non-isolated DC/DC converter such as Ćuk, switched capacitor, coupled inductor, and quasi Z-source converters are used for converting voltage up or down in a relatively low ratio [20,21]. Due to low cost, high efficiency, and lower ripples, DC/DC converters are famous in EV technology. However, present switching control techniques are not reliable enough for EVs. An isolated DC/DC converter is used when the ratio of output and input voltage is high. The buck–boost converter, push–pull converter (PPC), DC/DC resonant converter (RC), zero-voltage switching converter (ZVSC), and full-bridge boost DC/DC converter (FBC) are widely used converters in EV technology, with each having individual drawbacks.

EV technology is not only a revolution in the transportation industry, but also a roadmap to economic development [16]. The increasing EV industry has tremendously influenced the economy by creating jobs that meet sustainable economic growth, which is related to SDG8 [17]. Unlike other transportation technologies, EV technology totally depends on battery storage; thus, there is no need to burn coal, oil, or gas, which means that it is a technology that provides clean and green energy, which is the requirement of SDG7 [18]. Furthermore, EV technology is an environmentally friendly solution that emits zero carbon which meets the major requirement of SDG13 [19]. Moreover, EV technology can also integrate renewable energy sources (RESs). As a result, industries which produce the materials for generating energy from renewable energy sources will develop. This will represent industrial innovation that can fulfill another goal (SDG9) [20]. Smart cities are currently becoming popular, and EVs are the most precious requirement for smart cities related to SDG11 [21].

To date, many technologies related to EV energy storage have been proposed by many researchers throughout the world. Hannan et al. [14] presented a strong review with criticism of lithium-ion batteries in which they illustrated a brief history and performance and demand of lithium-ion batteries in the EV industry, as well as the effect of environment-related facts and issues. Another study by Tie et al. [22] studied alternative energy sources, energy storage systems, energy management and control, supervisory control, and algorithms. The authors developed a relationship between the EV industry and economic growth relating to SDGs, but they did not consider other SDGs with EV technology. Sujitha et al. [23] presented a review of RES-based EV charging systems in which various types of power converters were discussed. However, the authors delivered specific topologies and their working principle. Katoch et al. [24] conducted a detailed review on the thermal management system for EV batteries, such as air cooling, liquid cooling, direct refrigerant cooling, phase change material cooling, thermoelectric cooling, and heat pipe cooling. Manzetti et al. [11] illustrated a wide history of EV batteries from the beginning to the present. They mentioned that bio batteries could be a promising solution in green battery technology over metal–lithium batteries in upcoming days in the EV industry. Lipu et al. [5] reviewed various converter schemes and controller technologies in EV application; however, a comprehensive study based on other EV-based technologies was not mentioned. Although green chemistry, which meets one of the major SDGs (climate), was considered in this study, how this technology can complement the other SDGs is still needed to investigate. In summary, the relationship among EV technology, SDGs, and battery technology was not considered in any of the abovementioned studies. Therefore, further study is required to integrate the SDGs with the EV industries.

To bridge the existing shortcomings, this study highlights a detailed survey on prospects of EV and SDG integration. Furthermore, the study presents various technological advancements, issues, challenges, and future recommendations. This review provides the following contributions:

- This review critically examines the various battery storage systems, materials, characteristics, and performance. Additionally, the key components of the battery management system are outlined.
- The various technological advancements of EVs concerning the power electronics technology and charging strategies are discussed rigorously.
- The state-of-the-art methods, algorithms, controllers, and optimization schemes applied in EVs are explained thoroughly.
- The work establishes the relationship of EV energy management and technologies with sustainable development targeting various goals such as clean energy, sustainable cities, economic development, industry, infrastructure, and emission reduction.
- Lastly, this research illustrates the scope, opportunities, and future trends for the advancement of EVs. The analysis, key findings, and suggestions can be helpful in successfully integrating the EV technologies with SDG targets.

The remainder of the paper is divided into seven sections. Section 2 covers several battery energy storage systems and key components of battery management systems. The EV technologies concerning power electronics converters and charging features and technologies are presented in Section 3. In Section 4, the algorithms, methods, approaches, controllers, and optimizations employed in EVs are reported. The open issues and challenges are highlighted in Section 5. The impacts of EVs in achieving different goals of sustainable development are examined in Section 6. Lastly, the conclusions, along with the future trends, are provided in Section 7.

2. Battery Energy Storage and Management in EVs

This section broadly discusses and analyzes the various battery storage characteristics, features, and key components of battery management in EV applications.

2.1. Battery Storage Technology
2.1.1. Lead–Acid (Pb–Acid)

The lead–acid battery is considered as one of the oldest battery technologies to be used globally. Lead–acid batteries display a specific energy of 20–40 Wh/kg at 100% of the state of charge (SOC) of a lead–acid battery. It contributes a small cycle life due to the shedding of active material compared to other types of batteries such as nickel metal hydride. The low energy-to-weight ratio and low energy-to-volume ratio are considerable limitations. Furthermore, the lead–acid battery is not an environmentally sound technology due to the presence of lead and acid. Despite several drawbacks, the low manufacturing cost of around 100 USD/kWh makes it suitable for small-scale, light-performance vehicles [10,11].

2.1.2. Nickel-Cadmium (NiCd) Battery

Nickel–cadmium (NC) battery technology was employed in the 1990s, presenting high energy density. The NC battery technology was employed in applications such as power quality and energy reserves for telecommunication and portable services [25]. NC batteries provide a long lifecycle span of 1500 cycles compared with NiMH battery. However, the NC battery may cause damage due to deep discharge and a faster charging time. NiMH batteries possess a memory loss effect, which occurs due to the battery's frequent charging process before complete discharge. One of the major drawbacks of NC is the adoption of toxic metals such as cadmium during the manufacturing process. Cadmium causes adverse effects on the environment and human health. Equation (1) shows the electrochemical cell reaction of an NC battery, where Cd is used as an anode and NiO(OH) is used as a cathode [12].

$$Cd + 2NiO(OH) + 2H_2O \leftrightharpoons 2NiO(OH)_2 + Cd(OH)_2. \tag{1}$$

2.1.3. Lithium-Ion Battery

The lithium-ion battery is considered as one of the leading battery technologies used in EVs. The high energy density, greater efficiency, longer life cycle, and better performance at high temperatures are the well-known features of Li-ion batteries. Lithium maintains the lowest redox potential, which is around (−3.05 V), and the largest electrochemical equivalence of (3.86 Ah·g^{-1}) [25]. It is still recognized as suitable for battery-driven EVs. Moreover, lithium alone has the smallest reduction potential capacity of any element, which enables this battery to hold the topmost cell potential. The most important advantage of this battery relates to the recycling ability of the various components used. However, the material's unavailability and high cost per kWh (135 USD /kWh) represent significant drawbacks [12]. Generally, the performance of the lithium-ion battery relies upon various internal material properties. The selection of materials is crucial, specifically the positive electrode, which controls the battery characteristics such as power, safety, cost, and lifespan [26–29]. The positive electrode material can be classified into lithium cobalt oxide, lithium manganese oxide, lithium iron phosphate, lithium nickel manganese cobalt oxide, lithium nickel cobalt aluminum oxide, and lithium titanate, which are discussed below [13–15].

Lithium Cobalt Oxide—LiCoO$_2$

Goodenough announced the existence of layered transition metal oxides in 1980 [25]. They are considered among the most commonly implemented positive electrodes. Initially, Sony marketed lithium cobalt oxide (LCO) in 1991 and used cobalt oxide as a cathode which was the most commonly used material in lithium-ion battery technology. The theoretical capacity of LCO is approximately 274 mAh·g^{-1} with a high volumetric capacity of 1363 mAh·cm^{-3}. It constitutes high energy density, offers a moderate lifespan, and has considerable safety applicable for several electronic gadgets such as cameras, notebooks, and tablets [9,12,13]. Nevertheless, the LCO shows unsatisfactory behavior during its operating condition at the rate of high current of charge–discharge. Consequently, proper protections are required due to excessive heating and stress. In addition, the cost of cobalt is high due to its limited availability [30]. Hence, an alternative to cobalt cathode materials is preferred to raise the appropriateness of LCO in EVs.

Lithium Manganese Oxide—LiMn$_2$O$_4$

Lithium manganese oxide (LMO) is one of the most reviewed cathode materials for lithium-ion battery technology due to its easy accessibility to raw materials and low cost [31,32]. The Bellcore lab developed the LMO battery technology in 1994. The 3D spinel architecture of LMO helps to reduce the internal resistance and simultaneously increases the charge/discharge current flow. It exhibits decent specific power and energy density and can carry >50% more energy than nickel-based batteries. The theoretical capacity of LMO is about 148 mAh·g^{-1}. Pristine LMO ensures 95% delivery of its capacity, which is not possible in the case of LCO [33]. However, it has negative effects on its life cycle and performance. Moreover, LMO has extensive manganese breakdown in the electrolyte at high temperatures, which results in a high capacity loss. The capacity of LMO is approximately 33% lower than that of cobalt-based batteries [34–36]. Presently, the application of LMO is carried out in Nissan Leaf EV technology [14].

Lithium Iron Phosphate—LiFePO$_4$

The University of Texas investigated the application of phosphate as a cathode material and concluded that phosphate demonstrates better performance than LCO or LMO batteries at high temperatures and in overcharged states. Phosphates exhibit good thermal stability, operating in the temperature range of $-30\ ^\circ$C to $60\ ^\circ$C [37,38]. Lithium iron phosphate (LFP) can contribute with a nominal voltage of approximately 3.2 V and moderate power and energy density. In addition, the LFP battery has low cost, a long lifespan, an enhanced safety system, and high load-handling capability. The major drawbacks of LFP relate to poor lithium diffusion, poor electronic conductivity, and lower specific energy of 160 mAh/g compared to LCO and LMO battery technology. Furthermore, it requires a small particle size and carbon coating to enable performance at high current rates, resulting in a high processing cost [39].

Lithium Nickel Manganese Cobalt Oxide—Li(Ni, Mn, Co)O$_2$

Lithium nickel manganese cobalt oxide (LNMC) battery technology was first commodified in the year 2004. At present, battery industries are focusing on improving the cathode material by developing composite nickel–manganese–cobalt (NMC). These NMC electrode sheets are available in four different compositions, namely, NANOMYTE® BE-50E (NMC111), BE-52E (NMC532), BE-54E (NMC622), and BE-56E (NMC811). These different compositions possess unique outcomes. In terms of minimum capacity, BE-50E (NMC111), BE-52E (NMC532), BE-54E (NMC622), and BE-56E (NMC811) reveal 150 mAh/g, 155 mAh/g, 166 mAh/g, and 190 mAh/g, respectively, while the experimental outcomes were $\geq$155 mAh/g (2.7–4.3 V @ 0.1 C), $\geq$165 mAh/g (2.7–4.3 V @ 0.1 C), $\geq$175 mAh/g (3–4.3 V @ 0.1 C), and $\geq$200 mAh/g (3–4.3 V @ 0.1 C), respectively [40]. The cathode material of LNMC is developed by utilizing 33% nickel, 33% manganese, and 34% cobalt. The hybrid mixture of NMC draws out the low internal resistance effect of manganese and the high specific energy of nickel. Moreover, LNMC offers high power and energy density with improved lifespan and performance. At present, LNMC has high demand in EV applications for its low self-heating rate and long lifespan (1000–2000 cycles) [41–43]. It is suggested that LMNC battery characteristics could be altered by varying the combination of nickel, manganese, and cobalt for certain applications. The increment in manganese percentage leads to an enhancement of specific power, while the increment in nickel leads to an enhancement of specific energy. Presently, the BMW i3 is run by NMC-based lithium-ion batteries [13].

Lithium Nickel Cobalt Aluminum Oxide—Li(Ni, Co, Al)O$_2$

The nickel cobalt aluminum oxide (NCA) battery was commercially presented in 1999. The maximum utilization of nickel as a cathode material reduces the dependency of cobalt in LCO. It provides increased specific power, excellent specific energy of 200–250 Wh/kg, and a durable life cycle. In recent years, lithium nickel cobalt aluminum oxide (LNCA) battery technology has gained increasing attention in EV applications. Due to its high power and energy densities, automobile companies are concentrating on LNCA battery application in EV technology. However, further advancement is needed to improve its safety mechanism [37,44,45]. The automobile giant Tesla is currently utilizing LNCA battery technology to develop EVs [13].

Lithium Titanate—Li$_4$Ti$_5$O$_{12}$

Lithium titanate (LTO) has a spinel architecture and is configured using LMO, LNCA as a cathode material, and titanate as an anode material. The spinel configuration delivers a few advantages, such as structural firmness due to the zero strain effect and considerable reversibility [46]. LTO delivers high performance and a long lifespan. Furthermore, LTO operates safely at cold temperatures [47–49]. However, the power and energy density of LTO are lower compared to NMC- and NCA-based lithium-ion battery technology. The LTO recommends a constant active potential of around 1.55 V, but the electronic structure depicts insulating behavior with a bandgap of 2–3 eV [50]. The main obstacles that appear in LTO batteries are gas evolution, which leads to battery swelling, and low performance during charge/discharge due to low electrical conductivity [51]. Thus, further explorations are focused on improving these areas, including specific energy and cost reduction.

A detailed comparative study of lithium-ion battery materials, performance, and characteristics is shown in Figure 1 [52]. The figure clearly presents that LMO, LNMC, and LNCA batteries are the best depending upon the voltage, power, and energy categories. On the other hand, LFP and LTO batteries can be used when high lifecycle and safety are major concerns. Moreover, LTO is economically excellent and capable of delivering high performance.

Lithium-Ion Polymer

At the beginning of the 21st century, lithium-ion battery technology started to shift the paradigm from liquid electrolyte cells with metal housing to plastic casings. The battery technology was generally named as a lithium-ion polymer (LPO) battery [53]. The LPO battery technology is a secondary battery that consists of a polymer electrolyte in the liquid electrolyte utilized in usual lithium-ion batteries [54]. All Li-ion cells expand at high levels of state of charge (SOC) or overcharge due to slight vaporization of the electrolyte. This may result in delamination and bad contact of the internal layers of the cell, which in turn brings diminished reliability and overall cycle life of the cell. Lithium-ion polymer batteries have delivered satisfactory outcomes and have taken over nickel–metal hydride (NiMH) batteries for moveable electronic devices such as smartphones and laptops. They provide excellent high energy density (400 Wh·L^{-1}) compared to other types of batteries. The high power and energy density make them qualified candidates for EV and HEV applications [55]. They also contribute toward the extended cycle life of ordinary Li-ion batteries. The temperature should be kept at less than 50 °C to ensure the available cell capacity and utilize the full life span. However, functional instability materializes during limited battery discharge and overload conditions.

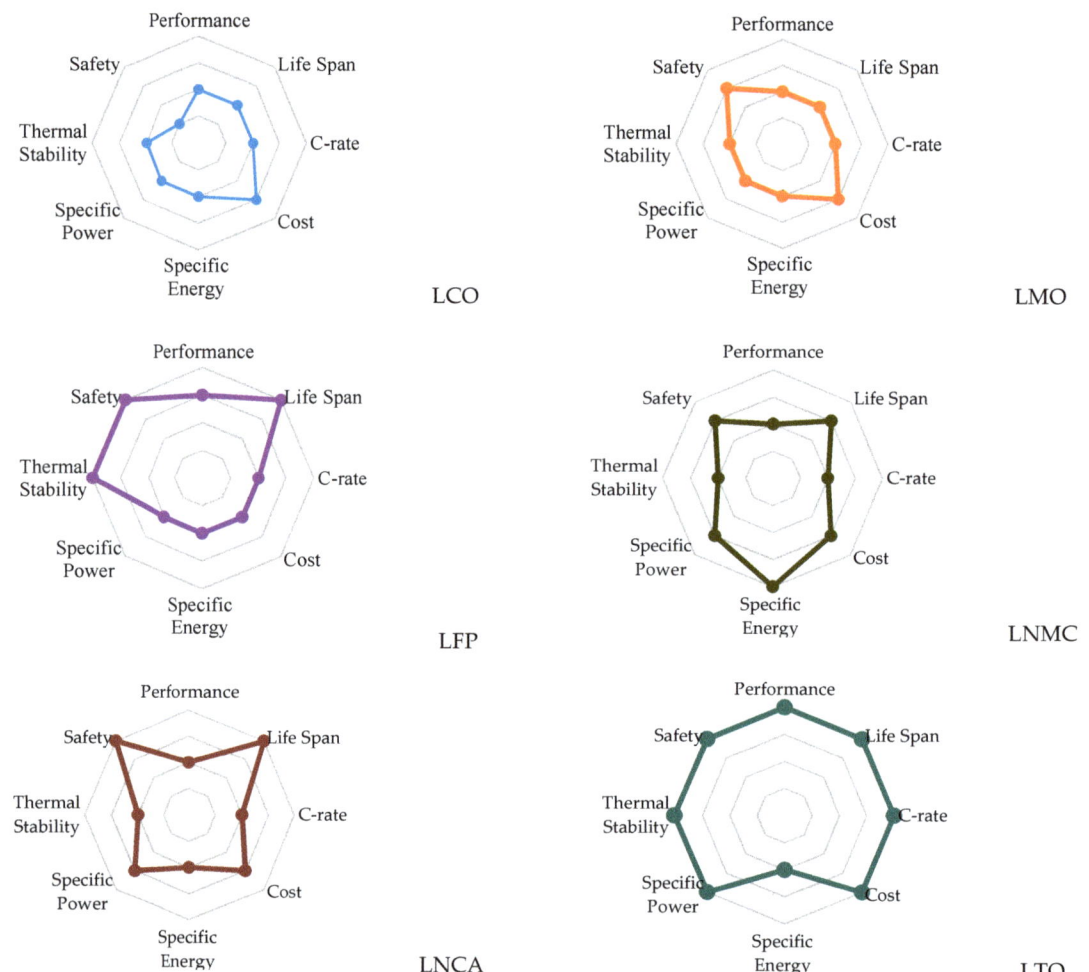

Figure 1. Performance comparison of different lithium-ion batteries according to indexed values.

Lithium-Ion Silicon

The initial work with lithium-ion silicon (LS) technology was initiated by Sharma and Seefurth in 1976 [56]. The preliminary research in LS battery technology was conducted by Dahn et al., as well as other teams from the year 1990 to 2000. At the same time, extensive research work was conducted between 2011 and 2015 [57]. The LS battery shows great potential due to its high capacity and long cycle life. However, some concerns related to LS battery technology need to be overcome, such as low Coulombic efficiency, lower real mass loading density, and high cost. Silicon anodes are among the most promising anode materials for lithium-ion batteries due to their various advantageous features, including the highest known capacity and relatively low working potential. However, the problem of extremely large volumetric change must be addressed before silicon anodes can be utilized in practical lithium batteries.

2.1.4. Sodium–Nickel Chloride (Na/NiCl$_2$)

The high-temperature sodium–nickel chloride (SNC) battery, also known as the ZEBRA (zero-emission battery research activity) battery, is manufactured from diluted sodium and nickel chloride. The solid ceramic simultaneously acts as an electrolyte and a separator at an optimal operating temperature of 270–350 °C. The specific energy of SNC is reported to be 125 Wh/kg with an energy efficiency of 92%, which is better than Pb–acid, NCd, and nickel–metal hydride (NMH) battery technologies [58]. The major concern of SNC is its operational safety due to the high operating temperature of 300 °C and its storage for longer periods. Furthermore, the high internal resistance and faster self-discharge cause low power capability for SNC batteries [59].

2.1.5. Fuel Cell

A fuel cell is an electrochemical device that uses two redox processes to transform the chemical energy of a fuel (typically hydrogen) and an oxidizing agent into electrical energy. Fuel cells require a constant supply of fuel and oxygen (often from the air) to sustain the chemical reaction, and they can constantly create electricity as long as fuel and oxygen are available. Sir William Grove created the first fuel cells in 1838. One century later, Francis Thomas Bacon created the hydrogen–oxygen fuel cell in 1932 [60]. Song et al. [61] examined the temperature effects on the performance of fuel cell-based hybrid EVs. Quan et al. [62] evaluated the fuel cell EV energy management strategies using model predictive control considering performance degradation in real time. Fuel cells offer a much more silent and a smoother alternative to conventional energy production that can greatly reduce CO_2 and harmful pollutant emissions. However, the fuel cell is expensive to manufacture due to the high cost of catalysts (platinum).

2.1.6. Supercapacitor

A supercapacitor (SC), sometimes known as an ultracapacitor, is a high-capacity capacitor that bridges the gap between electrolytic capacitors and rechargeable batteries. It has a capacitance value that is significantly higher than ordinary capacitors but with lower voltage restrictions. In comparison to electrolytic capacitors, it typically stores 10 to 100 times more energy per unit volume or mass, accepts and delivers charge considerably more quickly, and can withstand many more charge and discharge cycles than rechargeable batteries. Nguyen et al. [63] used the SC for energy storage in EV applications. Although the SC exhibits long life, it has some drawbacks, such as the generally lower amount of energy stored per unit weight compared to an electrochemical battery.

2.2. Battery Management System in EVs

The battery management system (BMS) can be defined as a system that assists in managing the battery operation via electronic, mechanical, and advanced technological systems [64]. An advanced BMS for EV applications is presented in Figure 2 [4]. The basic aims of BMS are cell/battery protection from being damaged and ensuring optimum operating conditions. The BMS ensures the proper supervision of the battery storage systems through control and continuous monitoring via various control techniques such as charge–discharge control, temperature control, cell potential, current, and voltage monitoring, thus enhancing the safety and lifetime of the energy management system (EMS) [65–67]. Nonetheless, the deep charge and discharge of the battery during long-distance traveling in EV are fundamental issues [68], potentially causing the failure of the battery or shock hazards due to the high discharge and thermal effect [69]. The BMS becomes effective in minimizing these difficulties by controlling the charge and discharge profile, as well as managing the thermal behavior of the battery packs. The state of charge (SOC), state of health (SOH), and remaining useful life (RUL) are the key parameters in the BMS for understanding the status of the battery. The BMS also protects the battery pack from high-voltage stress and short-circuit current by integrating controllers, actuators, and sensors [14]. The key components and operation of BMS applications in EV technology are explained below.

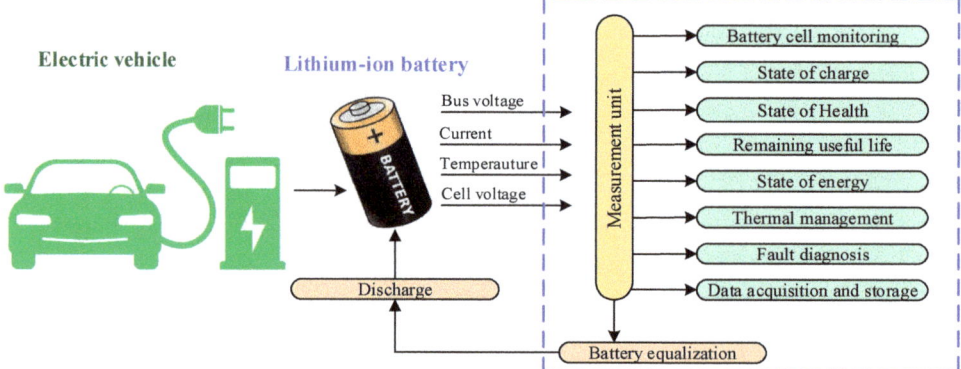

Figure 2. Various critical applications of BMS in EV technology.

2.2.1. Battery Cell Monitoring

The information on battery charging/discharging, health, temperature, and fault diagnosis is the foundation for completing the BMS duties [70,71]. Generally, a pack of battery cells is used in EVs [72]. The battery cell may react differently during the battery charging/discharging operation. As a result, continual battery cell monitoring is required to investigate the different states and performance indices [52]. The findings of the battery cell monitoring can help the system function better by managing, protecting, balancing, and controlling operations [73].

2.2.2. Voltage and Current Measurement

The battery cells are connected in series and parallel to the battery bank to acquire a sufficient amount of voltage and current. Hundreds of cells are linked in series in battery packs of electric automobiles, resulting in a large number of voltage measurement channels. When the cell voltage is measured, there is accumulated potential, and the combined potential of one cell differs from another. Hence, a suitable charge equalization must be provided in order to enhance EV autonomy [74]. An accurate battery cell measurement is required for the estimation of SOC and other battery states. One of the common voltage monitoring methods is the voltage divider technique which consists of a resistor and precise temperature-corrected voltage reference used to monitor the cell voltage. The other available methods are the optical coupling relay, optical coupling isolation amplifier, discrete transistor, and distributed measurement [64]. High-voltage current sensors are used to monitor the current of the battery module, which is later converted to a digital signal via analog to digital conversion (ADC). Finally, the voltage and current data are utilized to appropriately estimate the SOC, SOH, and RUL [75].

2.2.3. Data Acquisition

The data acquisition system (DAS) is used for measuring and estimating the parameters of the battery pack, such as current, voltage, temperature, and SOC [76,77]. This facilitates diagnosing the battery's health and identifying defective cells. It also investigates battery changes that would assist in delivering the status of battery aging, climate, and other factors. The DAS is an integral part of the BMS, which consists of a hardware device (microcontroller unit) and software. The DAS uses the ADC module for data conversion. A controlled area network (CAN) bus and serial communication interface (SCI) module are used to exchange information and communicate with the BMS [78–80]. A cloud-based DAS platform within the BMS to extract critical information such as battery current, voltage, and temperature is presented in Figure 3 [81].

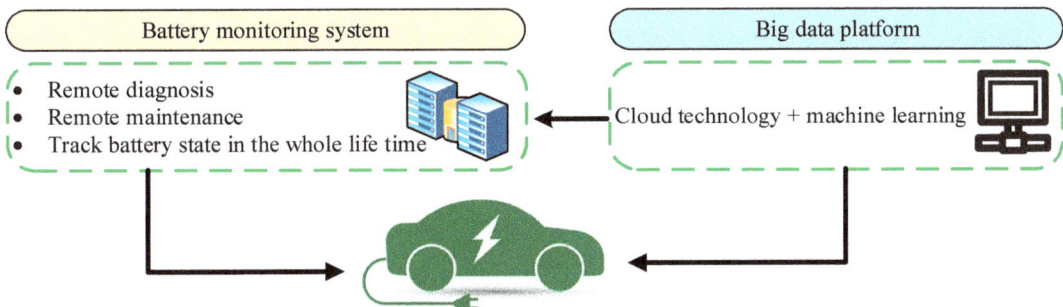

Figure 3. BMS-based cloud-integrated data acquisition framework for EV technology.

2.2.4. Battery State Estimation

The battery state estimation is critical for estimating battery charge and health. The SOC, SOH, RUL, state of function (SOF), state of power (SOP), state of energy (SOE), and state of safety (SOS) are some of the common battery states [82]. A framework for estimating SOC, SOH, RUL, SOF, SOP, SOE, and SOS for the BMS is depicted in Figure 4 [83].

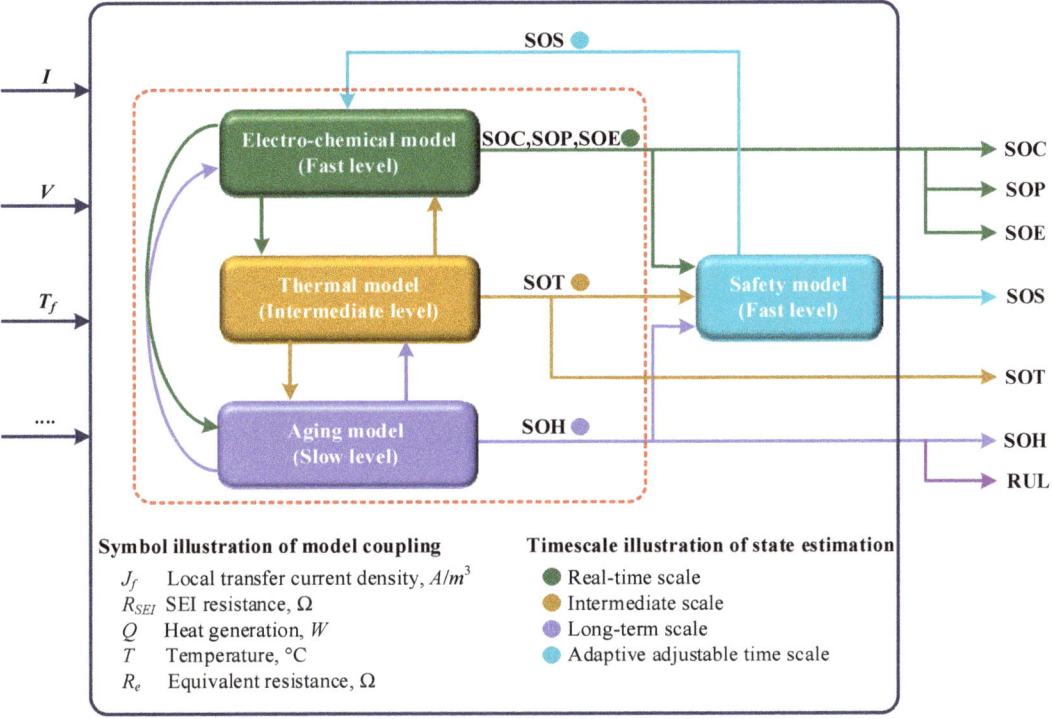

Figure 4. BMS-based battery SOC, SOH, RUL, SOF, SOP, SOE, and SOS estimation framework for EV applications.

State of Charge

SOC can be defined as the proportion of currently available capacity to maximal battery capacity [84]. It is not directly measurable from terminals; hence, a method must be created to predict the state from measured data [85]. The appropriate assessment of SOC is not only needed for battery protection from degradation but also for the highest level of energy management [86]. Several methods are available to estimate the SOC, such as the discharge test method, sliding mode observer method, neural network method, fuzzy logic method, impedance method, and internal resistance method, as well as Kalman filter (KF), machine learning, and deep learning approaches [52]. The Ampere-hour (Ah) and open-circuit voltage methods are also common methods to calculate the SOC [87,88]. The Ah technique becomes a simple choice for SOC calculation since charging or discharging current may be easily monitored. However, the accuracy of SOC estimation is not error-free, and the firmness of the initial state is complex. Furthermore, the estimation accuracy is improved in the open-circuit voltage (OCV) method, but the long resting time limits the rapid application of this method in EVs [89]. KF-based SOC estimation achieves accurate results but has complex mathematical computation and functional relationships [90]. Recently, machine learning and deep learning methods for SOC estimation have received wide attention due to their high accuracy, improved learning capability, better generalization performance, and convergence speed [91].

State of Health

The SOH of the battery can be defined as the available maximum capacity left by the cycling effect of charge–discharge [92]. The following equation can be used to estimate the SOH:

$$\text{SOH (\%)} = (Q_{actual} Q_{rated}) \times 100, \qquad (2)$$

where Q_{actual} is the actual capacity of the battery, and Q_{rated} is the rated capacity [93].

The SOH can be easily estimated from an understanding of capacity degradation and the internal resistance of the battery. A variety of methods have been developed to estimate battery SOH, which can be divided into three categories: model-free, model-based, and data-driven methods [93]. Electrochemical impedance spectroscopy (EIS) analysis is much more convenient compared with direct methods for capacity and internal resistance estimation in a model-free method [94]. On the other hand, model-based methods follow the equivalent circuit model and electrochemical model to estimate the capacity and internal resistance during battery operation. Similarly, the data-driven method uses the support vector machine (SVM) mechanism to estimate SOH by measuring the terminal voltage, current, and temperature [95].

Remaining Useful Life

Accurate and robust EV performance is subjected to the battery's remaining useful life (RUL). The battery's continuous charging and discharging process results in capacity degradation, which can deliver unacceptable outcomes such as major breakdown, economic loss, and safety issues [5,96]. Therefore, it is crucial to estimate the RUL of the battery toward the achievement of safe, accurate, robust, and reliable operation of EV technology [97]. When the battery is charged and discharged continuously, and its capacity remains 70% or 80% of the initial capacity, the battery needs replacement. Therefore, several model-based and data-driven-based techniques have been explored to predict the RUL of the battery. The model-based techniques rely on a mathematical model and detailed experiments; however, the technique requires a huge volume of data to estimate the battery degradation pattern. On the other hand, data-driven methods depend on battery historical data, which comprise various parameters such as voltage, current, impedance, capacity, and temperature. Data-driven methods predict the RUL by considering battery data and do not require complex mathematical models [96].

State of Function

The SOF is described as the capability of a battery that can finish a specific task. It narrates the performance of the battery in terms of meeting the power demand [98]. It can also be determined from the ratio of available useable energy to the maximum stored energy of the battery [99]. The SOF is estimated with the help of SOC, SOH, and temperature [64]. The SOF can be calculated from a few approaches, such as (adaptive) characteristic maps and equivalent circuit models, including the fuzzy logic control method [100]. The SOC, power pulse duration, power, voltage, and temperature are the characteristics needed in (adaptive) characteristic maps [98]. Additionally, KF and artificial neural network (ANN) algorithms are adopted in model-based methods for the accurate estimation of SOF. The parameters related to SOC, SOH, and C-rate of the battery are also employed in the fuzzy logic algorithm to estimate the battery SOF [64].

2.2.5. Battery Protection Strategies

Battery protection is one of the major tasks of BMS. Due to alterations in physical and chemical characteristics of the battery and frequent charge–discharge, voltage and charge deviance may occur in battery cells [101]. The overall battery performance and lifetime may be reduced because of the deviation of voltage and charge. Moreover, the deep discharge below the minimum SOC limit and overcharge of the battery beyond the C-rating may cause a critical situation for the battery [102]. Thus, a suitable protection system for the battery in EV applications is important. The proper maintenance of operating temperature is also a significant parameter for ensuring safety. The BMS provides temperature safety limits which stand between 0 °C and 60 °C for charge and between −20 °C and 60 °C for discharge [103]. It also provides deep discharge protection, overcharge protection, high-temperature protection, uplifted voltage protection, and power cutoff safety. However, BMS safety protocols should comply with the automobile International Organization for Standardization (ISO) 26262 [104]. A fully integrated, cost-effective, and low-power single-chip lithium-ion battery protection IC (BPIC) was proposed by Lee et al. [105], as shown in Figure 5.

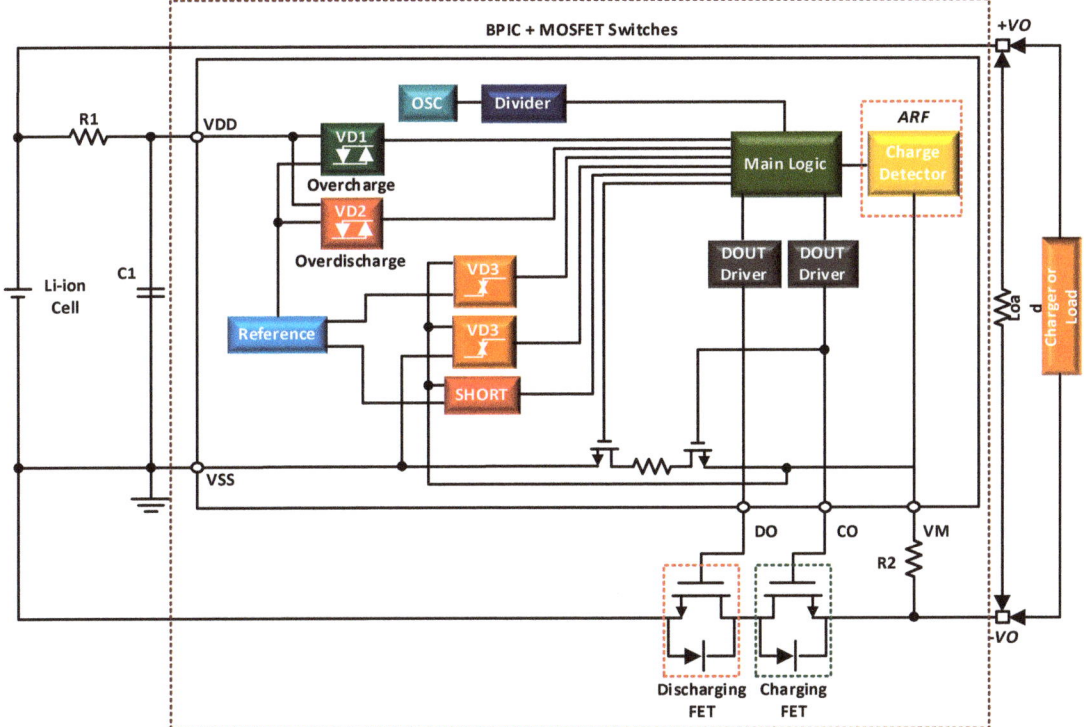

Figure 5. Block diagram of the lithium-ion battery protection circuit.

2.2.6. Battery Equalizer Control

The BMS can protect the battery from abnormalities that are caused by the under/overcharging of the battery through individual cell monitoring and charge equalization control [106,107]. The undercharging of the battery can deteriorate the lifetime, and overcharging of the battery can damage it completely. To enhance and maintain the constant performance of the battery, the equalization of voltage and charge of battery cells is critical [108]. Battery equalizer control can be broadly categorized into active and passive charge equalization controllers, as shown in Figure 6 [109].

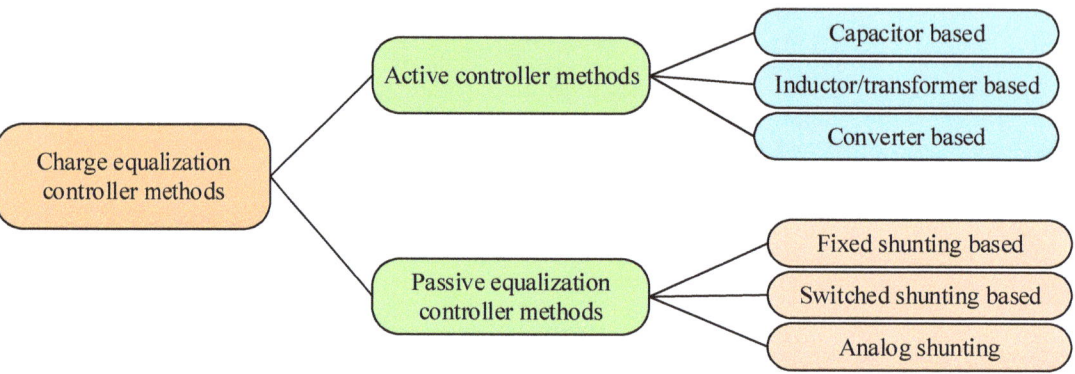

Figure 6. Charge equalization controllers for lithium-ion batteries.

The active charge equalization controller (CEC) works on the principle of transferring energy from cell to cell, cell to battery pack, or battery pack to cell [110]. The excess energy is collected from overcharged cell and delivered to the undercharged cell to equalize the charge and voltage. The active CEC can also be categorized into three types, namely, capacitor-based, inductor/transformer-based, and converter based. The energy transfer from cell to heat via a shunting resistor is the basic hypothesis of passive CEC, which can be distinguished into fixed shunting, switched shunting, and analog shunting. A fixed resistor is used in the fixed shunting method to bypass the current flows and control the voltage. Similarly, a controllable switch (relay) bypasses the release path from the overcharged battery in the switched shunting method. The most effective method among the three is the analog shunting method which uses a transistor instead of a resistor to complete the task of the current bypass from high-energy cells [109]. A constant current string-to-cell battery equalizer with open-loop current control was proposed by Wei et al. [111], as depicted in Figure 7.

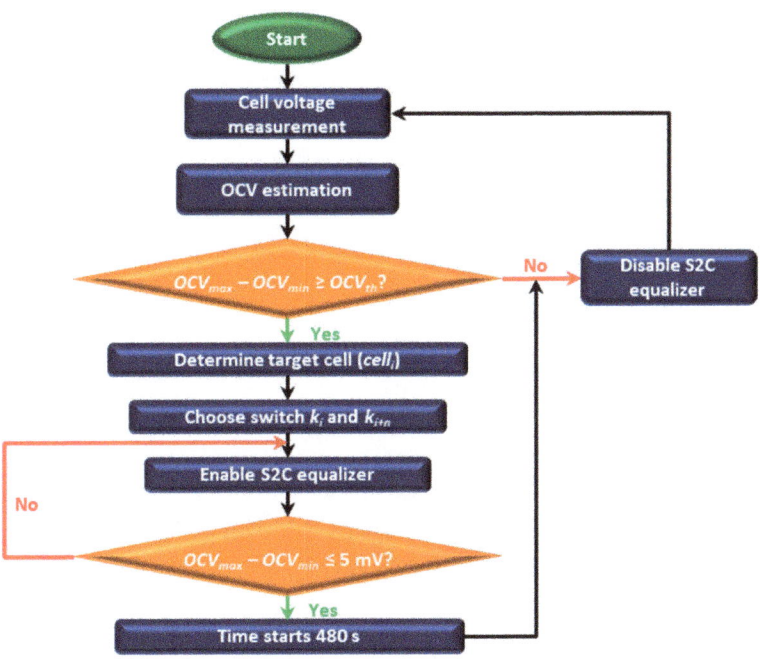

Figure 7. Flowchart of string-to-cell battery equalization algorithm.

2.2.7. Charge and Discharge Control

The battery charging/discharging determines the protection, performance, and durability. Incorrect charging drastically accelerates the battery's deterioration. Nonetheless, enhancing battery efficiency, reducing overheating, and prolonging the life cycle depends on controlled and quality charge and discharge. There are a few conventional but widely used charging techniques for resolving battery charging issues with a variety of aims and termination circumstances. The charging techniques can be classified into four types: constant-current (CC) charging, constant-voltage (CV) charging, constant-current/constant-voltage (CC-CV) charging, and multistage constant-current (MCC) charging [112]. A constant current rate is the main approach adopted by the CC technique to charge the battery. During the CC technique, a low current rate can lower the charging speed, which is not suitable for EV applications. The CV charging method works on the basis of a predefined constant voltage to charge the battery, eliminating the risk of overcharge and enhancing the battery cycle life. The charging speed and temperature variation are new modifications that have been added to this technique. The hybrid charging technique is CC-CV which works on interconnecting the principle of predefined current, a voltage of CC, and CV. In the beginning, the battery is charged with constant current (CC), and then the voltage is increased to a safe limit. In the end, the battery starts working in the CV phase and remains as such until the target capacity is obtained. Constant multistage series current is injected into the battery during the whole process of charging in the MCC charging technique. This highlights the basic difference between CC-CV and MCC. The speed of MCC charging is quite slow compared to the CC-CV technique. However, fuzzy logic technology has been incorporated with MCC to improve performance [95,113]. An orderly charge and discharge control process for EVs based on charging reliability indicators was developed by Li et al., as presented in Figure 8 [114].

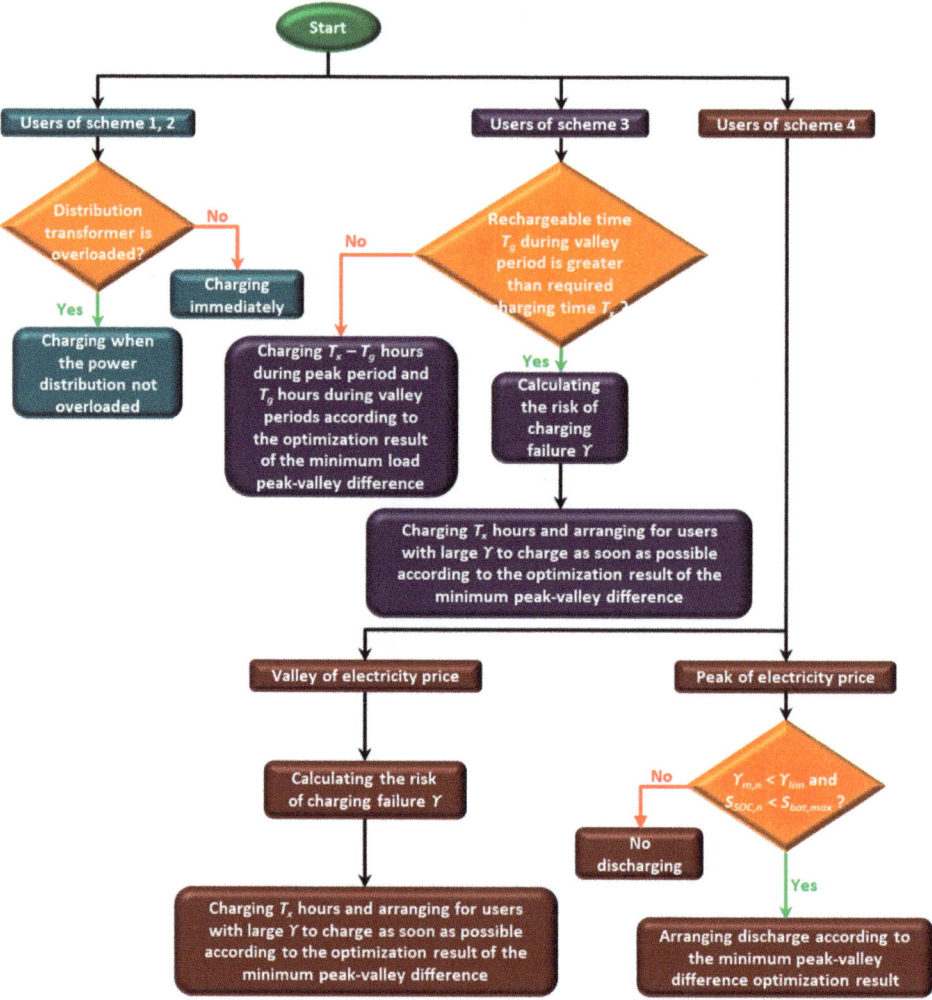

Figure 8. Orderly charge and discharge control process in EV applications.

2.2.8. Power/Energy Management Control

EV performance depends not only on energy storage but also on power and energy intelligent control strategies. In order to regulate power/energy flow efficiently in electric vehicles, the energy storage control system must be capable of dealing with high peak power when accelerating or decelerating [114]. Two basic types of common strategies are adopted for power and energy management (PEM) control [115]. A low-level component control strategy enhances the PEM performance and flexibility via a power transfer train mechanism that connects ESS, auxiliary ES, ICE, and generators altogether. A high-level supervisory control system works on the time-based data extraction process and balances the operations of different components. Various types of efficient PEMC systems have also been reported in several articles for HEV; among them, two major types are rule-based and optimization-based. The rule-based strategy can be classified into two types: deterministic rule-based and fuzzy logic. The real-time optimization and global optimization PEMC systems are types of optimization-based PEMC systems [22,116,117]. Based on the driving schedule, the powertrain model, and two neural networks, the energy management strategy based on the deep Q-learning method for a hybrid EV was proposed by Du et al., as shown in Figure 9 [118].

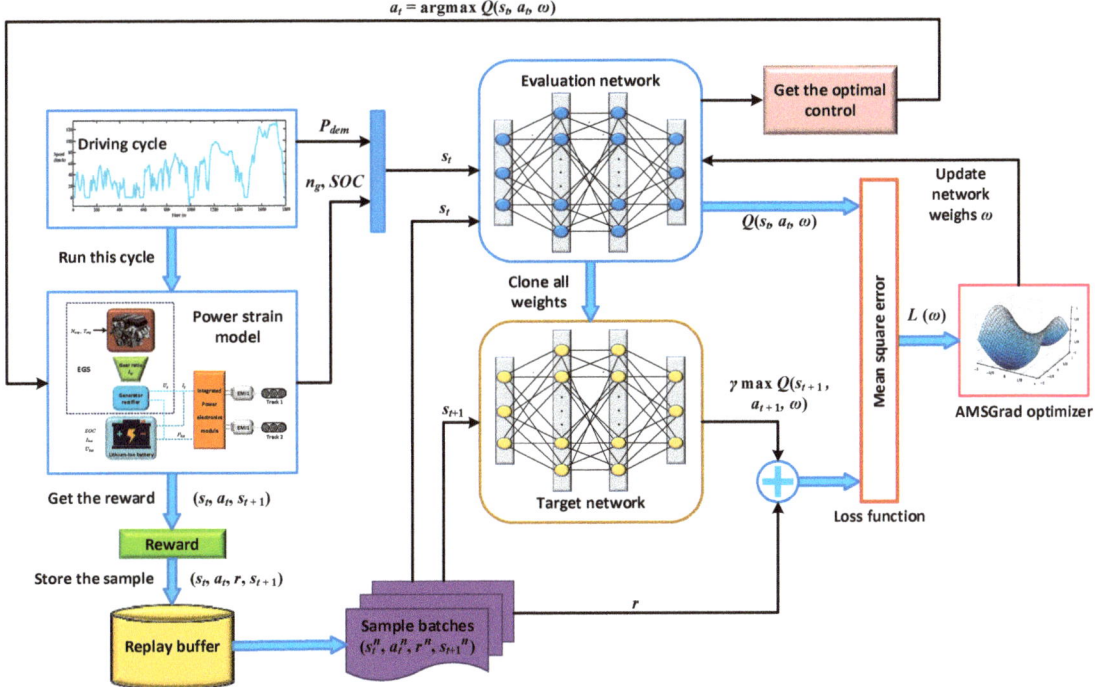

Figure 9. Advanced energy management strategy framework for EV.

2.2.9. Operating Temperature Control

The battery temperature significantly impacts several aspects of battery performance, including longevity, energy conversion efficiency, and safety [119]. The rapid charge–discharge cycle of the battery was identified as the main cause of the rising operating temperature of the battery, which reduces the battery performance [120]. A low operating temperature affects the electrolyte performance, and a high operating temperature causes thermal runway and safety issues. Temperatures of more than 40 °C and less than −10 °C cause capacity losses and performance degradation of the battery. Hence, the thermal management of a battery pack in an EV is a crucial aspect [121]. To ensure the operation at optimal operating temperature, a BTMS should perform crucial tasks such as heat removal from the battery by cooling, increasing heat when the temperature is too low, and providing suitable ventilation for exhaust gases. According to the heat transfer medium, the BTMS can be classified into air, liquid, and phase-change material (PCM) types [122]. The internal temperature estimation of a battery is another important issue that can prevent the battery from aging and explosion risk. The internal temperature estimation can be performed using micro-temperature sensors, EIS measurement, and a lumped-parameter battery thermal model [95,122]. In contrast to battery-based EV applications, fuel cell vehicles (FCVs) have shown huge potential toward decarbonization. They are more efficient than conventional internal combustion engine vehicles and produce no tailpipe emissions since they only emit water vapor and warm air. However, thermal management in FCV should be considered an important research area to be explored [60]. Accordingly, Hu et al. [123] developed an operating temperature tracking control framework to decouple the operating temperature from the complicated driving conditions of the FCV, as shown in Figure 10.

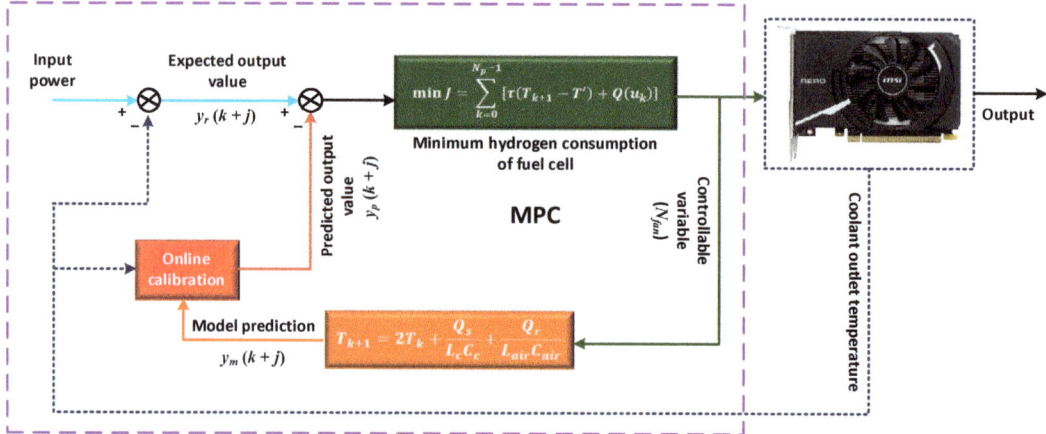

Figure 10. The optimal operating temperature tracking control framework of a fuel cell.

2.2.10. Fault Diagnosis

The unbalance, undercharge, overcharge, overcurrent, and extremely low or high temperatures are some critical issues suffered by battery storage systems [124]. Moreover, other types of faults related to data acquisition, networking, programming, etc. are experienced by BMSs in EV technology. The International Electrotechnical Commission (IEC) developed a BMS standard in 1995 that stipulates that BMSs for EVs must have battery fault diagnosis functions that can provide early warnings of battery aging and risk [64]. Analytical model-based, signal processing, knowledge-based, and data-driven methods are frequently used for fault diagnosis in EV applications [125]. The model-based method detects the faulty parameters with the help of a residual signal that is compared with a threshold to determine the fault. However, the diagnosis results can be affected by measurement and process noise. Time-domain analysis is a key tool to collect the test data for fault analysis in the signal processing-based method. Wavelet transform is a widely used technique in signal processing methods to carry out multiscale fault analysis for battery systems [126]. In addition, machine learning and expert systems are the methods used in a knowledge-based method for fault diagnosis, which can also be utilized to identify the battery lifetime. Moreover, the information entropy, local outlier factor, and correlation coefficient are the key tools to detect faulty data in the data-driven method for fault diagnosis [125,127]. A flowchart describing fault diagnosis of a lithium-ion battery system as proposed by Xiong et al. is shown in Figure 11 [125].

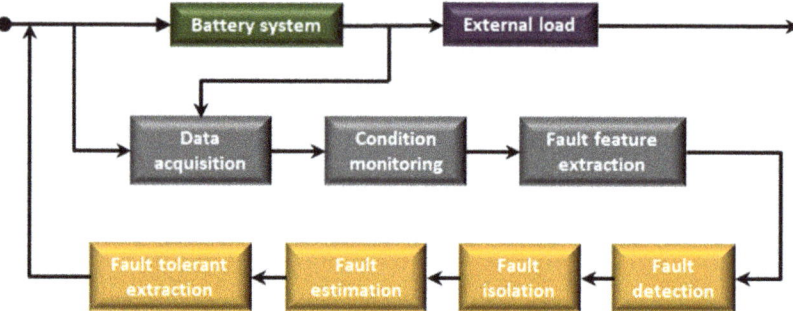

Figure 11. Fault diagnosis process of a battery system.

2.2.11. Communication and Networking

BMS communication with the EV and its external system is essential to protect the battery storage unit. The communication system can be established through wires or data links [128]. A simple BMS consists of a microcontroller unit, debugger, CAN bus, and host computer. For battery status monitoring, a monitoring IC such as AS8505 is used to communicate with the microcontroller through I/O lines via a CAN bus, which controls the cell data and monitors the balancing process [129]. The battery parameters such as voltage, current, and cell temperature are utilized to estimate the SOC, SOH, DOD, etc. in BMSs, which can also be used in IoT-based wireless communication systems with EVs to monitor battery health. The wireless communication technologies that can be employed to monitor battery comprise ZigBee communication, Wi-Fi communication, GSM communication, Bluetooth communication, GPRS communication, and GPS [130,131]. Furthermore, parameter identifier (PID) codes can also be used to collect critical parameters such as voltage, temperature, energy, power, SOC, SOH, DOD, and resistance. Additionally, PIDs use the CAN bus for data processing. Therefore data-driven personalized battery management schemes based on the platform of big data and cloud computing were introduced by Wang et al., as presented in Figure 12 [129]. A summary of BMS components, functions, algorithms, targets, and contributions is presented in Table 1.

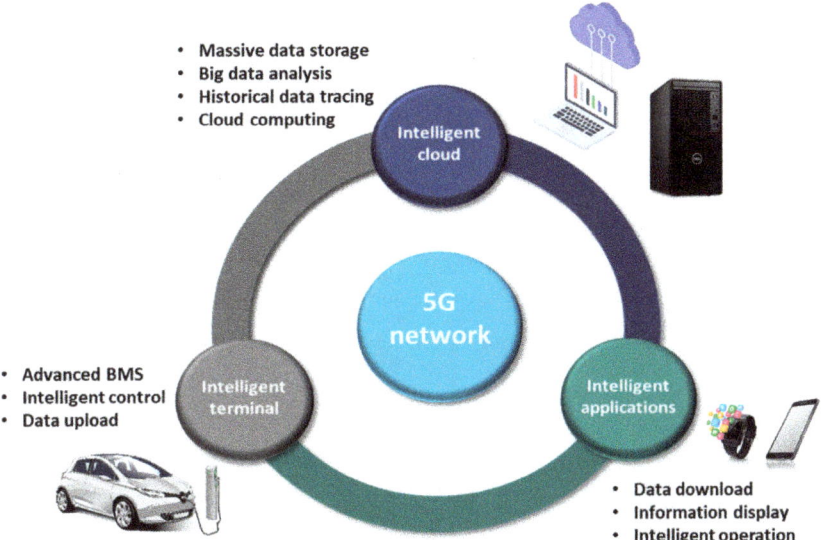

Figure 12. Advanced BMS architecture with 5G in EV technology.

Table 1. The various BMS components, functions, algorithms, targets, and outcomes in EV applications.

BMS Components	Functions	Algorithm/Methods	Target	Outcomes
Monitoring and data acquisition	■ Cell monitoring ■ Voltage and current monitoring ■ Data processing	■ Voltage and current measurement ■ Voltage divider technique ■ CAN bus ■ SCI	■ Current, voltage, and temperature monitoring ■ Communication	■ Better systems function by managing, protecting, balancing, and controlling operations ■ Enhanced EV autonomy ■ Information exchange

Table 1. Cont.

BMS Components	Functions	Algorithm/Methods	Target	Outcomes
State estimation	<ul><li>SOC</li><li>SOH</li><li>SOF</li></ul>	<ul><li>Ampere-hour (Ah)</li><li>Open-circuit voltage methods</li><li>Model-free, model-based, and data mining methods</li><li>Fuzzy logic algorithm</li></ul>	<ul><li>To minimize estimation error</li><li>To reduce the computational cost</li></ul>	<ul><li>Enhanced vehicle performance</li><li>Accurate estimation</li><li>Understanding of maximum capacity</li><li>Reduced capacity degradation</li><li>Identification of internal resistance</li></ul>
Control operation	<ul><li>Charge and discharge control</li><li>Power/energy management control</li><li>Operating temperature control</li></ul>	<ul><li>CC</li><li>CV</li><li>CC-CV</li><li>MCC</li><li>PEM</li></ul>	<ul><li>To enhance performance</li><li>To improve durability</li><li>To increase efficiency</li><li>To provide protection</li><li>To control energy flow</li><li>To ensure safety</li></ul>	<ul><li>Controlled operation in battery charging and discharging</li><li>Improved efficiency and safety</li><li>Increased life cycle</li></ul>
Fault diagnosis and protection	<ul><li>Battery protection</li><li>Unbalance</li><li>Undercharge</li><li>Overcharge</li><li>Overcurrent</li></ul>	<ul><li>Deep discharge protection and overcharge protection</li><li>Analytical model-based methods, signal processing-based methods, knowledge-based methods, and data-driven methods</li></ul>	<ul><li>Protection of battery due to physical and chemical alteration</li><li>Warnings of battery aging</li><li>Warnings of explosion</li></ul>	<ul><li>Improved overall battery performance</li><li>Increased battery lifetime</li><li>Protection from high temperature obtained</li><li>Protection from aging confirmed</li><li>Detection of faulty system</li></ul>
Communication and networking	<ul><li>Monitor and protect the battery</li></ul>	<ul><li>Microcontroller unit, Debugger, and CAN bus</li><li>Wireless</li><li>PIDs</li></ul>	<ul><li>To control the battery data and monitoring</li></ul>	<ul><li>Monitoring of the battery status using wired or wireless approach</li></ul>

3. Key Technological Progress of EVs

This section presents the various technological advancements of EVs concerning power electronics and charging systems in EV applications.

3.1. Power Electronics Technology

Power converter structures need to be dependable and lightweight for automotive applications with minimal electromagnetic interference and low current/voltage ripples to meet the automotive industry standards for high reliability and efficiency [132,133]. A proper interface between energy storage systems (ESSs) and power electronics converters is required for effective EV operation. There are numerous varieties of ESSs that are coupled to different types of power electronic converters in electric vehicles. AC/DC converters are typically used to charge ESSs through charging stations or grids. To accelerate the vehicle, ESSs transmit the necessary energy from a battery to the motor. However, the energy provided by ESSs is unreliable and suffers from significant voltage dropouts. As a result, DC/DC converters are crucial in transforming uncontrolled power flow into controlled/regulated power flow to support various electrical loads and auxiliary power supply in EVs [5]. The layout of the power conversion technique using various power electronics components is shown in Figure 13 [5]. The classification of power electronics technology in EV applications is shown in Figure 14.

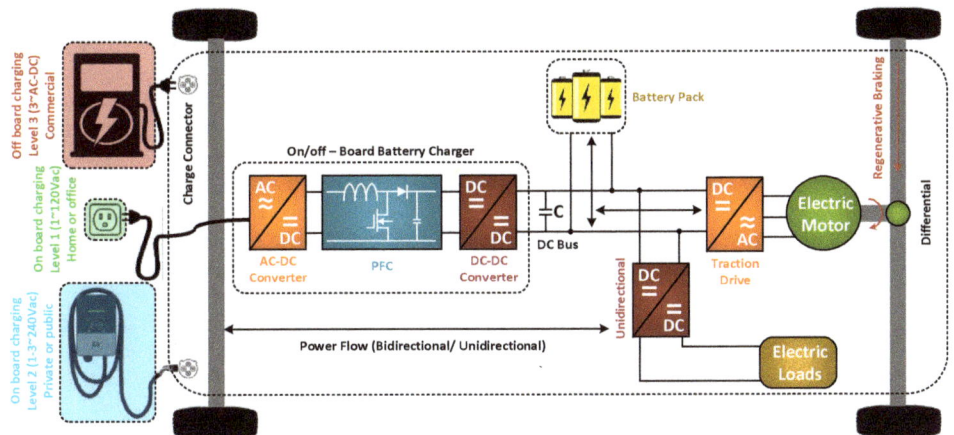

Figure 13. The EV drivetrain with converters connected with a charging system.

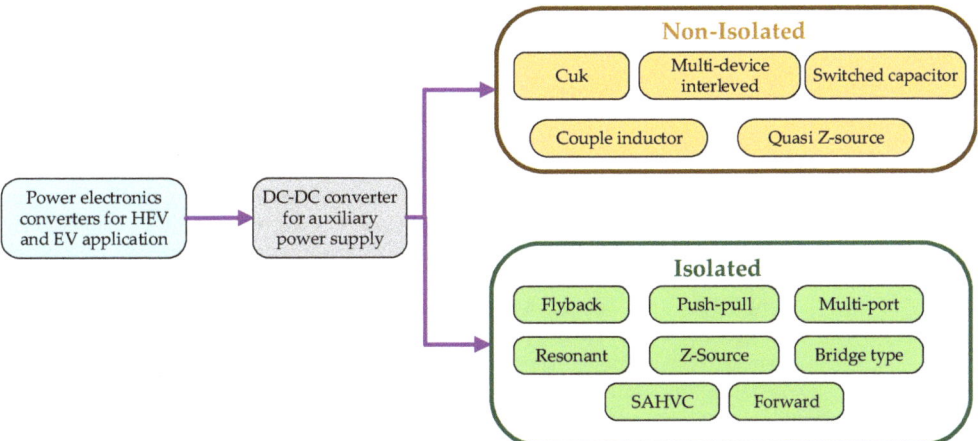

Figure 14. Power electronics converter types in EV applications.

3.1.1. DC/DC Converter: Non-Isolated

When voltage needs to be stepped up or down by a relatively small ratio, non-isolated converters are typically utilized. These types of converters are applicable where the presence of dielectric isolation is not a major issue [134,135]. The mid- and high-range vehicular types are more appropriate for utilizing non-isolated DC/DC converters [136]. A conventional boost DC/DC converter is usually employed where low DC voltage gain is required (<4%). There are five major types of commonly used DC/DC converters: multi-device interleaved, Ćuk, switched capacitor, coupled inductor, and quasi Z-source converters [5,137].

Because of their simple construction, low cost, high efficiency, lower ripples, and easy-to-use control method, multidevice interleaved (MDI) bidirectional DC/DC converters are frequently utilized in BEV and PHEV powertrains. They can maintain the constant magnitude of input current and output voltage ripple without including additional devices such as an inductor or capacitor. Some advantageous features of interleaved converters make them highly suitable for EV applications, such as enhanced heat dispersal, high energy density, reduced current stress, high efficiency, small size filter, and inherent ability to eliminate current ripples [138]. The interleaved topology suggested in [139] is structurally simple and has high modularity, resulting in reduced current stress on the switches and enhanced heat distribution. However, the current configuration of this converter can only be used for a low-power EV that can carry a maximum of two passengers. For high-power EVs, the structure needs to be extended using supercapacitors, making it structurally complex, bulky, and expensive. A two-phase bidirectional interleaved converter was proposed in [140] for EVs. This converter can be operated in both buck and boost mode with fast and low overshoot switching performance. A major drawback of this converter is that its operation is highly reliant on the switching control technique. In order to achieve the optimized performance from this converter, a highly complex control technique named optimal Bézier curve is required. The dynamic response of this converter with direct switching control is poor and may not be suitable for EVs. A two-phase hybrid mode interleaved converter was developed by [141] for EV fuel cells. This converter has the ability to simultaneously operate in continuous conduction mode and discontinuous conduction mode. With regard to load conditions and duty cycles, the boundaries between the two conducting modes are distinguished. Although this converter showed high-performance efficiency with low output voltage and current ripple, the dynamic performance of this converter has not been verified. Since the modes are distinguished on the basis of load conditions, transient conditions can greatly hamper the performance of this converter.

The Ćuk converter (CC) delivers flexibility toward regulating the output power as compared with input power. The Ćuk converter is developed by utilizing a single magnetic core and delivers important features such as low ripples/harmonics and high efficiency. The Ćuk converter also delivers high-performance efficiency compared with the DC/DC boost converter by controlling the current ripple through the L−C filter. A modified Ćuk converter for Toyota Prius was proposed in [142] with a basic proportional–integral (PI) controller for tuning and filtering. Since the conventional PI controller has various performance deficiencies such as sluggish transient response, high overshoot, manual tuning, and poor filtration, the real-time execution of this converter is questionable. Furthermore, a bridgeless modified Ćuk converter was developed in [143] to improve the power quality of the EV charger. The proposed converter's operation was verified under transient voltage conditions, and it successfully followed the IEC 6100-3-2 standard for reduced current harmonics. The only downside of this proposed topology is that it utilizes an additional flyback converter for current harmonic reduction, which is connected through a transformer. Thus, the implementation of this converter will be significantly expensive, and it will also have increased weight.

Additionally, In EVs, the switched-capacitor bidirectional converter (SCBC) uses synchronous rectification to conduct turn-on and turn-off actions. By utilizing switched capacitors, the SCBC can deliver stable voltage and current without magnetic coupling. Additionally, with the employment of power switches in SCBC, the application of additional components is minimized to improve the power conversion efficiency [144]. Some of the recently proposed SCBCs for EV applications can be found in [144–146]. Nonetheless, the SCBC suffers from various factors such as high harmonics and low efficiency at widespread input to output voltages. Zhang et al. [144] designed a switched capacitor-based converter for EV applications without the magnetic coupling that can deliver continuous inductor current and a stable switched-capacitor voltage through the switched capacitors. The performance of the proposed configuration was investigated using a 300 W prototype considering a wide voltage-gain range and variable low-voltage side (40 V to 100 V). The results demonstrated a maximum efficiency of the converter of 94.39% in step-up mode and 94.45% in step-down mode. Zhang et al. [147] developed a hybrid bidirectional DC/DC converter with switched capacitor-based converter for hybrid energy source-based EVs.

According to the experimental outcome obtained from a 400 W prototype, the authors validated the characteristics and theoretical analysis of the proposed converter.

On the other hand, the coupled inductor bidirectional converter (CIBC) demonstrates improved performance efficiency due to high voltage gain and low voltage stress compared to SCBC [148]. A reduced-component CIBC was proposed in [149] for EV charging applications. This topology can provide a wide range of voltage conversion by operating in both buck and boost mode. It also showed a high performance efficiency greater than 95%. Another CIBC for EVs was proposed in [150], and the operation was verified under transient conditions using simulation and experimental results. However, both of these CIBC topologies have inherent weaknesses due to the application of leakage inductance, resulting in resonance and voltage spikes [151]. This drawback can cause serious consequences in EV applications; therefore, the implementation of CIBC for EVs is still limited in the industry. Wu et al. [152] developed a couple inductor-based converter for EVs to enhance the voltage gain and decrease the switching voltage stress. An experimental model of the converter rated as 1 kW, 40–60 V to 400 V, was designed to validate its performance.

Lastly, due to several significant properties such as simple design, common ground, and wide range of voltage gain, the quasi Z-source bidirectional converter (QZBC) is employed in EV technology. Commonly, the EV application employs a conventional two-level QZBC topology. The QZBC replaces the conventional Z-source DC/DC boost converter by enhancing the output voltage gain, which is suitable for high step-up voltage conversion [153]. However, the employment of QZBC results in various drawbacks such as uneven input current and capacitance of high-voltage stress [135].

3.1.2. DC/DC Converter: Isolated

Isolated converters are essential where the output is completely separated from the input. In low- and medium-power vehicle applications, isolated DC/DC converters are commonly utilized [5]. Some of the important isolated DC/DC converters employed in BEV and PHEV applications are flyback, push–pull, multiport, resonant, zero-voltage switching (ZVS), dual-active bridge full bridge, ZVS full-bridge, and forward converters.

When a buck–boost converter splits an inductor into transformers, the result is a flyback converter (FBC). This is an isolated DC/DC converter that stores energy during the on state and transfers it to the off state. The application of FBCs can be carried out in low-power applications due to their various characteristics, such as low cost, high output voltage, and electrical isolation [154]. The constructional features of FBCs can obtain high gain and reduce the output current ripple and the leakage inductance [155]. A boosting multioutput FBC was proposed in [156] for EV application. This topology consists of three separate FBCs to provide multioutput voltage. Although this topology has high voltage gain and can be applied in high-power EV because of a parallel connection, it requires a transformer winding technique to decrease leakage inductance, which can drastically increase the cost and weight of this topology. A multiphase bidirectional FBC was developed for hybrid EVs in [157]. Due to its modularity, it is suitable for high-power applications while maintaining structural minimization and features such as high voltage gain, accurate operation during load fluctuations, decreased current ripple, and parallel-battery energy capacity. However, similar to the FBC in [156], it also utilizes a transformer, making it highly expensive and overweight.

The working principle of the push–pull converter (PPC) is based on the transformer operation, which transforms power from primary to secondary. A rectifier diode, bypass capacitor, power switches, and transformer circuit are the basic circuit components of the PPC configuration. The PPC demonstrates simple topology with high efficiency and results in low conduction loss due to low peak current. However, careful attention is required while operating the PPC due to the formation of a low impedance path and high current [158]. Some notable PPC topologies developed particularly for EV applications can be found in [159].

The multiport isolated converter (MPIC) performs the operation while considering several input sources and offers galvanic isolation. The performance efficiency and functionality are improved by feeding back the recovered power obtained during regenerative braking to the input sources [160,161]. A highly energy-efficient T-type MPIC was proposed in [162] for EVs. In order to handle multiple energy generation/storage units, the suggested converter unit has multiple input sources. Because the unit requires fewer switching components, the cost of the power electronics interface for EV implementation is greatly reduced. Although this MPIC has shown promising performance, it requires a complex multipurpose algorithm for accurate energy management in different modes of operation. Furthermore, another novel MPIC with the inherent ability to control multidirectional power flow was suggested in [163]. Unlike [162], this topology offers galvanic isolation by using a common magnetic link among the multiple input sources. It can be stated that MPIC topologies comprise several advantageous features compared to other converter topologies, especially for EV applications. Nonetheless, they are still in the early phase of development for EVs, and further research is required to optimize their cost and weight since they utilize transformers [136].

A DC–DC resonant converter (RC) is made of a resonant tank constructed with a combination of inductors and capacitors. The RC exhibits several benefits, including low switching loss, zero circulating currents, zero-voltage switching, and high efficiency. These features can be essential for EV applications, as demonstrated in [164,165]. However, RC exhibits various limitations in terms of transformer design complexity and high magnetizing current [165], which requires further improvement.

The zero-voltage switching converter (ZVSC) was designed on the basis of the dual half-bridge topology placed on both sides of the main transformer. Due to various strengths such as less circuitry topology, easy control, soft switching, and higher efficiency, the ZVSC is regarded as highly suitable for EV technology [166]. The topology of ZVSC can be adopted for both BEV and PHEV powertrains even though it has a power limitation >10 kW for automobiles [167,168]. The experimental verification of a 53.2 V, 2 kWh low-voltage and high-current lithium-ion battery energy storage system based on a 6 kW single-phase dual-active bridge (full-bridge) achieved efficiency as high as 96.9% [169]. A three-phase dual-active bridge with phase-shift modulation and burst mode switching was evaluated for battery energy storage systems to achieve high power density, high efficiency, and galvanic isolation [170].

The full-bridge boost DC/DC converter (FBC) is the most convenient converter topology that diminishes the voltage and current stresses on diodes and switches. The FBC operates in three stages: initially an inverter (DC/AC), then a high-frequency transformer (HFT), and finally a rectifier (AC/DC). This type of converter contributes higher step-up voltage due to HFT and galvanic isolation between input and load. An improved FBC was developed in [171] for efficient power conversion and distribution in EV charging. It has other valuable characteristics such as the minimized size of the EV charger and switching loss, faster operation, and economical performance. Moreover, another FBC was suggested in [172] with the phase-shift switching control technique. Even though FBCs have some effective characteristics for EV applications, they have a major performance deficiency; their maximum achievable efficiency is only around 91.5% [173].

Lastly, the forward converter (FC) works on the forward balancing technique, which has a fast balancing time and is easy to control. It consists of one magnetic core with one primary winding and multiple secondaries based on the desired application. The energy is transferred to the secondary when the switch is turned on. A few forward converter topologies have recently been developed for EV applications [174–176]. A detailed comparative study of different power electronics converters is shown in Table 2.

Table 2. Comparison of various power electronic converters in EV applications.

Converter Type	Converter Topologies	Strength	Weakness	Objectives	Outcomes	Refs.
Non-isolated	MDI	- Efficiency up to 97% - Low current stress - Reduced component size	- Complicated analysis during transient and steady state	Multiple input to a single output	- Enhancement of efficiency - Reduction in additional components	[138]
	CC	- Continuous input and output currents - Power factor improvement.	- Uncontrolled and undamped resonance	Reduction in energy loss	- Ripple-free constant output	[143]
	SCBC	- Improved power conversion efficiency - Cost-effective - Compact design	- High ripple current	High voltage gain and efficiency	- Efficiency greater than 90% - Stable voltage and current	[177]
	CIBC	- Operational flexibility - Small in size	- No consideration for voltage ripples	To reduce output current and inductor current ripples	- Increased efficiency by increasing the coupling coefficient	[178]
	QZBC	- Lower switch stress - Bidirectional operation - Enhanced output voltage gain	- Discontinuous input current	Maximum and minimum efficiency of 96.44% and 88.17%, respectively	- To obtain high voltage gain for step-up conversion	[153]
Isolated	FBC	- Applicable to higher-load-voltage situations - Ability to regulate multiple output voltages	- Ripple current	Attains lower leakage inductance to an acceptable limit	- To enable support of a wide input voltage range	[156]
	PPC	- Less filtering is required - Low conduction losses	- Protection required during switching	To change the voltage of the DC power supply	- Achieves low current and voltage on the primary side	[158]
	MPIC	- Low output voltage ripple current - Galvanic isolation bidirectional power flow	- High sensitivity corresponding to duty cycle under load changes - Difficult to achieve proper synchronization	To minimize the overall system losses	- Independent control of power flow	[163]
	RC	- High conversion ratio - High efficiency	- Complex integrated transformer	To minimize magnetic components and passive filters	- Low switching loss	[164]
	ZVSC	- Low EMI factor - Soft switching - Increased power density	- Poor fault-tolerant capability. - High gate current rating	To clamp the output diode bridge voltage	- Achieves zero-voltage switching under all load conditions	[167]
	Sinusoidal amplitude high-voltage bus converter (SAHVC)	- Flat output impedance up to 1MHz - Ensures noise-free operation	- Complex gate switching pattern - Not suitable for high-power conversion	Lowers voltage stress on the switching circuit	- High voltage bus conversion	[134]

Table 2. Cont.

Converter Type	Converter Topologies	Strength	Weakness	Objectives	Outcomes	Refs.
Isolated	Single-phase and three-phase DAB	- High power density - Bidirectional power transfer - Zero-voltage switching - Low voltage and current stresses	- Circulating current in the high frequency-transformer (three-phase) - Optimum efficiency is achieved only when the ratio of the DC-link voltages is equal to 1 using the phase-shift modulation switching method	-Galvanic isolation -Voltage matching	- High efficiency for low-voltage and high-current applications - Low DC voltage ripple	[169,170]
	FC	- Fast balancing time - Easy to control - High efficiency	- Nonuniform voltage of the secondaries	Voltage equalizing	- Reduces use of an inductor - The number of components is decreased	[174–176]

3.2. EV Charging Technology

EV charging is a major barrier to sustainable adoption in the global market. Charging entails injecting a suitable amount of electrical power from the grid into the battery. The length of time it takes to charge a battery is determined by the battery's capacity and the charger's power level. Three methods are frequently utilized for charging the battery of an electric vehicle (EV), i.e., conductive charging, inductive charging, and battery swapping [179,180]. Figure 15 shows the typical architecture of an electric vehicle charging system, where both the on-board and the off-board one are represented [181].

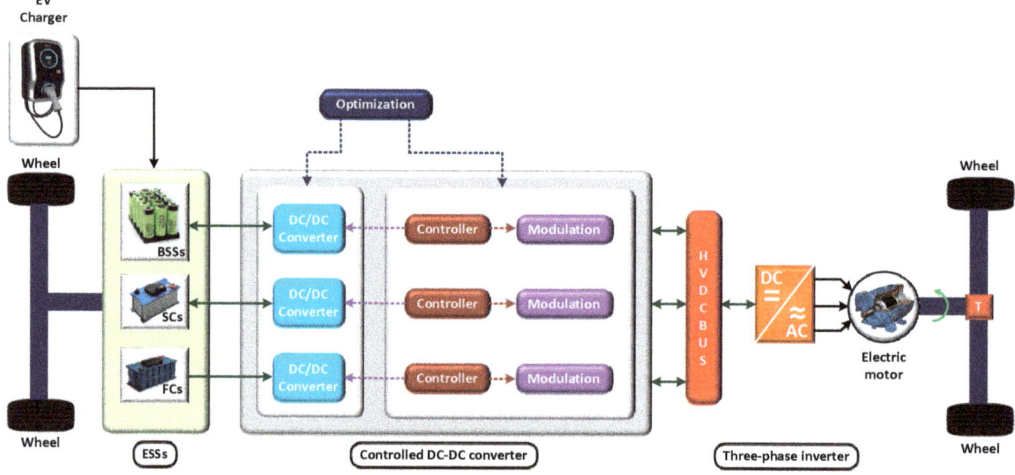

Figure 15. Charging system configuration for EV applications.

3.2.1. Conductive Charging

Conductive charging involves EV charging by connecting to the grid through a wire, allowing for a direct connection between the charger and the vehicle. This charging method comprises a rectifier (AC/DC) and converter (DC/DC) with power factor correction (PFC), which is categorized as an onboard and off-board charger. The construction of the rectifier and the DC/DC converter initially determines the topology of an on-board and off-board conductive charger [180,182]. The on-board charger is placed inside the EV, which is frequently utilized for slow charging. However, a fixed location is mandatory for an off-board charger, which is applicable for quick charging. The Nissan Leaf, Tesla Roadster, and Chevy Volt all are suitable EVs having conductive charging [183]. Figure 16 shows the conductive on-board and off-board charging infrastructures, as proposed by Khalid et al. [184].

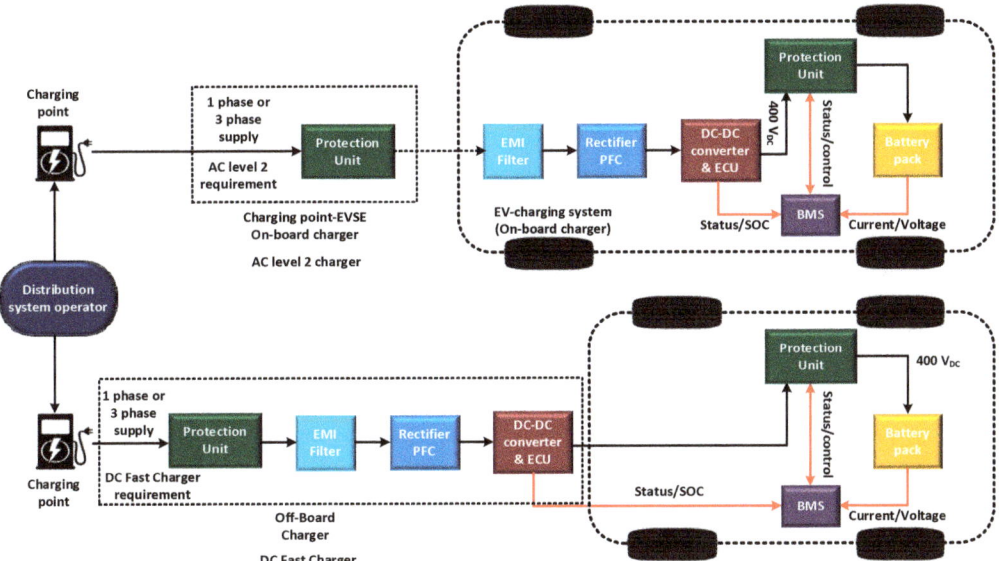

Figure 16. The on-board and off-board conductive charging infrastructures.

When it comes to charging electric vehicles, several standards are used. These requirements are mostly determined by the location in which the EV technology is embraced and employed. For instance, the charging of EVs in North America and the Pacific is based on SAE-J1772 specifications. Furthermore, the charging of EVs in China is based on GB/t 20234 standards, whereas the charging of EVs in Europe is based on IEC-62196 standards. The standards for North America, the Pacific, and China depend on the application of charging modes. On the other hand, the European standards are solely divided into categories on the basis of charging power, i.e., AC or DC.

The North American standard was developed in 1996 for electric connections of EVs, promoted by SAE International. The various charging mode standards and their implementation in several regions (the USA, Japan, Europe, and China) are depicted in Table 3 [185].

According to the survey, the North American SAE-J1772 standard is only compatible with the 120 V recharge mode, in contrast to the IEC-62196 and GB/T-20234 standards, which can operate at a greater voltage even in their lowest charging modes.

Moreover, the GB/T-20234 standard has a lower current intensity (10 A) than the other two standards, which have a current intensity of 16 A. However, the SAE-J1772 only supports a maximum intensity of 200 A in its most powerful modes, compared to 400 A for the IEC-62196 and 250 A for the GB/T-20234. In addition, the North American SAE-J1772 standard provides a reduced power of 1.9 kWh in comparison to the 2.5 kWh of the GB/T-20234 and the 3.8 kWh based on the AC power source. On the other hand, the IEC-62196 standard offers the power of 120 kW at 480 V AC which is much higher than the other two standards.

Table 3. Standard charging power ratings of various conductive Charging standards for EV applications.

	Charge Method	Volts	Maximum Current (Amps—Continuous)	Maximum Power
SAE-J1772	AC level 1	120 V AC	16 A	1.9 kW
	AC level 2	240 V AC	80 A	19.2 kW
	DC level 1	200–500 V DC maximum	80 A	40 kW
	DC level 2	200–500 V DC maximum	200 A	100 kW
IEC-62196	Single-phase	230–240 V AC	16 A	3.8 kW
	Three-phase	480 V AC	16 A	7.6 kW
	Single-phase	230–240 V AC	32 A	7.6 kW
	Three-phase	480 V AC	32 A	15.3 kW
	Single-phase	230–240 V AC	32–250 A	60 kW
	Three-phase	480 V AC	32–250 A	120 kW
GB/T-20234.2	AC charging	250 V and 440 V	10–63 A	27.7 kW
GB/T-20234.3	DC charging	750–1000 V	80–250 A	250 kW

For charging EVs in China, the Guobiao (GB) GB/T-20234 standard was adopted and promoted. This standard categorizes the charging modes between AC and DC. A detailed comparative analysis of various conductive charging standards for EV operation is presented in Tables 3 and 4.

Moreover, the International Electrotechnical Commission (IEC) established the IEC-62196 standard in 2001 as a global standard for charging an electric vehicle in Europe and China. The general guideline for the charging process and energy transferred pattern was introduced by the IEC-62196 standard, which was deduced from the IEC-61851 standard. The IEC-61851 administers a first classification of the type of charging based on its nominal power and the recharging time [186,187]. The EV users are offered four modes of charging the vehicle, as mentioned below. The different charging modes for EV operation are shown in Figure 17 [186].

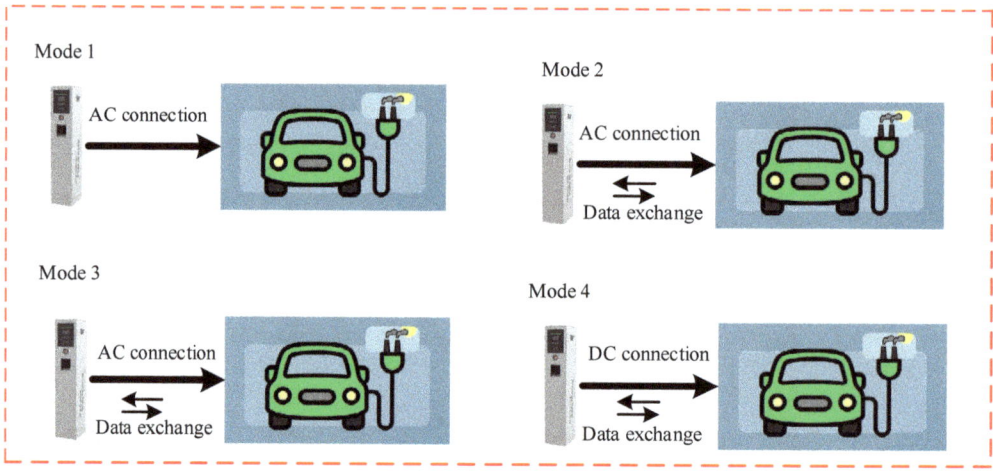

Figure 17. The various charging modes for EV applications.

Table 4. Comparative study of various conductive charging standards for EV operation.

Charging Standard	Country	Mode of Operation/ Classification		Features	Advantages	Disadvantages	Ref.
SAE-J1772	USA and Japan	AC level (single-phase)		Provides a physical connection Utilized at home, workplace, and public charging facilities	High output voltage regulation and high slew rate	Charging rate is limited by battery chemistry, infrastructure	[188]
		DC level					
IEC-62196	Europe and China	Mode-1	Single-phase	Only for domestic (household) use	The range of charging is high, i.e., recharge from 3 and 43 kW, and can support single phase up to 16 A and three phases up to 63 A	Can only be used with three-phase supply due to its specific design	[22]
			Three-phase				
		Mode-2	Single-phase	Overcurrent protection Over-temperature protection			
			Three-phase				
		Mode-3	Single-phase	Useable in public places or at home Utilizes EVSE			
			Three-phase				
		Mode-4	DC	The charger is part of the charging station, not part of the vehicle Utilizes an off-board charger			
GB/T-20234.2	China	AC charging		Conductive charging	Fast charging	-	[189]
GB/T-20234.3		DC charging					

- Mode-1 (slow charging). This mode is designed for domestic use purposes, frequently used in client houses. It provides the maximum current intensity of 16 A with a single-phase or three-phase power outlet facility, including neutral and earth conductors.
- Mode-2 (semi-fast charging). A similar charging approach is implemented in this mode with a slight modification in current intensity and user facility. This mode can handle the current intensity of a maximum of 32 A, and it also allows users to utilize the charging in public places.
- Mode-3 (fast charging). This mode contributes to a fast charging process with the help of current intensity from 32 A to 250 A. This model also adopts the specific power supply known as EV supply equipment (EVSE), which is utilized for recharging electric vehicles. This EVSE device accommodates a communication system that provides a communication advantage with the vehicles. Additionally, a control system to regulate energy flow, a monitoring system to observe the charging process, and a protection system are incorporated for protection with the EVSE.
- Mode-4 (ultrafast charging). According to the latest IEC-62196-3 standard, this model has a maximum charging power capacity of up to 400 kW. This standard also defines a direct connection between the EV and the DC supply network, having a maximum voltage of 1000 V and a current intensity of up to 400 A. An external charger is required in this mode, which provides protection, control, and communication between the vehicle and the recharging point [182].

3.2.2. Wireless Charging

Wireless power transfer (WPT) has been around for over two centuries. Nikola Tesla conducted tests at Colorado Springs, USA, in 1899 to see if electrical energy could be transmitted without wires. Wireless charging technology involves transferring electricity from one medium to another without the use of a contact medium. Electromagnetic radiation, electric coupling, and magnetic coupling are the three primary types of WPT systems. Moheamed et al. [190] classified the available WPT technologies into three categories.

Figure 18 shows a classification diagram for the different wireless power transmission technologies [190].

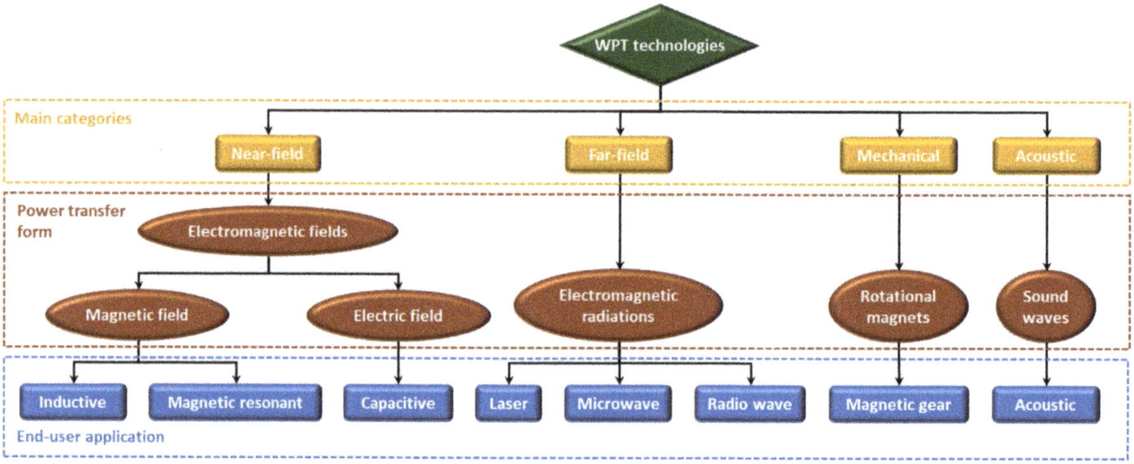

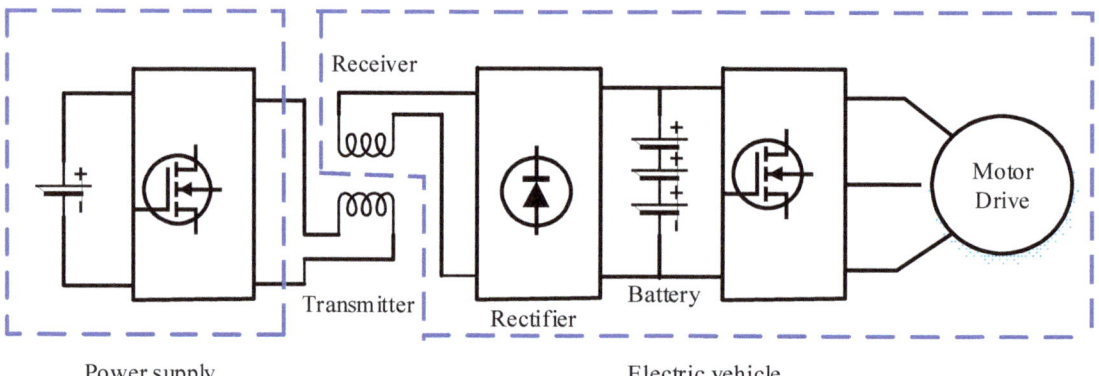

Figure 18. The categories of WPT technologies for EV applications.

WPT works in three stages: initial conversion of power supply, then resonance between coil to transfer power, and final charging of the battery. An input AC power supply is converted to high-frequency AC at the first stage. This high-frequency AC is utilized to generate an alternating magnetic field at the transmitter side (primary); as a result, AC voltage is induced at the receiver (secondary side coil). Finally, the AC voltage at the receiver is converted to DC to charge the battery. A magnetic resonant coupling and DC/DC converter can be incorporated at the secondary to improve the performance of the system. The converter system provides an efficiency of 90% under the frequency variation of 20 to 100 kHz. Figure 19 illustrates an outline of WPT for EV technology [180].

Figure 19. The layout of the WPT system in EV applications.

WPT is also a convenient source of charging because of its flexibility and comfort. Currently, there exist two wireless recharge modes, namely, capacitive power transfer (CPT) and inductive power transfer (IPT). However, IPT is the most often utilized since it can be applied to a wide range of gap lengths and power levels. In contrast, CPT, despite showing promising results with high power levels in terms of kilowatt-power-level applications, is only suitable for small gap power transfers.

An IPT system is electrically separated, and there is minimal wear and tear on mechanical components because no physical touch is necessary. The design of the magnetic structure is crucial in the IPT system for EV charging due to high-power applications. The magnetic coupling between the primary and secondary pads determines the power transfer capabilities of an IPT system, which is determined by the geometry, size, materials, and relative location of the magnetic couplers [191]. Recently, a 30 kW bus online electric vehicle (OLEV) IPT system was used at a bus stop, maintaining a charging height of 170 mm with an efficiency of 80% [192].

Moreover, the CPT technology is based on the notion of a capacitor's functioning. An air gap (d) between the conducting plates of a capacitor is generally filled with a dielectric substance for insulation. The direction of the electric field is reversed every half-cycle in an AC excitation, and the charge and discharge are alternately repeated. According to this method, the capacitor is thought to be carrying an AC. Power transmission via a metal barrier, system simplicity, minimal eddy current loss, and less electromagnetic interference (EMI) are all advantages of CPT technology [193,194].

Furthermore, depending on the situation, there are three different types of wireless recharges: (a) stationary charging, where the vehicle remains stationary or static during charging. For acceptable misalignment, the owner may just park the car in a location and leave it for charging with a set range, (b) opportunity charging, which occurs when the vehicle is stopped for a short period of time, and (c) dynamic charging, which occurs when the vehicle is moving along a dedicated charging lane. Utilizing this method, the charging of public transport (buses and taxis) is possible at the stops when passengers board and alight [180,195,196].

3.3. Battery Swapping

The battery swapping approach is one of the most time-efficient and hassle-free charging methods. In this method, the EV replaces the drained battery with a completely charged battery at a battery swapping station (BSS). Then, the BSS transfers the empty battery to the battery charging station (BCS) to recharge it. After the complete charge, the BCS transfers it back to the BSS for exchange in EVs. To complete the BSS process, a distribution transformer, AC/DC converters, battery chargers, vehicle batteries, robotic arms, charging racks, a maintenance system, a control system, and other types of equipment are required. One major advantage is that the battery swapping stations may execute bulk bidirectional power transfer with the grid. During peak demand, the fully charged batteries can inject electricity into the grid, while charging occurs during off-peak hours [179]. The battery swapping method is depicted in Figure 20 [197].

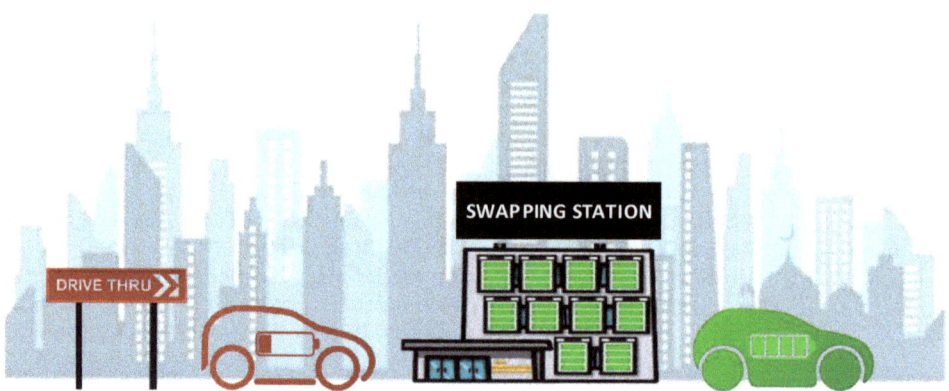

Figure 20. Battery swapping station layout for EV applications.

4. Intelligent Control Schemes, Optimization Algorithms, and Methods in EVs

4.1. EV Control Strategies

Improvements in fuel economy and carbon emission reduction can be achieved with an appropriate power split between the energy sources. Therefore, various control strategies have been applied, and their performance has been analyzed toward the achievement of the abovementioned objectives. Furthermore, the EV performance and battery state of charge are not compromised during the implementation of control strategies. Primarily, the control strategies employed in EV technology can be divided into two categories, i.e., offline control and online control strategies [198,199].

4.1.1. Offline Control Strategies

The offline control strategies in EV applications present incompetence toward delivering outcomes in the real-time world; nonetheless, their application is utilized to validate the performance of real-time controllers. In recent times, techniques such as linear programming (LP), dynamic programming (DP), genetic algorithm (GA), stochastic control (SC) strategy, and particle swarm optimization (PSO) have been employed as offline control techniques in EV applications.

- Linear programming (LP): A nonlinear fuel consumption model of HEV for a globally optimal solution can be estimated and resolved by linear programming. Convex optimization and linear matrix inequality techniques are used in LP to analyze the propulsion capabilities and minimize fuel consumption [200].
- Dynamic programming: The dynamic programming (DP) technique aims to figure out the optimal control policies based on multistage decision making without depending on the previous decision. The backward recursive method and the dynamic forward method are the common DP algorithms, as introduced by Bellman [201].
- Stochastic control (SC) strategy: The SC control technique is implemented to solve the optimization issues related to uncertainties. The formulation of the infinite-horizon stochastic dynamic optimization issue is conducted using this technique. Furthermore, the SC strategy delivers optimal control outcomes while considering diverse driving patterns. Liu et al. developed a hybrid power optimal control strategy by utilizing stochastic dynamic programming (SDP) to analyze the effects of harmonics on emissions from the engine. Additionally, Tate et al. developed two variants of SC strategy for parallel HEV application to analyze fuel consumption and tailpipe emissions. A two-stage stochastic programming method was proposed by Zeynali et al. [202] for a home energy management system including battery energy storage and EVs, as shown in Figure 21.

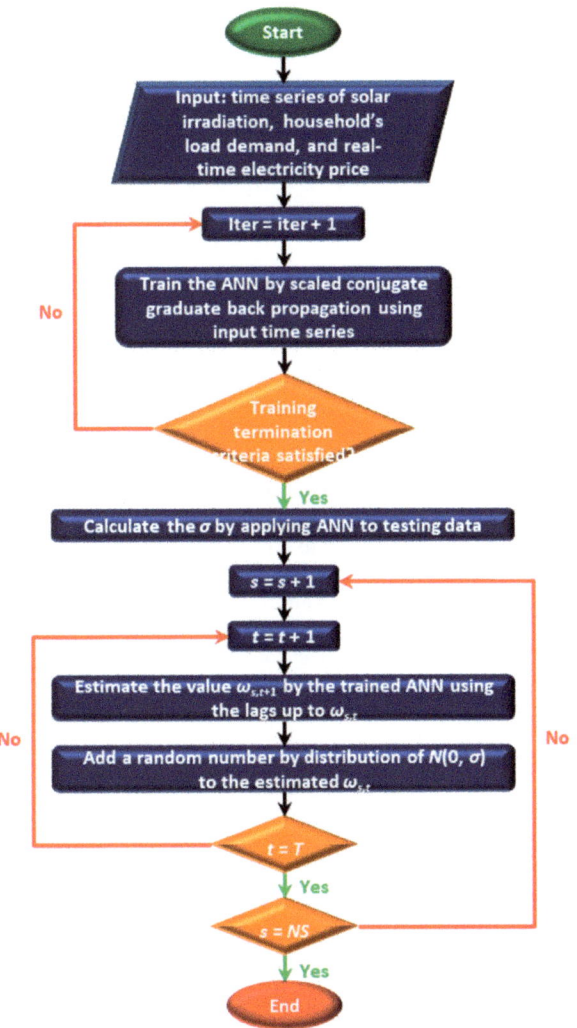

Figure 21. The flowchart diagram of the proposed scenario generation algorithm.

4.1.2. Online Control Strategies

Rule-based (RB) control strategies are implemented according to a set of predefined rules which rely on some intuition, heuristics, or the human experience without a priori knowledge of a predetermined driving cycle. The control strategies can be simply expressed as an if–then type of control rule, which determines the shutdown of the engine or the power discharge at a specific point in time. Alsharif et al. [203] developed a simple rule-based control strategy that can control the power flow from grid to EV and EV to the grid. The control objectives were formulated to minimize two objective functions, including loss of power supply probability (LPSP) and cost of electricity (COE). Several data inputs were considered to perform the operation, such as mean wind speed (m/s), mean solar irradiance (W/m^2), max solar irradiance (W/m^2), mean ambient temperature (°C), mean energy demand (kW), peak energy demand (kW), and min energy demand (kW). Even though the RB energy management system (EMS) is simple and can be implemented in real time on vehicle engines, it has some drawbacks. The first is that it lacks optimality while needing prior knowledge of the driving cycle. Furthermore, a substantial amount of calibration work is necessary to ensure that the performance is within a reasonable range for each driving cycle. Deterministic and fuzzy logic EMSs represent rule-based approaches [204,205]. The rules may be drawn from experience in a deterministic RB-EMS, in which the major energy sources are regulated to function primarily within ideal operating circumstances. This approach works in a high-efficiency region to improve fuel economy and decrease energy transmission loss. Frequency-decoupling control is another deterministic rule for power splitting, in which low-frequency power is provided by energy sources with slow dynamics. However, peak and/or high-frequency power is provided by energy sources with rapid dynamics.

The fuzzy logic (FL) approach translates human thinking and experience into a set of if–then rules. Input quantization, fuzziness, fuzzy reasoning, inverse fuzziness, and output quantization are the five steps of this FL conversion process. This route provides the advantage of wholesomeness and easy tuning, which facilitates the independent adaptation of the control strategy. The FL also offers control of efficient engine operation and coordinates the parallel HEV subsystems. Optimized fuzzy rules control, adaptive fuzzy logic control, and predictive fuzzy logic control are the types of FL control strategies. An optimization method is followed by optimized fuzzy rules control to meet the target of reduction in fuel consumption, minimization of emission, improvement of driving performance, and maintenance of the SoC. The adaptive FL strategy works on an adaptive algorithm to enhance self-adaptation so that the HEV powertrain can accommodate the unknown tire dynamics, changing road surface, and vehicle loading. Moreover, the predictive FL control strategy is aimed at understanding the future states of vehicles and performing real-time control tasks. A fuzzy logic-based EMS controller was proposed by Mohd Sabri et al., as shown in Figure 22 [206]. The controller mechanism appropriately distributes the power via the hybrid train while achieving the minimum fuel consumption as its objective. The inputs to the controller are current vehicle speed, ICE speed demand, current ICE speed, current SOC of ESS, and total trip distance.

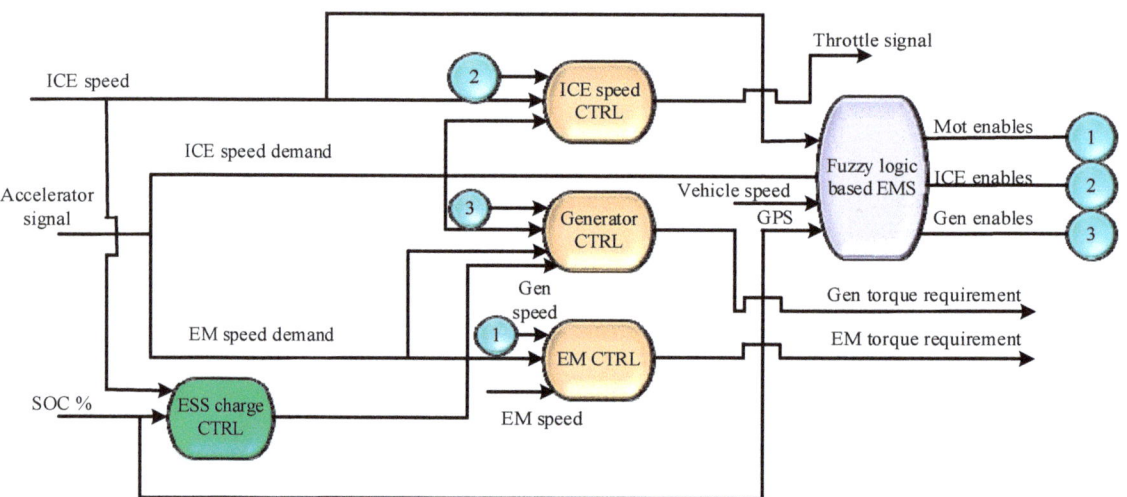

Figure 22. Fuzzy logic-based EMS controller block in EV applications.

4.2. EV Optimization Strategies

The neural network (NN) method works on the basis of periodicity and predictable operation with an optimal charging strategy that helps to estimate the energy demand and optimize the charging cost of EV. Compared to NN, the nonoptimal strategy is expensive and not suitable for the health of the battery. The NN can optimize the charging price, as well as estimate the daily energy demand, with the help of the energy predictor model named nonlinear autoregressive network with exogenous inputs (NARX) as a function of the meteorological data and previous energy data used [207]. In this case, the inputs to the NN model comprise SOC (0 . . . t), temperature, and charging time slots, while the output of the model is SOC (t + 1). Two different constraints, SOC and temperature, are applied to minimize the charging costs of EVs.

Pontryagin's minimal principle (PMP) is the most well-known technique for solving the optimal control problem, which is widely used in adaptive forms to develop real-time optimization-based EMS. Delprat et al. [208] presented a PMP application for optimizing the EMS of a parallel HEV. The constraints addressed in this work were the limitations of the motor, the engine, and the battery while fulfilling the objective toward minimizing fuel consumption during the driving interval. The global optimization issue represented by DP has been converted into an instantaneous Hamiltonian optimization problem through PMP, which was developed using a variational approach [209]. German et al. [63] proposed a new approach of PMP to develop real-time EMS in EVs, in which no additional adaptation of the co-state variable is required for real-time applications. The objective function was formulated to reduce battery degradation by considering the objective constraint as the set of system dynamical models.

The simulated annealing (SA) algorithm is utilized in derivative-free algorithms for EMS control. Kirkpatrick [210] invented the SA in 1983 on the basis of the metal annealing process. The SA method uses a stochastic technique to find a solution, where the solution candidates are selected, and the improvements are assessed on the basis of the objective function. The SA was utilized by Chen et al. [211] to discover the best engine-on power and maximum current coefficient, while the PMP was used to determine the battery current commands. The inputs to the model were battery power/current and SOC. The objective of the proposed study was to minimize fuel consumption by satisfying the constraints of driving power and battery SOC. Trovao et al. [212] also used the SA to find the best energy distribution between the battery and the SC for short-term power management. An SA algorithm-based control scheme was proposed by Song et al., as shown in Figure 23 [213].

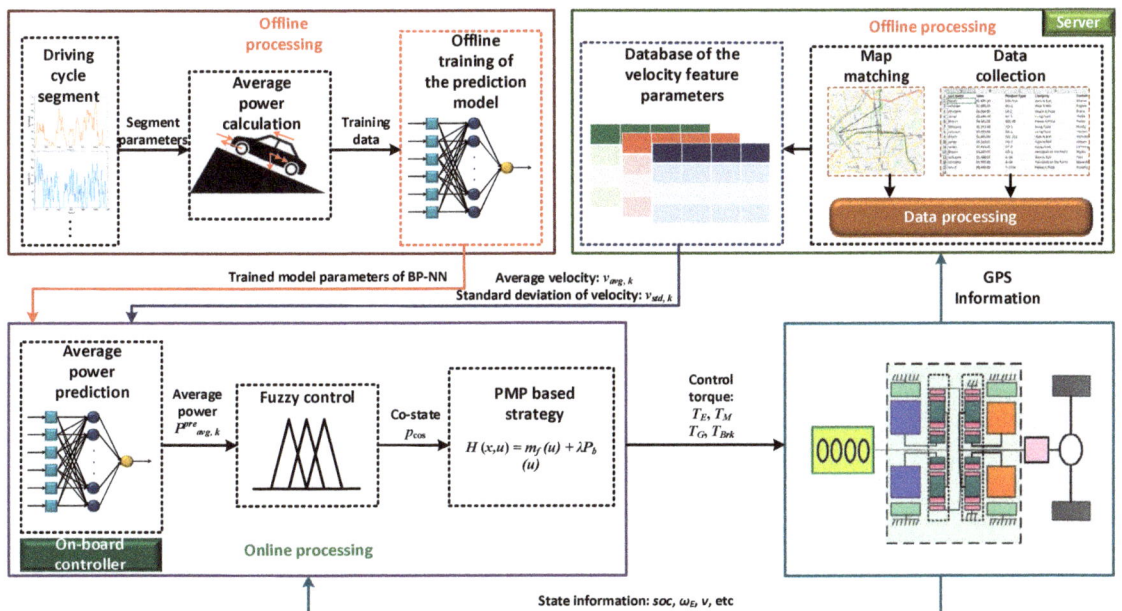

Figure 23. The control scheme of the proposed strategy in EV applications.

The genetic algorithm (GA) is another stochastic search approach, influenced by natural selection and genetic evolution. GA is a reliable and viable global optimization method with a large search space that may be used to solve complicated engineering optimization problems with nonlinear, multimodal, and nonconvex objective functions [199,204]. Chen et al. [211] utilized the GA to minimize the fuel consumption of a power-split PHEV by achieving the optimal engine-on power threshold and the QP to obtain the best battery current at high speed. The GA provides the advantage of multi-objective optimization such as fuel consumption, energy cost, the health of the battery, and emission. For instance, Piccolo et al. [214] presented a GA-based control strategy for optimizing fuel consumption and carbon emission terms for EV applications. Altundogan et al. [215] applied GA to find the optimal location of an EV charging station in urban areas. Li et al. [216] employed GA to address the fleet allocation issues of EV considering demand uncertainty.

The particle swarm optimization (PSO) technique is a meta-heuristic optimization technique utilized for searching a large area of a candidate solution. The PSO technique is inspired by bird flocking, where the optimization is carried out with suitable iterations for a given candidate solution. In recent times, the PSO technique has been employed in EV applications. PSO algorithms have been widely utilized for the optimized design of electromechanical systems, SC, and fuel cell size, in addition to energy control [217]. Zhang et al. [218] used the PSO-based multi-objective, multi-constraint optimization model to address the load dispatch issue in the microgrid. The results illustrated that the orderly charging–discharging method decreased cost and load variance by 13.4% and 78.8%, respectively. Yin and Ming [219] introduced a PSO-based charging and discharging scheduling strategy considering cost and environmental protection. The proposed approach reduces the user charging costs and improves the safe operation of the power grid. Similarly, Wang et al. [220] proposed a multi-objective PSO-based scheduling strategy for the orderly charging and discharging of EVs. Sadeghi et al. [221] developed multi-objective PSO to determine the optimal sizing of hybrid renewable energy systems in the presence of EVs. A comparative analysis of various methods, controllers, and optimization techniques for EV operation is shown in Table 5.

Table 5. The control strategies and optimization schemes are applied for EV applications.

Operation		Methods	Objectives	Benefits	Shortcomings	Achievements	Refs.
Control	Offline	LP	Minimization of fuel cost	Fuel consumption minimization Understanding the propulsion capabilities	Depends on prior knowledge.	Successful in automotive energy management	[222]
		DP	Reduction in emission	Computation efficiency can be improved Prior knowledge is not required	Computational burden	Improved fuel economy Multistage optimization	[201]
	Online	RB	Optimization of the energy flow management	Easy control strategies	Human skills are required Calibration work is needed	Real-time implementation of the vehicle engine	[205]
		FL	Energy cost and battery health.	Independent adaptation of the control strategy	Human thinking and experience are required It cannot guarantee optimal performance	Reduction in fuel consumption Minimization of emission Maintenance of the SOC	[204]

Table 5. Cont.

Operation	Methods	Objectives	Benefits	Shortcomings	Achievements	Refs.
Optimization	NN	Cost minimization	Able to predict the energy requirement	Meteorological data are required	Estimation of energy demand Optimization of the charging cost	[207]
	PMP	Minimizing battery degradation	Real-time optimization	Feedback controller is required	Optimization of EMS	[63]
	SA	Minimizing the fuel-consumption	Short- and long-term power management	Cannot guarantee a globally optimal solution	Optimal engine-on power Maximum current coefficient	[212]
	GA	Ensuring power demand between the electric motor and internal combustion engine.	Improvement of the overall vehicle environmental impact	Crossover probability effect on algorithm	Optimal fuel consumption Minimized emissions	[214]
	PSO	Multi-objective, multi-constraint optimization model providing load dispatch for a microgrid	The impact of EV charging on the power system is improved by enhancing safety and reducing cost	Slow convergence rate and easy to fall into local optimum in high-dimensional space	The orderly charging–discharging method decreases cost and load variance by 13.4% and 78.8%, respectively	[218]

5. Open Issues, Challenges, and Limitations

The future EVs have to be more sustainable in order to compete with conventional gasoline-powered vehicles, and they need smart optimization, controller, and management systems to maintain the charge level of the battery storage system. Simple structure, environmentally friendliness, lack of noise, and high efficiency are some value-added features of EVs. EV also offers noninterrupted acceleration and instantaneous high torques [223]. However, there are several areas where EVs fall short, which are covered in this section. As EVs are still under development, this study covers some crucial aspects and difficulties in attaining sustainable development.

5.1. Battery Storage Technology

There are several concerns regarding battery storage technology, including aging, charging current, and health degradation. Fast charging causes high current flow, leading to temperature rise, which affects the battery performance and shortens the battery life. Series and parallel combinations of small batteries are necessary to create a battery system with a safe structure, competitive cost, and high capacity in a compact form with air ventilation. In the future, the controller should split the batteries in such a way that some of them can charge from the source while others deliver power to the motor. Some cells have been divided into more segments that can enable fast charging [224]. The latest battery materials, including hydrogen and fuel cells, can be suggested with desirable characteristics and a reasonable price.

5.2. Battery Balancing and Temperature Issues

A pack of batteries is needed, which is made up of multiple batteries connected in series and parallel [225]. The performance of the battery pack as a whole is difficult to monitor since batteries can charge and discharge at different rates and operate under different conditions due to other operational states in terms of temperature, state of charge, and state of health. The battery management system must monitor the charge rate across the whole pack down to the cell level to ensure efficient battery-pack performance and prolonged battery life [4]. Charge balancing and thermal management are the two main responsibilities of the battery management system. Passive balancing is not helpful when batteries are discharged because of the limitation of weakening cells. Excess energy is dissipated as heat through an external resistor, and a cooling system is needed for these reasons [101]. Active balancing is necessary and more efficient in balancing a cell's energy because it redistributes the energy among cells rather than dissipating and wasting it. Power electronic devices are used to move energy from strong to weak cells to maximize the available energy and increase the module's capacity. Additionally, the thermal management of each battery has to operate in an acceptable and safe temperature range, and failing to do so will cause performance degradation or irreversible damage.

5.3. Motor Drive Technology

The induction motor is the most popular choice for EVs, which uses three-phase AC power input, providing a four-pole magnetic field. The induction motor speed is dependent on the frequency of the AC power supply; thus, by varying the frequency, the speed drive wheel will increase or decrease in place of the transmission gearbox, making the EV simple and dependable. The motor can spin from 0 to 1800 RPM with a single gear as compared to a regular combustion engine. The induction motor has a good starting torque when the car goes down a hill. Furthermore, there is no energy loss in the rotor; however, it is not efficient for a long and high-speed drive. This problem is caused by the back electromagnetic force in the rotor, which is a reverse voltage to the stator's supply voltage. Therefore, a higher speed results in a higher back electromagnetic force, which can affect motor performance. Moreover, this high-power magnet results in magnetic eddy current losses, thus increasing the motor's heat. In Tesla Model 3, a new motor called IPMsynRM uses a permanent magnetic and reluctance design to solve these issues.

5.4. Power Electronics Technology

Power electronic devices are a key technology for control in almost all EV applications because they can convert energy to run motors, batteries, and generators. The power electronics technology is used in two levels of EVs that require high-power electric energy to rotate the electric motors and energy management for other applications such as charging the battery. Power electronic components such as silicon-based power MOSFETs and IGBTs are used as power electronic switches in the power train system of automotive electrical and electronic systems to reduce the overall size [226]. The power electronics devices require powerful thermal management because, when operating at high temperatures, the power electronics devices can be defective and fail, in which case the EV would not function, necessitating a major operation to inspect the faulty parts and replace them. This could happen when the cooling system is not working efficiently due to fan or compressor problems. In power electronics, the reliability issues of power, semiconductors, and capacitors stand out because of the different stress factors and field return data. Antiferroelectric ceramics are needed to keep the capacitance stable under a voltage bias and to maintain performance at high operating temperatures [227]. The integration of capacitors, cooling, and active devices ensures that power electronics are safe and reliable.

5.5. EV Charging Technology

The increasing number of EVs has raised several issues based on the level of charger types, yet the main issue is the recharge time [228]. During fast charging, the charging procedure is interrupted frequently, which wastes passenger time. Consequently, there are challenges with the technology, cost, safety, sustainability, and environment [228]. Public EV charging station systems have problems such as being expensive to build and needing more charging stations in almost all parking spots along highways. Charging during times of high demand costs more, and the electricity load is problematic for utility providers.

Scalability is another leading challenge that EV chargers have to deal with. Wireless inductive charging is still in a standby situation and is waiting for advanced technology to make it possible for EVs to charge spontaneously without the need for cables [229].

5.6. Intelligent Control and Optimization Schemes

Intelligent controllers such as artificial intelligence and machine learning are always used in advanced technology, and they are the key to most improvements made in the last few years in EV applications. Yet, they can have a serious problem if the training process is not executed accurately or if too much or too little data are taken into account [230]. To solve the abovementioned concerns, various optimization schemes can be employed before setting up these smart controllers, especially during unexpected conditions. Regarding the optimization issues associated with EVs, they require a different level of optimization in many applications in EVs, ranging from the wheel size to the battery management system and controller for both batteries and motor [231].

5.7. EV Aerodynamic Mechanical Design and Materials

The selection of appropriate aerodynamic and mechanical design and materials is a key research area to be explored. Aerodynamic efficiency is a big factor during the manufacturing process, but these issues primarily affect the driver. The design and materials, including wheel size and material, vehicle body shape and size, battery sets, and motor size, are the main factors that could make the vehicle light and less resistant to the airflow [232].

5.8. Safety Design of EVs

A range of issues associated with power system security and safety in EVs need to be addressed effectively. Since the EV does not have an engine in the front of the vehicle, which always absorbs the shock of a crash, the driver and the passengers will be right up against the next car in the event of an accident [232]. When the battery is damaged by a severe accident, the high voltage may affect the driver and passengers.

5.9. Availability of Charging Stations

Charging station availability issues can be solved by the fast charging and popularity of EVs. However, this issue is vital because the user needs to identify the nearest charging station before traveling. The lack of charging stations may limit the proliferation of EVs and have a negative impact on society. The EV must be charged conveniently and quickly to ensure the EV owner's comfort. The EV charging infrastructure needs to accommodate exponential market growth and a wide range of charging use-cases.

5.10. V2G Concept Challenges

Vehicle-to-grid (V2G) interactions have never been so easy, but building an integrated system that can host a large number of EVs for the benefit of both parties poses many challenges and issues. Communication platforms, for example, network bandwidth and the radius of EV information, are needed for aggregation and network latency. Legislation and agreements are some of the challenges [233]. The governments' regulatory issues and electric grid upgrades also play an important role in this context. From this point of view, the EV revolution faces important regulatory and technological challenges which require close collaboration among different levels of the same government. EV gird interconnection has some main issues such as the Doppler effect of changing frequency waveforms, adjacent interference of power singles, multipath fading (which may cause signal attenuation and distortion), interference from other EVs or other sources, access delay of mobility between peers, and network stabilization time [234]. Detailed requirements for V2G communication such as network latency, network bandwidth, and actual radius information need to be investigated. Security threats need to be met in a smart EV charging service using an authentication protocol to guarantee a safe integration of power grid data [233].

5.11. Battery Environmental Issues

EVs are environmentally friendly vehicles because they can reduce emissions that contribute to climate change. However, they can be harmful when their batteries die and are not landfilled or recycled properly [235,236], or if the source of charge is nonrenewable resources. Although energy storage is a complex system having several factors, including state coupling, input coupling, environmental sensitivity, life degradation, and added characteristics [237], the majority of materials can be recycled.

6. EVs on the Road to Achieving Sustainable Development Goals

The United Nations (UN) established 17 Sustainable Development Goals (SDGs) in 2015, with the goal of providing a common vision for good living and a tranquil atmosphere for the globe and its inhabitants [16]. According to the SuM4All's Global Roadmap of Action (GRA) toward sustainable mobility, global GHG emissions from the transportation industry must be reduced from 8 billion tons of CO_2 to 2–4 billion tons by 2050, with net-zero emissions in the following decades [238]. As a result, the transportation industry is a key participant in the fight against climate change to achieve the SDGs. With the advancements in technology in the transport sector, electric vehicles (EVs) have been introduced. They are predicted to play a large role in lowering overall road transport-related emissions caused by internal combustion engine vehicles (ICEVs), as outlined in Figure 24 [239]. The environmental effect of ICEVs is mostly determined by the fossil fuel utilized in the combustion vehicles, but the EV's impact is determined by the energy utilized to generate electricity. In 2019, the worldwide EV fleet emitted approximately 51 million tons of carbon dioxide, which is almost half of what an ICE-powered fleet of the same size would have emitted, totaling 53 Mt CO_2-eq of averted emissions [240].

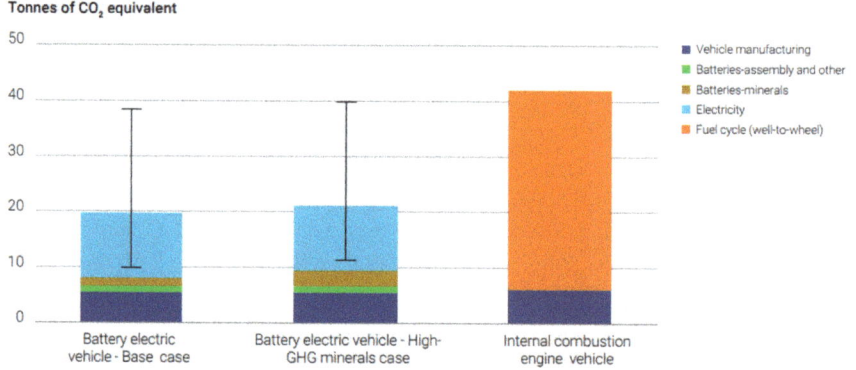

Figure 24. CO_2 emissions of a typical mid-size automobile in terms of the powertrain (2018) over a 10 year lifetime.

The transportation industry supports almost all SDGs, including those related to social and economic growth, improving access to services, enhancing agricultural production, and linking commodities to markets. Sustainable development requires sustainable, secure, and clean transportation that is available to all, and EVs fulfill all these features.

Electric vehicles have recently been linked to renewable energy, batteries, and other uses, all of which have improved environmental implications and can help achieve the relevant SDGs. As demonstrated in Table 6 and Figure 25, EVs have an impact on three aspects of sustainability (social, economic, and environmental), as well as a significant association with seven of the 17 SDGs.

Table 6. The correlation between EVs and SDGs.

Sector	SDGs	Objective	EVs on the Road to Achieving SDGs	Relevant Research that Supports the Correlation
Social	SDG 3: Good health and wellbeing	Reduce pollution-related illnesses	Unlike internal combustion engine (ICE) vehicles, electric vehicles (EVs) emit no pollution. As a result, EVs have been promoted as part of a larger global solution to bad air quality and the healthful life of city residents.	[241–249]
	SDG 11: Sustainable cities and communities	Improve inclusive and long-term urban planning and management	EVs are being utilized in the development of smart cities, which implies that all of the municipality services, such as local infrastructure and transportation, have been combined into a single, fully functional system. As a result, everyone benefits from a sustainable transportation system.	[21,250–253]
	SDG 7: Affordable and clean energy	Ensure that everyone has access to energy that is affordable, dependable, and contemporary	With the adoption of several functional activities, for example, optimum scheduling and energy optimization associated with EVs, affordable energy and reduced power consumption can be accomplished.	[17,254–263]
		Maximize the worldwide percentage of renewable energy in the energy mix by a significant amount	EVs can be used with a variety of renewable energy sources to produce a cost-effective alternative to fossil fuels.	
		Global energy efficiency improvement rate	EVs use distributed generation, energy efficiency, and energy storage to deliver contemporary, sustainable, and efficient energy.	
Economic	SDG 8: Decent work and economic growth	Encourage measures to promote productive activity and good employment creation	The success of the EV market, along with its numerous functions, particularly in the fields of renewable energy, electric buses, and trains, plays a part in economic growth and job creation in production, marketing, and supply.	[18,239,264–270]
	SDG 9: Industry, innovation, and infrastructure	Create high-quality, sustainable, dependable, and robust infrastructure to strengthen the economy	Electric vehicles are transforming the transportation sector into one that is adaptive, robust, and sustainable to changing global climatic circumstances while also promoting economic growth.	[3,260,271–275]
	SDG 12: Responsible consumption and production	Create a program framework for the sustainable use of resources	In the context of the virtual power plant, smart grid, distributed power production, and microgrid, energy management in EVs ensures the effective utilization of supply and load.	[276–283]
Environmental	SDG 13: Climate action	Take quick action to combat climate change's impacts	Carbon emissions can be reduced by combining various renewable energy sources with EV batteries to combat climate change.	[17,50–60]

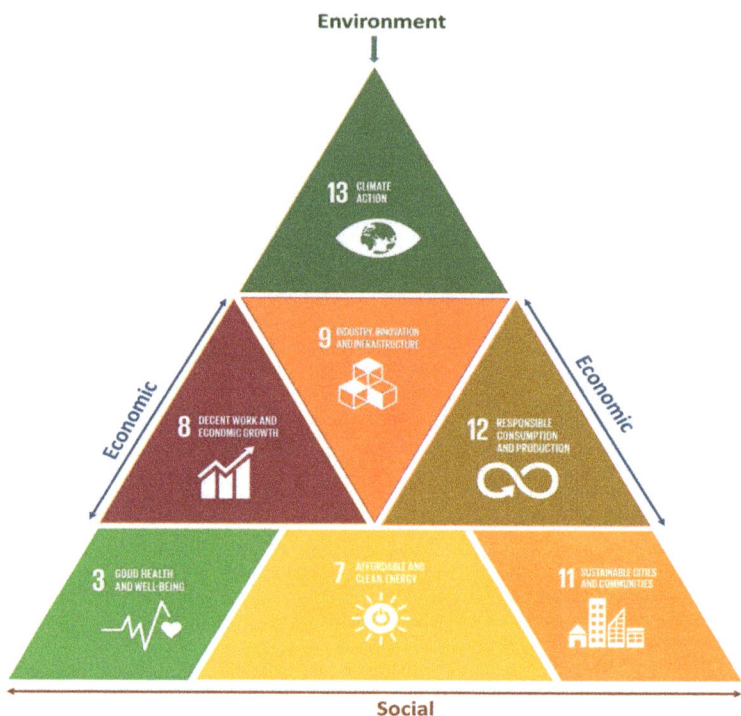

Figure 25. Association of EVs with SDGs.

6.1. Social Impact of EVs

Within the society category, the construction and implementation of EVs have a beneficial influence on the achievement of multiple SDGs. According to the categorization of the 17 SDGs into three categories, EVs can be related to three of the SDGs, as discussed below.

6.1.1. SDG3

We looked into a number of studies to validate the link between EVs and their social impact on SDG targets. EVs, for example, can provide thermal comfort [241], air quality comfort [242,243], temperature control [244], humidity control [245], and heat radiation [246], all of which are connected to SDG 3.9 aimed at excellent health and wellbeing. In the area of human health effects, however, the vehicle categories exhibit a distinct proportion of impact location. Human physiological consequences of ICVs arise when driving due to tailpipe pollution, which accounts for 94% of all ICV human health implications [247]. Public health effects from BEVs are related to the energy-generating subphase. Electrification of automobiles, on the other hand, benefits the human health effect sector since the consequences are reduced from ICV to BEV. According to research, one of the primary issues that healthcare institutions confront, particularly in rural and distant places, is the lack of reliable energy services [248]. Only 34% of hospitals have dependable energy connections in the Sub-Saharan African nations investigated. However, a study of two nations found that energy access has improved little over time. This crucial issue must be addressed in realistic initiatives to improve healthcare delivery in Sub-Saharan Africa [249]. This impacts a developing country's high maternal and infant mortality rates. However, a high-reliability power source may be established using a standalone hybrid energy system based-EV, improving the efficacy of healthcare services in rural/remote places.

6.1.2. SDG11

EVs can also assist in socioeconomic development through effective energy management [250], which is linked to SDG 11.3. Substantial published data exist for SDG 11 regarding the favorable function of environmentally friendly electric buses, electric vehicles, and solar cars [251]. They are also being utilized in the development of smart cities, which implies that all of the municipality services, for example, infrastructure and transportation, have been combined into a single, fully functional system. As a result, everyone benefits from a sustainable transportation system [21]. As batteries play an important part in EV technology, with today's technology, the efficiency and density of battery energy storage (BES) have aided in the development of inexpensive electric cars that emit little pollution. People will use electric vehicles as their major means of mobility when high-performance batteries are developed [21,252]. Batteries of various forms and sizes are considered to be among the most efficient power storage strategies, and studies on different battery technologies and applications can be found in the literature; nevertheless, the damaging outcomes of large usage on environmental and human health (Targets 11.5 and 11.6) continue to be a major problem. The BESS has a short lifespan since it is made up of large amounts of diverse raw materials, including metals and nonmetals. As a result, massive amounts of pollutants (e.g., emissions of greenhouse gases, poisonous gases, and toxic materials) can be created in the battery sector in many phases, including production, mining, shipping, application, storing, repair, recycling, and dumping [253].

6.1.3. SDG7

As electric vehicles are powered by electricity, they are strongly related to SDG7 on reasonable and sustainable energy. EVs can provide power generation [254], high energy efficiency [255] through power management [256], optimal scheduling [257], and renewable energy integration [258], all of which are related to Targets 7.1, 7.2, and 7.3. In order to meet Target 7.2, which calls for a significant increase in the proportion of renewable energy in the universal energy composition, considerable information is available showing that the EV, which is considered among the BES technologies, has aided in growing the market for renewable energy in the universal energy composition and can help achieve this goal [259]. In the case of SDG 7 (affordable and clean energy), there is evidence that the use and development of EV-BES will enable the attainment of all goals (100%) under this target. Many studies advocate the use of EV-BES in the pursuit of a contemporary energy service that is dependable, clean, and inexpensive. The BES, which can be connected with RE and electric vehicles (EVs), has grown in popularity as a way to replace the usage of conventional energy resources since it is more flexible in terms of storing and delivering electricity, making it a more economical option [17]. With the lowering cost of renewable energy and the rising scarcity of fossil fuels, a long-term solution is required in which renewable energy paired with EVs may assure universal access to power [259]. According to a study in China [260], a hybrid renewable resource-based charging station for EVs is economically and technically feasible and guarantees clean and dependable energy supply. The high initial financing required for BESs in EVs is a major drawback; however, studies have shown that the cost can be compensated during its operation [261]. For instance, to meet Target 7.2, which calls for a significant increase in the proportion of clean energy sources, there is significant proof that the EV, which is involved in storage systems and power systems, has aided in creating a market for renewable energy in the world energy pack and can help achieve this goal [262]. Despite the ongoing COVID-19 epidemic, research forecasted a 5% growth in the standalone (with storage system) renewables market in the power industry in 2020 relative to 2019. This is in line with a rise in BESSs needed for stationary and transportation applications [240].

6.2. Economic Influences of EVs

The three SDGs in the economic category (SDGs 8, 9, and 12) are concerned with employment, industries, growth of the economy, and technology and infrastructural facilities to achieve the objective.

6.2.1. SDG8

During the production of the different components of EVs, such as batteries, sensing devices, smart appliances, and electric motors, jobs can be created, aligning with Target 8.3 [18,263]. The extensive and diverse uses of EVs have been identified to provide economic prospects through the use of categorical employment from domestic resources and firms. In view of SDG8, development and suitable work, recent research shows that the success in the EV market along with its numerous functions, particularly in the fields of renewable energy, electric buses, and trains, has played a part in economic growth and job creation in production, marketing, and supply. EVs can help reach Target 8.3 by helping to create new employment. According to a 2019 forecast, battery-powered and plug-in hybrid cars might account for the bulk of new vehicle sales by 2040 [264]. In 2020, the German government announced a stimulus package involving 2.5 billion EUR to enhance electric car charging infrastructure and to encourage more e-mobility research and development, such as battery solutions. The UK government is committed to reducing 68% of CO_2 emissions by 2030. To meet the target, 43% of vehicles in the UK need to be electric [239]. In recent times, the US government announced a 3 billion USD initiative to enhance electric car battery production [265]. According to research published in [266], the electric automobile market is increasing rapidly in Flanders, Belgium. Using data from a large-scale poll conducted in 2011 and a choice-based conjoint study, it was estimated that battery electric vehicles would account for roughly 5% of new vehicle sales by 2020, while plug-in hybrid electric vehicles would account for around 7%. These percentages might rise to 15% and 29% by 2030, respectively. However, the rate of adoption of electric cars is influenced by their purchasing cost. The EV market also fulfills Target 8.6 by creating work possibilities for youngsters, lowering the youth unemployment rate, and helping to raise per capita income [267]. Job possibilities offered will also assist local children in obtaining more schooling and a greater employment rate since most projects would employ locals [268,269], favorably impacting Targets 8.1 and 8.3.

6.2.2. SDG9

EVs can contribute to the achievement of Target 9.1, which aims at the creation of dependable, high-quality, and long-lasting infrastructures to promote social wellbeing and economic growth [260,270]. For instance, an investigation in Sub-Saharan Africa found that using climate-resilient EV-BESs for photovoltaic house systems in remote regions has improved local infrastructure and aided public community, environment, and financial stability [271,272]. Creating a publicly recharging infrastructure system is a critical component for increasing the use of electric vehicles. As a result, GHG emissions associated with conventional fuel automobiles are reduced while also enhancing the metropolitan environment. The availability of charging stations (CS) and their characteristics is a major recognized obstacle to the adoption of electric vehicles among residents. The introduction of publicly CS systems helps to improve the EV customer journey by enhancing the accessibility of infrastructural facilities. The best way to encourage the formation and future growth of a sustainable local CS system is an apparent problem for city areas and local officials [273]. A case study in Canada [274] showed that British Columbia (BC) adopted EV technology as it is a favorable site for EVs since massive hydropower accounts for 85% of this region's electricity. EVs are ideally suitable for metropolitan environments, where the majority of housing structures in BC are situated, because of concerns with respect to their driving range and ability to cut local emissions.

6.2.3. SDG12

The deployment of EV integrated virtual power plants (VPPs) [275], distributed power production [276,277], smart grids [278], and microgrids [279] can help achieve effective energy usage between supply and load, which is relevant to Target 12.1. Unifying a significant quantity of EVs into VPPs will have ecological advantages, including energy efficiency and pollution minimization, safety advantages connected to the electrical grid's steady functioning, and economic incentives to VPPs and automobile stakeholders. More EVs will be included in VPPs in the long run because of these advantages. In addition to VPPs, electric vehicles in distribution networks are becoming controlled commodities with vehicle-to-grid (V2G) technology, enabling them to perform various auxiliary operations (e.g., maximum energy harvesting, voltage stabilization, and frequency regulation). Moreover, evidence exists to demonstrate the favorable impact of EV-BESs on SDG12, particularly Target 12.5, which calls for considerably reducing waste creation through minimization, recycling, and reuse. According to research, the BES of an EV is likely reused or discarded, or its components will be retrieved depending on the application [280]. Recycling the materials used in lithium-ion batteries (lithium, cobalt, nickel, and aluminum) can reduce the power density by 10–53%, and the construction price of lithium-oxygen batteries can be reduced to 1510 MJ/kWh from 1870 MJ/kWh, resulting in minimal GHG emissions [281]. According to Dewulf et al., recycling battery elements can reduce fossil source usage and, hence, minimize waste production [282].

Although there is a bright employability perspective, there will be beneficiaries and sufferers in the switch to the electric-powered transportation industry. There will be a loss of jobs in the oil business, at petrol stations, and perhaps in the car repair and mechanic sector because of the worldwide adaptation of EVs, since EVs require considerably lower maintenance compared to traditional petrol and diesel automobiles. There will not be practically as many employments needed to make batteries, electrical motors, and power electronic components as compared to the manufacture of petrol engines, exhaust systems, pollution control systems, fuel monitoring systems, gearboxes, and automotive components. Thousands of jobs may be lost in Germany alone by 2030 as a result of the switch to EVs, according to projections [268]. In order to transition to electrification, Volkswagen AG's parent company, which owns the premium brand Audi in Germany, stated last year that it would eliminate 7500 positions internationally. However, production, R&D, and battery manufacturing will all directly create employment in the EV sector. Installation and servicing of the equipment used in EVs will create indirect employment. Moreover, the workforce responsible for building petrol or combustion engine components might be reassigned to manufacture EV motor parts and battery packs. However, it is likely that there will be fewer such positions in the EV sector than currently employed in ICEV production lines [267].

Additionally, electric cars are less expensive to maintain and drive, leading to an impact on both direct and indirect employment. Each dollar saved on petrol, oil changes, and engine components can be reinvested in the domestic economy. The US Energy Information Administration estimates that more than 80% of the price of a gallon of petrol leaves the domestic economy straightaway [283,284]. More money will remain in the community and strengthen the local economy if fuel costs are reduced. According to research conducted by the California Electric Transportation Coalition, every dollar conserved on petrol that is put toward the cost of other domestic products and commodities leads to the creation of 16 jobs statewide [196,285]. Residents of New York City drive far less than those in typical US metro areas, keeping 19 billion USD annually within the city's economy [286].

6.3. Environmental Effects of EVs

Batteries have recently been linked to renewable energy, electric cars, and other uses, all of which have improved environmental implications and can help achieve the relevant SDGs. The advantages and disadvantages of EVs related to the environment are discussed below.

SDG13

The advantages of EVs for the climate (SDG13) can be summarized as their capacity to minimize emissions from conventional sources of energy utilized as a major element of RE to contain power [287]. Furthermore, with the advancing feature of EV solutions, many nations are incorporating initiatives aimed at environmental issues into their policy proposals [288]. The Intergovernmental Panel on Climate Change (IPCC) projected in 2018 that electric vehicles, electric motorcycles, and electric transport must displace fossil fuel-powered passenger vehicles by 2035–2050 in order to keep global warming below 1.5 °C. Targets 13.1 and 13.2 address environment-related problems and incorporate clean environment activities into social policies, schemes, and management. According to strategies reported in Europe [289], New Zealand [290], China [263], and the United States [291], by 2050, the use of EVs combined with RE will contribute to a 70% reduction in CO_2 emissions in the electricity sector. Many studies advocate the use of EVs in the pursuit of a contemporary energy service that is dependable and clean. EVs, which are connected with BESs, have grown in popularity as a way to replace the usage of traditional power sources since EVs are feasible in delivering and storing electricity, representing a more economical option [17]. With the lowering cost of renewable energy and the rising shortage of natural fuels, a sustainable strategy is required in which renewable energy and BESSs paired with EVs may ensure universal access to power [292]. Different studies, however, concurred that EVs have environmental drawbacks [293,294]. In certain cases, they employ dangerous and combustible ingredients, and they need substantial energy to make them, which results in extensive GHG emissions [295]. As a result, the BESS may function as a deterrent to Target 13.1's goal of reducing climate-related dangers. Because the BESS contains elements with severe environmental implications, it may have a detrimental influence [296]. EVs appear to be a viable climate answer because, if we can make our networks carbon-free, automobile emissions will decline dramatically. The excellent news for EVs is that many nations are now focusing on decarbonizing their power infrastructures. During the past few years, utility companies in the United States have abandoned hundreds of coal facilities in favor of a combination of solar, wind, and natural gas electricity with fewer emissions. Therefore, researchers have discovered [297] that EVs have become greener in general, and they will only grow greener.

7. Conclusions and Future Trends

Currently, global warming has triggered several studies in the field of energy toward curbing carbon emissions. A major portion of carbon emissions comes from the automobiles industries, which is currently undergoing drastic improvements by developing EVs. However, the development of EV technology requires several factors to consider. Therefore, in this review paper, battery storage and management, along with several EV technologies emphasizing power electronics converters, charging infrastructure, and methods, algorithms, controllers, and optimization, were reviewed toward achieving SDGs. Firstly, various battery storage technologies along with components of battery management were discussed. The analysis revealed that each battery technology features different performance characteristics such as specific power, specific energy, and thermal stability. Therefore, the battery features should be considered before their application in EV technology. Secondly, several EV-based technologies, such as power electronics technology and charging strategies, were critically reviewed. The investigation showed that power converters act as the key technology to control and optimize EV operations. With regard to charging strategies, more efficient technologies to limit the charging time, thermal loss, and appropriate thermal management should be developed. Thirdly, state-of-the-art methods, algorithms, and optimization approaches were investigated. It was reported that each algorithm and optimization technique for EV application delivers satisfactory outcomes. However, the computational complexity and high training time still need to be addressed. Fourthly, the open issues, limitations, and research gaps were identified. It was found that appropriate hybridization of various technologies such as battery storage systems, power converters, charging strategies, optimization approaches, and algorithms should be explored toward the efficient development of EV technology. Lastly, various SDG targets associated with EV application were explored and analyzed. On the basis of the issues and challenges, this article provides the following future trends and suggestions:

- For the power capacity of commercial and industrial energy storage systems, battery storage technology appears promising. The majority of EVs are powered by lithium-ion batteries. Fast charging shortens battery life and reduces performance because of the high current and temperature produced. In the future, the controller should split the batteries in such a way that some of them can charge from any source while others deliver power to the motor. Therefore, further study is suggested for designing controllers for improved performance and accuracy in EV technology.
- EVs cannot be powered by a single battery; instead, a battery pack comprises multiple modules connected together in series and parallel. The battery pack's performance is difficult to monitor at the pack level since batteries might function under different conditions. In order to balance a cell's energy, active balancing is required, which is more effective because it redistributes the energy among cells rather than letting it go to waste. Power electronic devices are employed to transmit energy from strong to weak cells and maximize the amount of energy available, which also increases the module's capacity. Henceforth, in-depth investigation is needed to deliver better active balancing between the battery pack by utilizing the converter circuits.
- Powerful thermal management is needed for the power electronics equipment since it may malfunction and fail while working at high temperatures. The power electronic devices are not entirely developed, and the thermal management is questionable because the EV industry is not totally mature and has various difficulties. Additionally, condition monitoring adds complexity and potential threats to the vehicle. Therefore, comprehensive exploration is needed to study the thermal management of power electronics devices.
- The major difficulty with EVs is the long recharge times; however, there are other problems related to the degree of charger types. High voltage, power, and energy transmission are needed for EV charging. Consequently, there is a difficulty with technology, cost, safety, sustainability, and the environment. Public rapid EV charging systems are dealing with problems such as being generally expensive to establish and all of the parking on the highways needing more chargers. Henceforth, considerable work needs to be accomplished to develop an appropriate charging system for EV applications.
- Intelligent controllers such as AI and ML are constantly used in cutting-edge technology, and they have been at the heart of the majority of advancements in recent years across a variety of applications. EV optimization problems require a varied level of optimization in various applications, starting with wheel size and extending to optimizing the battery management system and controller for both batteries and motor. Therefore, extensive optimization and AI techniques need to be explored for EV applications.
- EVs can lower emissions contributing to climate change, making them eco-friendly automobiles. However, when their batteries run out and are not properly disposed of or recycled, or when their power source is a nonrenewable resource, they can be dangerous. The battery energy storage system is complex with respect to state coupling, input coupling, environmental sensitivity, life-cycle deterioration, and ad-

ditional characteristics. Henceforth, further investigation is needed to study the life cycle of batteries and its associated factors.
- The power electronics converter technology is important toward controlling, stabilizing, and providing the conversion to operate motors, battery storage, and generators, as well as optimizing the EV operations for effective outcomes. At present, the power electronics converter technology is undergoing a drastic technological shift to develop lightweight converters, which depict less electromagnetic interference and fewer ripples to meet automotive industry standards. Therefore, further investigation is needed toward developing power converters with appropriate characteristics.
- Battery state estimation (e.g., SOC, SOH, and RUL) holds significant importance in EV technology. State estimation is important toward battery protection and energy management in EV applications. Various state-of-the-art technologies and methods, such as model-based, data-driven-based, and hybrid-based, have been applied to estimate the various battery states. However, in cases where the battery state is not appropriately estimated, system failure and economic loss could result. Additionally, inappropriate estimation may lead to early replacement of batteries, delay in battery replacement, and explicit failure events. Therefore, further exploration is necessary to develop a suitable estimation technique.
- The development of clean technology and SDG for EV applications can be achieved with the significant involvement of battery storage technology. Nonetheless, the participation and profitability of battery technology in the existing global energy market have not been explored comprehensively. Therefore, the development and analysis of various battery technologies in EV applications should be further studied.
- The performance, accuracy, and robustness of the EVs can be conducted by implementing the Internet of things (IoT) technology, which consists of sensors, data processors, and cloud technology. With IoT-based EV technology, EV data in the form of voltage, current, temperature, etc. can be stored and analyzed on the cloud platform. Henceforth, further examination to develop an effective IoT-based EV technology should be conducted.

Overall, the constructive discussion, analysis, concerns, and recommendations can provide decision makers with useful opportunities and directions for the adaptation of SDGs in the EV industries. In conclusion, proper selection and consideration of battery storage technology, battery management systems, power electronics technology, EV charging technology, and environmental issues of EVs can help in SDG integration.

Author Contributions: Conceptualization, M.S.H.L. and A.A.M.; methodology, M.S.H.L. and A.A.M.; validation, M.S.H.L., A.A.M., S.A. and M.S.M.; formal analysis, S.A., M.S.M., K.H. and S.T.M.; investigation, A.A.M., K.H. and T.R.; resources, M.S.H.L.; data curation, A.A.M., S.A. and M.S.M.; writing—original draft preparation, A.A.M., S.A., M.S.M., K.H. and T.R.; writing—review and editing, M.H.M., M.R.S., A.A., M.G.M.A. and N.M.L.T.; visualization, S.A. and S.T.M.; supervision, M.S.H.L.; project administration, M.S.H.L. and N.M.L.T.; funding acquisition, N.M.L.T. All authors read and agreed to the published version of the manuscript.

Funding: This work was supported by the Ministry of Science and Technology under the National Key R&D Program of China (Grant 2021YFE0108600).

Institutional Review Board Statement: Not applicable.

Informed Consent Statement: Not applicable.

Data Availability Statement: Not applicable.

Acknowledgments: The authors acknowledge the financial support provided by the Ministry of Science and Technology under the National Key R&D Program of China (Grant 2021YFE0108600). Special thanks to Department of Electrical and Electronic Engineering, Green University of Bangladesh for providing collaborative support.

Conflicts of Interest: The authors declare no conflict of interest.

Abbreviations

ADC	Analog-to-digital conversion
ANN	Artificial neural network
BCS	Battery charging station
BMS	Battery management system
BPIC	Battery protection IC
BSS	Battery swapping station
CAN	Controlled area network
CC	Constant current
CEC	Charge equalization controller
CIBC	Coupled inductor bidirectional converter
CPT	Capacitive power transfer
CS	Charging stations
CV	Constant voltage
DAS	Data acquisition system
DP	Dynamic programming
EIS	Electrochemical impedance spectroscopy
EMI	Electromagnetic interference
EMS	Energy management system
EV	Electric vehicle
FBC	Full-bridge boost DC/DC converter
FL	Fuzzy logic
GA	Genetic algorithm
GB	Guobiao Standards
GHG	Greenhouse gas
GRA	Global roadmap of action
HFT	High-frequency transformer
ICEV	Internal combustion engine vehicle
IEC	International Electrotechnical Commission
IPCC	Intergovernmental Panel on Climate Change
IPT	Inductive power transfer
ISO	International Organization for Standardization
KF	Kalman filter
LCO	Lithium cobalt oxide
LFP	Lithium iron phosphate
LMO	Lithium manganese oxide
LNMC	Lithium nickel manganese cobalt oxide
LNCA	Lithium nickel cobalt aluminum oxide
LTO	Lithium titanate oxide
LP	Linear programming
MCC	Multistage constant current
MPIC	Multiport isolated converter
NaNiCl	Sodium–nickel chloride
NCA	Nickel cobalt aluminum oxide
NiMG	Nickel–metal hydride
NMC	Nickel–manganese–cobalt
OCV	Open-circuit voltage
OLEV	Online electric vehicle
PCM	Phase-change material
PEM	Power and energy management
PFC	Power factor correction
PI	Proportional–integral
PID	Parameter identifiers
PMP	Pontryagin's minimal principle
PPC	Push–pull converter
PSO	Particle swarm optimization

QZBC	Quasi Z-source bidirectional converter
RC	Resonant converter
RES	Renewable energy sources
RUL	Remaining useful life
SA	Simulated annealing
SC	Stochastic control
SCBC	Switched-capacitor bidirectional converter
SCI	Serial communication interface
SDG	Sustainable Development Goals
SOC	State of charge
SOF	State of function
SOP	State of power
SOE	State of energy
SOS	State of safety
SVM	Support vector machine
UN	United Nations
VPP	Virtual power plants
V2G	Vehicle to the grid
WPT	Wireless power transfer
ZEBRA	Zero-emission battery research activity
ZVSC	Zero-voltage switching converter

References

1. Barkh, H.; Yu, A.; Friend, D.; Shani, P.; Tu, Q.; Swei, O. Vehicle Fleet Electrification and Its Effects on the Global Warming Potential of Highway Pavements in the United States. *Resour. Conserv. Recycl.* **2022**, *185*, 106440. [CrossRef]
2. Wang, Y.; Zhou, G.; Li, T.; Wei, X. Comprehensive Evaluation of the Sustainable Development of Battery Electric Vehicles in China. *Sustainability* **2019**, *11*, 5635. [CrossRef]
3. Debnath, R.; Bardhan, R.; Reiner, D.M.; Miller, J.R. Political, Economic, Social, Technological, Legal and Environmental Dimensions of Electric Vehicle Adoption in the United States: A Social-Media Interaction Analysis. *Renew. Sustain. Energy Rev.* **2021**, *152*, 111707. [CrossRef]
4. Hossain Lipu, M.S.; Hannan, M.A.; Karim, T.F.; Hussain, A.; Saad, M.H.M.; Ayob, A.; Miah, M.S.; Indra Mahlia, T.M. Intelligent Algorithms and Control Strategies for Battery Management System in EVs: Progress, Challenges and Future Outlook. *J. Clean. Prod.* **2021**, *292*, 126044. [CrossRef]
5. Lipu, M.S.H.; Faisal, M.; Ansari, S.; Hannan, M.A.; Karim, T.F.; Ayob, A.; Hussain, A.; Miah, M.S.; Saad, M.H.M. Review of Electric Vehicle Converter Configurations, Control Schemes and Optimizations: Challenges and Suggestions. *Electronics* **2021**, *10*, 477. [CrossRef]
6. Asekomeh, A.; Gershon, O.; Azubuike, S.I. Optimally Clocking the Low Carbon Energy Mile to Achieve the Sustainable Development Goals: Evidence from Dundee's Electric Vehicle Strategy. *Energies* **2021**, *14*, 842. [CrossRef]
7. Omahne, V.; Knez, M.; Obrecht, M. Social Aspects of Electric Vehicles Research—Trends and Relations to Sustainable Development Goals. *World Electr. Veh. J.* **2021**, *12*, 15. [CrossRef]
8. Hannan, M.A.; How, D.N.T.; Lipu, M.S.H.; Mansor, M.; Ker, P.J.; Dong, Z.Y.; Sahari, K.S.M.; Tiong, S.K.; Muttaqi, K.M.; Mahlia, T.M.I.; et al. Deep Learning Approach towards Accurate State of Charge Estimation for Lithium-Ion Batteries Using Self-Supervised Transformer Model. *Sci. Rep.* **2021**, *11*, 19541. [CrossRef] [PubMed]
9. How, D.N.T.; Hannan, M.A.; Hossain Lipu, M.S.; Ker, P.J. State of Charge Estimation for Lithium-Ion Batteries Using Model-Based and Data-Driven Methods: A Review. *IEEE Access* **2019**, *7*, 136116–136136. [CrossRef]
10. Mahmoudzadeh Andwari, A.; Pesiridis, A.; Rajoo, S.; Martinez-Botas, R.; Esfahanian, V. A Review of Battery Electric Vehicle Technology and Readiness Levels. *Renew. Sustain. Energy Rev.* **2017**, *78*, 414–430. [CrossRef]
11. Manzetti, S.; Mariasiu, F. Electric Vehicle Battery Technologies: From Present State to Future Systems. *Renew. Sustain. Energy Rev.* **2015**, *51*, 1004–1012. [CrossRef]
12. Hanifah, R.A.; Toha, S.F.; Ahmad, S. Electric Vehicle Battery Modelling and Performance Comparison in Relation to Range Anxiety. *Procedia Comput. Sci.* **2015**, *76*, 250–256. Available online: https://www.18650batterystore.com/blogs/news/lithium-ion-battery-raw-material-costs-continue-rise-in-2022 (accessed on 31 July 2022). [CrossRef]
13. Zubi, G.; Dufo-López, R.; Carvalho, M.; Pasaoglu, G. The Lithium-Ion Battery: State of the Art and Future Perspectives. *Renew. Sustain. Energy Rev.* **2018**, *89*, 292–308. [CrossRef]
14. Hannan, M.A.; Hoque, M.M.; Hussain, A.; Yusof, Y.; Ker, P. State-of-the-Art and Energy Management System of Lithium-Ion Batteries in Electric Vehicle Applications: Issues and Recommendations. *IEEE Access* **2018**, *6*, 19362–19378. [CrossRef]
15. Zhang, R.; Xia, B.; Li, B.; Cao, L.; Lai, Y.; Zheng, W.; Wang, H.; Wang, W.; Zhang, R.; Xia, B.; et al. State of the Art of Lithium-Ion Battery SOC Estimation for Electrical Vehicles. *Energies* **2018**, *11*, 1820. [CrossRef]

16. Moyer, J.D.; Hedden, S. Are We on the Right Path to Achieve the Sustainable Development Goals? *World Dev.* **2020**, *127*, 104749. [CrossRef]
17. Martinez-bolanos, J.R.; Edgard, M.; Udaeta, M.; Luiz, A.; Gimenes, V.; Oliveira, V. Economic Feasibility of Battery Energy Storage Systems for Replacing Peak Power Plants for Commercial Consumers under Energy Time of Use Tariffs. *J. Energy Storage* **2020**, *29*, 101373. [CrossRef]
18. Lau, Y.; Wu, A.Y.; Yan, M.W. A Way Forward for Electric Vehicle in Greater Bay Area: Challenges and Opportunities for the 21st Century. *Vehicles* **2022**, *4*, 420–432. [CrossRef]
19. Fonseca, L.M.; Domingues, J.P.; Dima, A.M. Mapping the Sustainable Development Goals Relationships. *Sustainability* **2020**, *12*, 3359. [CrossRef]
20. Cao, J.; Chen, X.; Qiu, R.; Hou, S. Electric Vehicle Industry Sustainable Development with a Stakeholder Engagement System. *Technol. Soc.* **2021**, *67*, 101771. [CrossRef]
21. Wen, J.; Zhao, D.; Zhang, C. An Overview of Electricity Powered Vehicles: Lithium-Ion Battery Energy Storage Density and Energy Conversion Efficiency. *Renew. Energy* **2020**, *162*, 1629–1648. [CrossRef]
22. Tie, S.F.; Tan, C.W. A Review of Energy Sources and Energy Management System in Electric Vehicles. *Renew. Sustain. Energy Rev.* **2013**, *20*, 82–102. [CrossRef]
23. Sujitha, N.; Krithiga, S. RES Based EV Battery Charging System: A Review. *Renew. Sustain. Energy Rev.* **2017**, *75*, 978–988. [CrossRef]
24. Bamrah, P.; Kumar Chauhan, M.; Singh Sikarwar, B.; Singh Katoch, S. A Detailed Review on Electric Vehicles Battery Thermal Management System CFD Analysis of Battery Thermal Management System. *IOP Conf. Ser. Mater. Sci. Eng.* **2020**, *912*, 042005. [CrossRef]
25. Maiyalagan, T.; Elumalai, P. *Rechargeable Lithium-Ion Batteries: Trends and Progress in Electric Vehicles*; CRC Press: Boca Raton, FL, USA, 2020. [CrossRef]
26. Nitta, N.; Wu, F.; Lee, J.T.; Yushin, G. Li-Ion Battery Materials: Present and Future. *Mater. Today* **2015**, *18*, 252–264. [CrossRef]
27. Mishra, A.; Mehta, A.; Basu, S.; Malode, S.J.; Shetti, N.P.; Shukla, S.S.; Nadagouda, M.N.; Aminabhavi, T.M. Electrode Materials for Lithium-Ion Batteries. *Mater. Sci. Energy Technol.* **2018**, *1*, 182–187. [CrossRef]
28. Miranda, D.; Gören, A.; Costa, C.M.; Silva, M.M.; Almeida, A.M.; Lanceros-Méndez, S. Theoretical Simulation of the Optimal Relation between Active Material, Binder and Conductive Additive for Lithium-Ion Battery Cathodes. *Energy* **2019**, *172*, 68–78. [CrossRef]
29. Yuan, H.; Song, W.; Wang, M.; Chen, Y.; Gu, Y. Lithium-Ion Conductive Coating Layer on Nickel Rich Layered Oxide Cathode Material with Improved Electrochemical Properties for Li-Ion Battery. *J. Alloys Compd.* **2019**, *784*, 1311–1322. [CrossRef]
30. Wang, K.; Wan, J.; Xiang, Y.; Zhu, J.; Leng, Q.; Wang, M.; Xu, L.; Yang, Y. Recent advances and historical developments of high voltage lithium cobalt oxide materials for rechargeable Li-ion batteries. *J. Power Sources* **2020**, *460*, 228062. [CrossRef]
31. Liu, H.; Li, M.; Xiang, M.; Guo, J.; Bai, H.; Bai, W.; Liu, X. Effects of crystal structure and plane orientation on lithium and nickel co-doped spinel lithium manganese oxide for long cycle life lithium-ion batteries. *J. Colloid Interface Sci.* **2021**, *585*, 729–739. [CrossRef] [PubMed]
32. Li, S.; Xue, Y.; Cui, X.; Geng, S.; Huang, Y. Effect of Sulfolane and Lithium Bis(Oxalato)Borate-Based Electrolytes on the Performance of Spinel LiMn2O4 Cathodes at 55 °C. *Ionics* **2016**, *22*, 797–801. [CrossRef]
33. Marincaș, A.H.; Ilea, P. Enhancing Lithium Manganese Oxide Electrochemical Behavior by Doping and Surface Modifications. *Coatings* **2021**, *11*, 456. [CrossRef]
34. Chen, K.; Yu, Z.; Deng, S.; Wu, Q.; Zou, J.; Zeng, X. Evaluation of the Low Temperature Performance of Lithium Manganese Oxide/Lithium Titanate Lithium-Ion Batteries for Start/Stop Applications. *J. Power Sources* **2015**, *278*, 411–419. [CrossRef]
35. Chen, Q.; Luo, L.; Wang, L.; Xie, T.; Dai, S.; Yang, Y.; Li, Y.; Yuan, M. Enhanced Electrochemical Properties of Y2O3-Coated-(Lithium-Manganese)-Rich Layered Oxides as Cathode Materials for Use in Lithium-Ion Batteries. *J. Alloys Compd.* **2018**, *735*, 1778–1786. [CrossRef]
36. Varsano, F.; Alvani, C.; La Barbera, A.; Masi, A.; Padella, F. Lithium Manganese Oxides as High-Temperature Thermal Energy Storage System. *Thermochim. Acta* **2016**, *640*, 26–35. [CrossRef]
37. Golubkov, A.W.; Scheikl, S.; Planteu, R.; Voitic, G.; Wiltsche, H.; Stangl, C.; Fauler, G.; Thaler, A.; Hacker, V. Thermal Runaway of Commercial 18650 Li-Ion Batteries with LFP and NCA Cathodes—Impact of State of Charge and Overcharge. *RSC Adv.* **2015**, *5*, 57171–57186. [CrossRef]
38. Zhao, C.; Yin, H.; Ma, C. Quantitative Evaluation of LiFePO4 Battery Cycle Life Improvement Using Ultracapacitors. *IEEE Trans. Power Electron.* **2016**, *31*, 3989–3993. [CrossRef]
39. Zhang, W.J. Structure and Performance of LiFePO4 Cathode Materials: A Review. *J. Power Sources* **2011**, *196*, 2962–2970. [CrossRef]
40. Lithium Nickel Manganese Cobalt Oxide (NMC) Tapes I NEI Corporation. Available online: https://www.neicorporation.com/products/batteries/cathode-anode-tapes/lithium-nickel-manganese-cobalt-oxide/ (accessed on 18 August 2022).
41. Maheshwari, A.; Heck, M.; Santarelli, M. Cycle Aging Studies of Lithium Nickel Manganese Cobalt Oxide-Based Batteries Using Electrochemical Impedance Spectroscopy. *Electrochim. Acta* **2018**, *273*, 335–348. [CrossRef]
42. Liu, Z.; Ivanco, A.; Onori, S. Aging Characterization and Modeling of Nickel-Manganese-Cobalt Lithium-Ion Batteries for 48V Mild Hybrid Electric Vehicle Applications. *J. Energy Storage* **2019**, *21*, 519–527. [CrossRef]

43. Liu, S.; Xiong, L.; He, C. Long Cycle Life Lithium Ion Battery with Lithium Nickel Cobalt Manganese Oxide (NCM) Cathode. *J. Power Sources* **2014**, *261*, 285–291. [CrossRef]
44. Kim, H.; Lee, K.; Kim, S.; Kim, Y. Fluorination of Free Lithium Residues on the Surface of Lithium Nickel Cobalt Aluminum Oxide Cathode Materials for Lithium Ion Batteries. *Mater. Des.* **2016**, *100*, 175–179. [CrossRef]
45. Joulié, M.; Laucournet, R.; Billy, E. Hydrometallurgical Process for the Recovery of High Value Metals from Spent Lithium Nickel Cobalt Aluminum Oxide Based Lithium-Ion Batteries. *J. Power Sources* **2014**, *247*, 551–555. [CrossRef]
46. Li, Z.; Li, J.; Zhao, Y.; Yang, K.; Gao, F.; Li, X. Influence of Cooling Mode on the Electrochemical Properties of Li4Ti5O12 Anode Materials for Lithium-Ion Batteries. *Ionics* **2016**, *22*, 789–795. [CrossRef]
47. Rashid, M.; Sahoo, A.; Gupta, A.; Sharma, Y. Numerical Modelling of Transport Limitations in Lithium Titanate Anodes. *Electrochim. Acta* **2018**, *283*, 313–326. [CrossRef]
48. Liu, W.; Wang, Y.; Jia, X.; Xia, B. The Characterization of Lithium Titanate Microspheres Synthesized by a Hydrothermal Method. *J. Chem.* **2013**, *2013*, 497654. [CrossRef]
49. Wang, L.; Wang, Z.; Ju, Q.; Wang, W.; Wang, Z. Characteristic Analysis of Lithium Titanate Battery. *Energy Procedia* **2017**, *105*, 4444–4449. [CrossRef]
50. Park, J.S.; Margez, C.L.; Greszler, T.A. Effect of Particle Size and Electronic Percolation on Low-Temperature Performance in Lithium Titanate-Based Batteries. *ACS Omega* **2019**, *4*, 21048–21053. [CrossRef]
51. Yuan, T.; Tan, Z.; Ma, C.; Yang, J.; Ma, Z.F.; Zheng, S. Challenges of Spinel Li$_4$Ti$_5$O$_{12}$ for Lithium-Ion Battery Industrial Applications. *Adv. Energy Mater.* **2017**, *7*, 1601625. [CrossRef]
52. Hossain Lipu, M.S.; Hannan, M.A.; Hussain, A.; Ayob, A.; Saad, M.H.M.; Karim, T.F.; How, D.N.T. Data-Driven State of Charge Estimation of Lithium-Ion Batteries: Algorithms, Implementation Factors, Limitations and Future Trends. *J. Clean. Prod.* **2020**, *277*, 124110. [CrossRef]
53. Li, M.; Lu, J.; Chen, Z.; Amine, K. 30 Years of Lithium-Ion Batteries. *Adv. Mater.* **2018**, *30*, 1800561. [CrossRef] [PubMed]
54. Arya, A.; Sharma, A.L. Polymer Electrolytes for Lithium Ion Batteries: A Critical Study. *Ionics* **2017**, *23*, 497–540. [CrossRef]
55. Chacko, S.; Chung, Y.M. Thermal Modelling of Li-Ion Polymer Battery for Electric Vehicle Drive Cycles. *J. Power Sources* **2012**, *213*, 296–303. [CrossRef]
56. Sharma, R.A.; Seefurth, R.N. Thermodynamic Properties of the Lithium-Silicon System. *J. Electrochem. Soc.* **1976**, *123*, 1763. [CrossRef]
57. Zuo, X.; Zhu, J.; Müller-Buschbaum, P.; Cheng, Y.J. Silicon Based Lithium-Ion Battery Anodes: A Chronicle Perspective Review. *Nano Energy* **2017**, *31*, 113–143. [CrossRef]
58. Marcondes, A.; Scherer, H.F.; Salgado, J.R.C.; De Freitas, R.L.B. Sodium-Nickel Chloride Single Cell Battery Electrical Model-Discharge Voltage Behavior. In *Proceedings of the Workshop on Communication Networks and Power Systems, Brasilia, Brazil, 3–4 October 2019*; pp. 1–4. [CrossRef]
59. Budde-Meiwes, H.; Drillkens, J.; Lunz, B.; Muennix, J.; Rothgang, S.; Kowal, J.; Sauer, D.U. A Review of Current Automotive Battery Technology and Future Prospects. *Proc. Inst. Mech. Eng. Part D J. Automob. Eng.* **2013**, *227*, 761–776. [CrossRef]
60. Zhang, G.; Jiao, K. Multi-Phase Models for Water and Thermal Management of Proton Exchange Membrane Fuel Cell: A Review. *J. Power Sources* **2018**, *391*, 120–133. [CrossRef]
61. Song, Z.; Pan, Y.; Chen, H.; Zhang, T. Effects of Temperature on the Performance of Fuel Cell Hybrid Electric Vehicles: A Review. *Appl. Energy* **2021**, *302*, 117572. [CrossRef]
62. Quan, S.; Wang, Y.X.; Xiao, X.; He, H.; Sun, F. Real-Time Energy Management for Fuel Cell Electric Vehicle Using Speed Prediction-Based Model Predictive Control Considering Performance Degradation. *Appl. Energy* **2021**, *304*, 117845. [CrossRef]
63. Nguyen, B.H.; German, R.; Trovao, J.P.F.; Bouscayrol, A. Real-Time Energy Management of Battery/Supercapacitor Electric Vehicles Based on an Adaptation of Pontryagin's Minimum Principle. *IEEE Trans. Veh. Technol.* **2018**, *68*, 203–212. [CrossRef]
64. Lu, L.; Han, X.; Li, J.; Hua, J.; Ouyang, M. Lithium-Ion Battery State of Function Estimation Based on Fuzzy Logic Algorithm with Associated Variables. *J. Power Sources* **2013**, *226*, 272–288. [CrossRef]
65. Xiong, R.; Li, L.; Tian, J. Towards a Smarter Battery Management System: A Critical Review on Battery State of Health Monitoring Methods. *J. Power Sources* **2018**, *405*, 18–29. [CrossRef]
66. Eghtedarpour, N.; Farjah, E. Distributed Charge/Discharge Control of Energy Storages in a Renewable-Energy-Based DC Micro-Grid. *IET Renew. Power Gener.* **2014**, *8*, 45–57. [CrossRef]
67. Cen, J.; Jiang, F. Li-Ion Power Battery Temperature Control by a Battery Thermal Management and Vehicle Cabin Air Conditioning Integrated System. *Energy Sustain. Dev.* **2020**, *57*, 141–148. [CrossRef]
68. Rahimi-Eichi, H.; Ojha, U.; Baronti, F.; Chow, M.-Y. Battery Management System: An Overview of Its Application in the Smart Grid and Electric Vehicles. *IEEE Ind. Electron. Mag.* **2013**, *7*, 4–16. [CrossRef]
69. Zhang, P.; Liang, J.; Zhang, F. An Overview of Different Approaches for Battery Lifetime Prediction. *IOP Conf. Ser. Mater. Sci. Eng.* **2017**, *199*, 012134. [CrossRef]
70. Wang, P.; Zhang, X.; Yang, L.; Zhang, X.; Yang, M.; Chen, H.; Fang, D. Real-Time Monitoring of Internal Temperature Evolution of the Lithium-Ion Coin Cell Battery during the Charge and Discharge Process. *Extrem. Mech. Lett.* **2016**, *9*, 459–466. [CrossRef]
71. Vincent, T.A.; Gulsoy, B.; Sansom, J.E.H.; Marco, J. A Smart Cell Monitoring System Based on Power Line Communication-Optimization of Instrumentation and Acquisition for Smart Battery Management. *IEEE Access* **2021**, *9*, 161773–161793. [CrossRef]

72. Hussain, S.; Nengroo, S.H.; Zafar, A.; Kim, H.-J.; Alvi, M.J.; Ali, M.U. Towards a Smarter Battery Management System for Electric Vehicle Applications: A Critical Review of Lithium-Ion Battery State of Charge Estimation. *Energies* **2019**, *12*, 446. [CrossRef]
73. Affanni, A.; Bellini, A.; Franceschini, G.; Guglielmi, P.; Tassoni, C. Battery Choice and Management for New-Generation Electric Vehicles. *IEEE Trans. Ind. Electron.* **2005**, *52*, 1343–1349. [CrossRef]
74. Alvarez-Diazcomas, A.; Estévez-Bén, A.A.; Rodríguez-Reséndiz, J.; Martínez-Prado, M.A.; Carrillo-Serrano, R.V.; Thenozhi, S. A Review of Battery Equalizer Circuits for Electric Vehicle Applications. *Energies* **2020**, *13*, 5688. [CrossRef]
75. Hou, Z.Y.; Lou, P.Y.; Wang, C.C. State of Charge, State of Health, and State of Function Monitoring for EV BMS. In Proceedings of the IEEE International Conference on Consumer Electronics, Las Vegas, NV, USA, 8–10 January 2017; pp. 310–311. [CrossRef]
76. Meah, K.; Hake, D.; Wilkerson, S. Design, Build, and Test Drive a FSAE Electric Vehicle. *J. Eng.* **2020**, *2020*, 863–869. [CrossRef]
77. Lelie, M.; Braun, T.; Knips, M.; Nordmann, H.; Ringbeck, F.; Zappen, H.; Sauer, D.U. Battery Management System Hardware Concepts: An Overview. *Appl. Sci.* **2018**, *8*, 534. [CrossRef]
78. Svendsen, M.; Winther-Jensen, M.; Pedersen, A.B.; Andersen, P.B.; Sørensen, T.M. Electric Vehicle Data Acquisition System. In Proceedings of the IEEE International Electric Vehicle Conference, IEVC, Florence, Italy, 17–19 December 2014; pp. 1–6.
79. Qiang, J.; Yang, L.; Ao, G.; Zhong, H. Battery Management System for Electric Vehicle Application. In Proceedings of the 2006 IEEE International Conference on Vehicular Electronics and Safety, Shanghai, China, 13–15 December 2006; pp. 134–138. [CrossRef]
80. Dong, T.; Wei, X.; Dai, H. Research on High-Precision Data Acquisition and SOC Calibration Method for Power Battery. In Proceedings of the IEEE Vehicle Power and Propulsion Conference, Harbin, China, 3–5 September 2008; pp. 1–5.
81. Yang, S.; Zhang, Z.; Cao, R.; Wang, M.; Cheng, H.; Zhang, L.; Jiang, Y.; Li, Y.; Chen, B.; Ling, H.; et al. Implementation for a Cloud Battery Management System Based on the CHAIN Framework. *Energy AI* **2021**, *5*, 100088. [CrossRef]
82. Maures, M.; Mathieu, R.; Briat, O.; Capitaine, A.; Delétage, J.Y.; Vinassa, J.M. An Incremental Capacity Parametric Model Based on Logistic Equations for Battery State Estimation and Monitoring. *Batteries* **2022**, *8*, 39. [CrossRef]
83. Hu, X.; Feng, F.; Liu, K.; Zhang, L.; Xie, J.; Liu, B. State Estimation for Advanced Battery Management: Key Challenges and Future Trends. *Renew. Sustain. Energy Rev.* **2019**, *114*, 109334. [CrossRef]
84. Hannan, M.A.; How, D.N.T.; Hossain Lipu, M.S.; Ker, P.J.; Dong, Z.Y.; Mansur, M.; Blaabjerg, F. SOC Estimation of Li-Ion Batteries with Learning Rate-Optimized Deep Fully Convolutional Network. *IEEE Trans. Power Electron.* **2021**, *36*, 7349–7353. [CrossRef]
85. Hannan, M.A.; How, D.N.T.; Mansor, M.B.; Hossain Lipu, M.S.; Ker, P.; Muttaqi, K. State-of-Charge Estimation of Li-Ion Battery Using Gated Recurrent Unit with One-Cycle Learning Rate Policy. *IEEE Trans. Ind. Appl.* **2021**, *57*, 2964–2971. [CrossRef]
86. Kalikatzarakis, M.; Geertsma, R.D.; Boonen, E.J.; Visser, K.; Negenborn, R.R. Ship Energy Management for Hybrid Propulsion and Power Supply with Shore Charging. *Control Eng. Pract.* **2018**, *76*, 133–154. [CrossRef]
87. Xiong, X.; Wang, S.L.; Fernandez, C.; Yu, C.M.; Zou, C.Y.; Jiang, C. A Novel Practical State of Charge Estimation Method: An Adaptive Improved Ampere-Hour Method Based on Composite Correction Factor. *Int. J. Energy Res.* **2020**, *44*, 11385–11404. [CrossRef]
88. Dong, G.; Wei, J.; Zhang, C.; Chen, Z. Online State of Charge Estimation and Open Circuit Voltage Hysteresis Modeling of LiFePO4 Battery Using Invariant Imbedding Method. *Appl. Energy* **2016**, *162*, 163–171. [CrossRef]
89. Zhang, M.; Fan, X. Review on the State of Charge Estimation Methods for Electric Vehicle Battery. *World Electr. Veh. J.* **2020**, *11*, 23. [CrossRef]
90. Rzepka, B.; Bischof, S.; Blank, T. Implementing an Extended Kalman Filter for SoC Estimation of a Li-Ion Battery with Hysteresis: A Step-by-Step Guide. *Energies* **2021**, *14*, 3733. [CrossRef]
91. How, D.N.T.; Hannan, M.A.; Lipu, M.S.H.; Sahari, K.S.M.; Ker, P.J.; Muttaqi, K.M. State-of-Charge Estimation of Li-Ion Battery in Electric Vehicles: A Deep Neural Network Approach. *IEEE Trans. Ind. Appl.* **2020**, *56*, 5565–5574. [CrossRef]
92. Lin, H.; Kang, L.; Xie, D.; Linghu, J.; Li, J. Online State-of-Health Estimation of Lithium-Ion Battery Based on Incremental Capacity Curve and BP Neural Network. *Batteries* **2022**, *8*, 29. [CrossRef]
93. Lipu, M.S.H.; Hannan, M.A.; Hussain, A.; Hoque, M.M.; Ker, P.J.; Saad, M.H.M.; Ayob, A. A Review of State of Health and Remaining Useful Life Estimation Methods for Lithium-Ion Battery in EVs: Challenges and Recommendations. *J. Clean. Prod.* **2018**, *205*, 115–133. [CrossRef]
94. Galeotti, M.; Cinà, L.; Giammanco, C.; Cordiner, S.; Di Carlo, A. Performance Analysis and SOH (State of Health) Evaluation of Lithium Polymer Batteries through Electrochemical Impedance Spectroscopy. *Energy* **2015**, *89*, 678–686. [CrossRef]
95. Liu, K.; Li, K.; Peng, Q.; Zhang, C. A Brief Review on Key Technologies in the Battery Management System of Electric Vehicles. *Front. Mech. Eng.* **2019**, *14*, 47–64. [CrossRef]
96. Ansari, S.; Ayob, A.; Lipu, M.S.H.; Hussain, A.; Saad, M.H.M. Multi-Channel Profile Based Artificial Neural Network Approach for Remaining Useful Life Prediction of Electric Vehicle Lithium-Ion Batteries. *Energies* **2021**, *14*, 7521. [CrossRef]
97. Ansari, S.; Ayob, A.; Lipu, M.S.H.; Hussain, A.; Saad, M.H.M. Data-Driven Remaining Useful Life Prediction for Lithium-Ion Batteries Using Multi-Charging Profile Framework: A Recurrent Neural Network Approach. *Sustainability* **2021**, *13*, 13333. [CrossRef]
98. Shen, P.; Ouyang, M.; Lu, L.; Li, J.; Feng, X. The Co-Estimation of State of Charge, State of Health, and State of Function for Lithium-Ion Batteries in Electric Vehicles. *IEEE Trans. Veh. Technol.* **2018**, *67*, 92–103. [CrossRef]

99. Ouyang, J.; Xiang, D.; Li, J. State-of-Function Evaluation for Lithium-Ion Power Battery Pack Based on Fuzzy Logic Control Algorithm. In Proceedings of the IEEE 9th Joint International Information Technology and Artificial Intelligence Conference, Chongqing, China, 11–13 December 2020; pp. 822–826.
100. Wang, D.; Yang, F.; Gan, L.; Li, Y. Fuzzy Prediction of Power Lithium Ion Battery State of Function Based on the Fuzzy C-Means Clustering Algorithm. *World Electr. Veh. J.* **2019**, *10*, 1. [CrossRef]
101. Gabbar, H.A.; Othman, A.M.; Abdussami, M.R. Review of Battery Management Systems (BMS) Development and Industrial Standards. *Technologies* **2021**, *9*, 28. [CrossRef]
102. Křivík, P. Methods of SoC Determination of Lead Acid Battery. *J. Energy Storage* **2018**, *15*, 191–195. [CrossRef]
103. Li, K.; Yan, J.; Chen, H.; Wang, Q. Water Cooling Based Strategy for Lithium Ion Battery Pack Dynamic Cycling for Thermal Management System. *Appl. Therm. Eng.* **2018**, *132*, 575–585. [CrossRef]
104. Marcos, D.; Garmendia, M.; Crego, J.; Cortajarena, J.A. Functional Safety Bms Design Methodology for Automotive Lithium-based Batteries. *Energies* **2021**, *14*, 6942. [CrossRef]
105. Lee, S.; Jeong, Y.; Song, Y.; Kim, J. A Single Chip Li-Ion Battery Protection IC with Low Standby Mode Auto Release. In Proceedings of the International SoC Design Conference, Jeju, Korea, 3–6 November 2014; pp. 38–39.
106. Ju, F.; Deng, W.; Li, J. Performance Evaluation of Modularized Global Equalization System for Lithium-Ion Battery Packs. *IEEE Trans. Autom. Sci. Eng.* **2016**, *13*, 986–996. [CrossRef]
107. Hannan, M.; Hoque, M.; Ker, P.; Begum, R.; Mohamed, A. Charge Equalization Controller Algorithm for Series-Connected Lithium-Ion Battery Storage Systems: Modeling and Applications. *Energies* **2017**, *10*, 1390. [CrossRef]
108. Carter, J.; Fan, Z.; Cao, J. Cell Equalisation Circuits: A Review. *J. Power Sources* **2020**, *448*, 227489. [CrossRef]
109. Hoque, M.M.; Hannan, M.A.; Mohamed, A.; Ayob, A. Battery Charge Equalization Controller in Electric Vehicle Applications: A Review. *Renew. Sustain. Energy Rev.* **2016**, *75*, 1363–1385. [CrossRef]
110. Diao, W.; Xue, N.; Bhattacharjee, V.; Jiang, J.; Karabasoglu, O.; Pecht, M. Active Battery Cell Equalization Based on Residual Available Energy Maximization. *Appl. Energy* **2018**, *210*, 690–698. [CrossRef]
111. Wei, Z.; Peng, F.; Wang, H. An LCC-Based String-to-Cell Battery Equalizer with Simplified Constant Current Control. *IEEE Trans. Power Electron.* **2022**, *37*, 1816–1827. [CrossRef]
112. Hua, A.C.C.; Syue, B.Z.W. Charge and Discharge Characteristics of Lead-Acid Battery and LiFePO4 Battery. In Proceedings of the Nternational Power Electronics Conference, Sapporo, Japan, 21–24 June 2010; pp. 1478–1483.
113. Wang, S.C.; Chen, Y.L.; Liu, Y.H.; Huang, Y.S. A Fast-Charging Pattern Search for Li-Ion Batteries with Fuzzy-Logic-Based Taguchi Method. In Proceedings of the IEEE Conference on Industrial Electronics and Applications, Auckland, New Zealand, 15–17 June 2015; pp. 855–859.
114. Li, S.; Huang, T. Optimization Control for Orderly Charge and Discharge Control Strategy of Electric Vehicles Based on Reliable Index of Charging. In Proceedings of the iSPEC 2020 IEEE Sustainable Power and Energy Conference: Energy Transition and Energy Internet, Chengdu, China, 23–25 November 2020; pp. 2200–2205.
115. Sulaiman, N.; Hannan, M.A.; Mohamed, A.; Majlan, E.H.; Wan Daud, W.R. A Review on Energy Management System for Fuel Cell Hybrid Electric Vehicle: Issues and Challenges. *Renew. Sustain. Energy Rev.* **2015**, *52*, 802–814. [CrossRef]
116. Zheng, C.; Li, W.; Liang, Q. An Energy Management Strategy of Hybrid Energy Storage Systems for Electric Vehicle Applications. *IEEE Trans. Sustain. Energy* **2018**, *9*, 1880–1888. [CrossRef]
117. Zhang, P.; Yan, F.; Du, C. A Comprehensive Analysis of Energy Management Strategies for Hybrid Electric Vehicles Based on Bibliometrics. *Renew. Sustain. Energy Rev.* **2015**, *48*, 88–104. [CrossRef]
118. Du, G.; Zou, Y.; Zhang, X.; Liu, T.; Wu, J.; He, D. Deep Reinforcement Learning Based Energy Management for a Hybrid Electric Vehicle. *Energy* **2020**, *201*, 117591. [CrossRef]
119. Li, A.; Yuen, A.C.Y.; Wang, W.; Chen, T.B.Y.; Lai, C.S.; Yang, W.; Wu, W.; Chan, Q.N.; Kook, S.; Yeoh, G.H. Integration of Computational Fluid Dynamics and Artificial Neural Network for Optimization Design of Battery Thermal Management System. *Batteries* **2022**, *8*, 69. [CrossRef]
120. Akinlabi, A.A.H.; Solyali, D. Configuration, Design, and Optimization of Air-Cooled Battery Thermal Management System for Electric Vehicles: A Review. *Renew. Sustain. Energy Rev.* **2020**, *125*, 109815. [CrossRef]
121. Xia, G.; Cao, L.; Bi, G. A Review on Battery Thermal Management in Electric Vehicle Application. *J. Power Sources* **2017**, *367*, 90–105. [CrossRef]
122. Kim, J.; Oh, J.; Lee, H. Review on Battery Thermal Management System for Electric Vehicles. *Appl. Therm. Eng.* **2019**, *149*, 192–212. [CrossRef]
123. Hu, D.; Wang, Y.; Li, J.; Yang, Q.; Wang, J. Investigation of Optimal Operating Temperature for the PEMFC and Its Tracking Control for Energy Saving in Vehicle Applications. *Energy Convers. Manag.* **2021**, *249*, 114842. [CrossRef]
124. Xiong, R.; Sun, W.; Yu, Q.; Sun, F. Research progress, challenges and prospects of fault diagnosis on battery system of electric vehicles. *Appl. Energy* **2020**, *279*, 115855. [CrossRef]
125. Kaplan, H.; Tehrani, K.; Jamshidi, M. A fault diagnosis design based on deep learning approach for electric vehicle applications. *Energies* **2021**, *14*, 6599. [CrossRef]
126. Yao, L.; Xiao, Y.; Gong, X.; Hou, J.; Chen, X. A Novel Intelligent Method for Fault Diagnosis of Electric Vehicle Battery System Based on Wavelet Neural Network. *J. Power Sources* **2020**, *453*, 227870. [CrossRef]

127. Li, D.; Zhang, Z.; Liu, P.; Wang, Z.; Zhang, L. Battery Fault Diagnosis for Electric Vehicles Based on Voltage Abnormality by Combining the Long Short-Term Memory Neural Network and the Equivalent Circuit Model. *IEEE Trans. Power Electron.* **2021**, *36*, 1303–1315. [CrossRef]
128. Tudoroiu, N. (Ed.) *Battery Management Systems of Electric and Hybrid Electric Vehicles*; MDPI: Basel, Switzerland, 2021; ISBN 978-3-0365-1060-6.
129. Wang, Y.; Tian, J.; Sun, Z.; Wang, L.; Xu, R.; Li, M.; Chen, Z. A Comprehensive Review of Battery Modeling and State Estimation Approaches for Advanced Battery Management Systems. *Renew. Sustain. Energy Rev.* **2020**, *131*, 110015. [CrossRef]
130. Sivaraman, P.; Sharmeela, C. IoT-Based Battery Management System for Hybrid Electric Vehicle. In *Artificial Intelligent Techniques for Electric and Hybrid Electric Vehicles*; Wiley: Hoboken, NJ, USA, 2020; pp. 1–16.
131. Chon, A.; Lee, I.; Lee, J.; Lee, M. Wireless Battery Management System (WiBMS). In Proceedings of the Large Lithium Ion Battery Technology and Application Symposium, LLIBTA 2015—Held at AABC 2015, Mainz, Germany, 26–27 January 2015; pp. 315–320.
132. Habib, S.; Khan, M.M.; Abbas, F.; Ali, A.; Faiz, M.T.; Ehsan, F.; Tang, H. Contemporary Trends in Power Electronics Converters for Charging Solutions of Electric Vehicles. *CSEE J. Power Energy Syst.* **2020**, *6*, 911–929. [CrossRef]
133. Pahlevani, M.; Jain, P.K. Soft-Switching Power Electronics Technology for Electric Vehicles: A Technology Review. *IEEE J. Emerg. Sel. Top. Ind. Electron.* **2020**, *1*, 80–90. [CrossRef]
134. Chakraborty, S.; Vu, H.N.; Hasan, M.M.; Tran, D.D.; El Baghdadi, M.; Hegazy, O. DC-DC Converter Topologies for Electric Vehicles, Plug-in Hybrid Electric Vehicles and Fast Charging Stations: State of the Art and Future Trends. *Energies* **2019**, *12*, 1569. [CrossRef]
135. Mumtaz, F.; Yahaya, N.Z.; Meraj, S.T.; Singh, B.; Kannan, R.; Ibrahim, O. Review on Non-Isolated DC-DC Converters and Their Control Techniques for Renewable Energy Applications. *Ain Shams Eng. J.* **2021**, *12*, 3747–3763. [CrossRef]
136. Bairabathina, S.; Balamurugan, S. Review on Non-Isolated Multi-Input Step-up Converters for Grid-Independent Hybrid Electric Vehicles. *Int. J. Hydrog. Energy* **2020**, *45*, 21687–21713. [CrossRef]
137. Krithika, V.; Subramani, C. A Comprehensive Review on Choice of Hybrid Vehicles and Power Converters, Control Strategies for Hybrid Electric Vehicles. *Int. J. Energy Res.* **2018**, *42*, 1789–1812. [CrossRef]
138. Lai, C.M.; Lin, Y.C.; Lee, D. Study and Implementation of a Two-Phase Interleaved Bidirectional DC/DC Converter for Vehicle and DC-Microgrid Systems. *Energies* **2015**, *8*, 9969–9991. [CrossRef]
139. de Melo, R.R.; Tofoli, F.L.; Daher, S.; Antunes, F.L.M. Interleaved Bidirectional DC–DC Converter for Electric Vehicle Applications Based on Multiple Energy Storage Devices. *Electr. Eng.* **2020**, *102*, 2011–2023. [CrossRef]
140. Wang, F.; Luo, Y.; Li, H.; Xu, X. Switching Characteristics Optimization of Two-Phase Interleaved Bidirectional DC/DC for Electric Vehicles. *Energies* **2019**, *12*, 378. [CrossRef]
141. Wen, H.; Su, B. Hybrid-Mode Interleaved Boost Converter Design for Fuel Cell Electric Vehicles. *Energy Convers. Manag.* **2016**, *122*, 477–487. [CrossRef]
142. Balachander, K.; Amudha, A.; Ramkumar, M.S.; Emayavaramban, G.; Divyapriy, S.; Nagaveni, P. Design and Analysis of Modified CUK Converter for Electric Hybrid Vehicle. *Mater. Today Proc.* **2021**, *45*, 1691–1695. [CrossRef]
143. Kushwaha, R.; Singh, B. A Modified Bridgeless Cuk Converter Based EV Charger with Improved Power Quality. In Proceedings of the 2019 IEEE Transportation Electrification Conference and Expo, Detroit, MI, USA, 19–21 June 2019; pp. 1–6.
144. Zhang, Y.; Gao, Y.; Zhou, L.; Sumner, M. A Switched-Capacitor Bidirectional DC-DC Converter with Wide Voltage Gain Range for Electric Vehicles with Hybrid Energy Sources. *IEEE Trans. Power Electron.* **2018**, *33*, 9459–9469. [CrossRef]
145. Janabi, A.; Wang, B. Switched-Capacitor Voltage Boost Converter for Electric and Hybrid Electric Vehicle Drives. *IEEE Trans. Power Electron.* **2020**, *35*, 5615–5624. [CrossRef]
146. Ahmed, A. *Series Resonant Switched Capacitor Converter for Electric Vehicle Lithium-Ion Battery Cell Voltage Equalization*; Concordia University: Montreal, QC, Canada, 2012.
147. Zhang, Y.; Liu, Q.; Gao, Y.; Li, J.; Sumner, M. Hybrid Switched-Capacitor/Switched-Quasi-Z-Source Bidirectional DC-DC Converter with a Wide Voltage Gain Range for Hybrid Energy Sources EVs. *IEEE Trans. Ind. Electron.* **2019**, *66*, 2680–2690. [CrossRef]
148. Farakhor, A.; Abapour, M.; Sabahi, M. Study on the Derivation of the Continuous Input Current High-Voltage Gain DC/DC Converters. *IET Power Electron.* **2018**, *11*, 1652–1660. [CrossRef]
149. Ayachit, A.; Hasan, S.U.; Siwakoti, Y.P.; Abdul-Hak, M.; Kazimierczuk, M.K.; Blaabjerg, F. Coupled-Inductor Bidirectional DC-DC Converter for EV Charging Applications with Wide Voltage Conversion Ratio and Low Parts Count. In Proceedings of the 2019 IEEE Energy Conversion Congress and Exposition, ECCE, Baltimore, MD, USA, 29 September–3 October 2019; pp. 1174–1179.
150. Gonzalez-Castano, C.; Vidal-Idiarte, E.; Calvente, J. Design of a Bidirectional DC/DC Converter with Coupled Inductor for an Electric Vehicle Application. In Proceedings of the IEEE International Symposium on Industrial Electronics, Edinburgh, UK, 19–21 June 2017; pp. 688–693.
151. Liu, H.; Hu, H.; Wu, H.; Xing, Y.; Batarseh, I. Overview of High-Step-Up Coupled-Inductor Boost Converters. *IEEE J. Emerg. Sel. Top. Power Electron.* **2016**, *4*, 689–704. [CrossRef]
152. Wu, H.; Sun, K.; Chen, L.; Zhu, L.; Xing, Y. High Step-Up/Step-Down Soft-Switching Bidirectional DC-DC Converter with Coupled-Inductor and Voltage Matching Control for Energy Storage Systems. *IEEE Trans. Ind. Electron.* **2016**, *63*, 2892–2903. [CrossRef]

153. Zhu, X.; Zhang, B. High Step-up Quasi-Z-Source DC–DC Converters with Single Switched Capacitor Branch. *J. Mod. Power Syst. Clean Energy* **2017**, *5*, 537–547. [CrossRef]
154. Taneri, M.C.; Genc, N.; Mamizadeh, A. Analyzing and Comparing of Variable and Constant Switching Frequency Flyback DC-DC Converter. In Proceedings of the 2019 4th International Conference on Power Electronics and their Applications, ICPEA, Elazig, Turkey, 25–27 September 2019; pp. 1–6.
155. Kanthimathi, R.; Kamala, J. Analysis of Different Flyback Converter Topologies. In Proceedings of the 2015 International Conference on Industrial Instrumentation and Control, ICIC, Pune, India, 28–30 May 2015; pp. 1248–1252.
156. Sangeetha, J.; Santhi Mary Antony, A.; Ramya, D. A Boosting Multi Flyback Converter for Electric Vehicle Application. *Res. J. Appl. Sci. Eng. Technol.* **2015**, *10*, 1133–1140. [CrossRef]
157. Bhattacharya, T.; Umanand, L.; Giri, V.S.; Mathew, K. Multiphase Bidirectional Flyback Converter Topology for Hybrid Electric Vehicles. *IEEE Trans. Ind. Electron.* **2009**, *56*, 78–84. [CrossRef]
158. Deshmukh, S.H.; Sheikh, A.; Giri, M.M.; Tutakne, D.D.R. High Input Power Factor High Frequency Push-Pull DC/DC Converter. *IOSR J. Electr. Electron. Eng.* **2016**, *11*, 42–47. [CrossRef]
159. Veeresh, H.; Kusagur, A. Design and Development of Push Pull DC-DC Converter by ZCS/ZVS to Electrical Vehicle (EVs) Applications. *Int. J. Mod. Trends Eng. Res.* **2015**, *2*, 1024–1031.
160. Forouzesh, M.; Siwakoti, Y.P.; Gorji, S.A.; Blaabjerg, F.; Lehman, B. Step-Up DC-DC Converters: A Comprehensive Review of Voltage-Boosting Techniques, Topologies, and Applications. *IEEE Trans. Power Electron.* **2017**, *32*, 9143–9178. [CrossRef]
161. Al-Jafeary, I.M.K.; Tan, N.M.L.; Mansur, B.M.; Buticchi, G.A. A Multi-Pulse Phase-Modulation/Duty Cycle Control for a Triple Active Bridge Converter. In Proceedings of the IEEE International Symposium on Industrial Electronics (ISIE 2022), Anchorage, AK, USA, 1–3 June 2022; pp. 1–6.
162. Savrun, M.M.; İnci, M.; Büyük, M. Design and Analysis of a High Energy Efficient Multi-Port Dc-Dc Converter Interface for Fuel Cell/Battery Electric Vehicle-to-Home (V2H) System. *J. Energy Storage* **2022**, *45*, 103755. [CrossRef]
163. Miah, M.S.; Shahadat, M.; Lipu, H.; Meraj, T.; Hasan, K.; Ansari, S.; Jamal, T.; Masrur, H.; Elavarasan, R.M.; Hussain, A.; et al. Optimized Energy Management Schemes for Electric Vehicle Applications: A Bibliometric Analysis towards Future Trends. *Sustainability* **2021**, *13*, 12800. [CrossRef]
164. Kim, D.H.; Kim, M.S.; Nengroo, S.H.; Kim, C.H.; Kim, H.J. Llc Resonant Converter for Lev (Light Electric Vehicle) Fast Chargers. *Electronics* **2019**, *8*, 362. [CrossRef]
165. Ahn, S.H.; Gong, J.W.; Jang, S.R.; Ryoo, H.J.; Kim, D.H. Design and Implementation of Enhanced Resonant Converter for EV Fast Charger. *J. Electr. Eng. Technol.* **2014**, *9*, 143–153. [CrossRef]
166. Lin, F.; Wang, Y.; Wang, Z.; Rong, Y.; Yu, H. The Design of Electric Car DC/DC Converter Based on the Phase-Shifted Full-Bridge ZVS Control. *Energy Procedia* **2016**, *88*, 940–944. [CrossRef]
167. Peng, F.Z.; Li, H.; Su, G.J.; Lawler, J.S. A New ZVS Bidirectional DC-DC Converter for Fuel Cell and Battery Application. *IEEE Trans. Power Electron.* **2004**, *19*, 54–65. [CrossRef]
168. Pahlevaninezhad, M.; Das, P.; Drobnik, J.; Jain, P.K.; Bakhshai, A. A Novel ZVZCS Full-Bridge DC/DC Converter Used for Electric Vehicles. *IEEE Trans. Power Electron.* **2012**, *27*, 2752–2769. [CrossRef]
169. Tan, N.M.L.; Abe, T.; Akagi, H. Design and Performance of a Bidirectional Isolated DC-DC Converter for a Battery Energy Storage System. *IEEE Trans. Power Electron.* **2012**, *27*, 1237–1248. [CrossRef]
170. Syahira, N.; Sharifuddin, M.; Tan, N.M.L.; Akagi, H. Evaluation of a Three-Phase Bidirectional Isolated DC-DC Converter with Varying Transformer Configurations Using Phase-Shift Modulation and Burst-Mode Switching. *Energies* **2020**, *13*, 2836. [CrossRef]
171. Aswathi, C.; Lakshmiprabha, K. An Improved Full Bridge DC-DC Converter for Electric Vehicles. *Int. J. Technol. Eng. Syst.* **2014**, *6*, 89–94.
172. Sharma, A.; Agarwal, V. Performance Analysis of Full Bridge DC-DC Converter for Electrical Vehicle Application. 2021. Available online: https://ssrn.com/abstract=3808509 (accessed on 13 July 2022).
173. Kabalo, M.; Blunier, B.; Bouquain, D.; Miraoui, A. State-of-the-Art of DC-DC Converters for Fuel Cell Vehicles. In Proceedings of the IEEE Vehicle Power and Propulsion Conference, VPPC 2010, Lille, France, 1–3 September 2010; pp. 1–6.
174. Moghaddam, A.F.; Van den Bossche, A. Forward Converter Current Fed Equalizer for Lithium Based Batteries in Ultralight Electrical Vehicles. *Electronics* **2019**, *8*, 408. [CrossRef]
175. Joseph, P.K.; Devaraj, E. Design of Hybrid Forward Boost Converter for Renewable Energy Powered Electric Vehicle Charging Applications. *IET Power Electron.* **2019**, *12*, 2015–2021. [CrossRef]
176. Rehlaender, P.; Grote, T.; Schafmeister, F.; Boecker, J. Analytical Modeling and Design of an Active Clamp Forward Converter Applied as a Single-Stage On-Board DC-DC Converter for EVs. In Proceedings of the International Exhibition and Conference for Power Electronics, Intelligent Motion, Renewable Energy and Energy Management, Nuremberg, Germany, 7–9 May 2019; IEEE: Nuremberg, Germany, 2019; pp. 1–8.
177. Zhang, Y.; Zhang, W.; Gao, F.; Gao, S.; Rogers, D.J. A Switched-Capacitor Interleaved Bidirectional Converter with Wide Voltage-Gain Range for Super Capacitors in EVs. *IEEE Trans. Power Electron.* **2020**, *35*, 1536–1547. [CrossRef]
178. Salehahari, S.; Babaei, E. A New Hybrid Multilevel Inverter Based on Coupled-Inductor and Cascaded H-Bridge. In Proceedings of the International Conference on Electrical Engineering/Electronics, Computer, Telecommunications and Information Technology, Chiang Mai, Thailand, 28 June–1 July 2016; pp. 1–6.

179. Xian, Z.; Wang, G. Optimal Dispatch of Electric Vehicle Batteries between Battery Swapping Stations and Charging Stations. In Proceedings of the IEEE Power and Energy Society General Meeting, Boston, MA, USA, 17–21 July 2016; pp. 1–6.
180. Ahmad, A.; Khan, Z.A.; Saad Alam, M.; Khateeb, S. A Review of the Electric Vehicle Charging Techniques, Standards, Progression and Evolution of EV Technologies in Germany. *Smart Sci.* **2018**, *6*, 36–53. [CrossRef]
181. Brenna, M.; Foiadelli, F.; Leone, C.; Longo, M. Electric Vehicles Charging Technology Review and Optimal Size Estimation. *J. Electr. Eng. Technol.* **2020**, *15*, 2539–2552. [CrossRef]
182. Dericioglu, Ç.; Yirik, E.; Ünal, E.; Cuma, M.U.; Onur, B.; Tumay, M. A Review of Charging Technologies for Commercial Electric Vehicles. *Int. J. Adv. Automot. Technol.* **2018**, *2*, 61–70. [CrossRef]
183. Shareef, H.; Islam, M.M.; Mohamed, A. A Review of the Stage-of-the-Art Charging Technologies, Placement Methodologies, and Impacts of Electric Vehicles. *Renew. Sustain. Energy Rev.* **2016**, *64*, 403–420. [CrossRef]
184. Khalid, M.R.; Alam, M.S.; Sarwar, A.; Jamil Asghar, M.S. A Comprehensive Review on Electric Vehicles Charging Infrastructures and Their Impacts on Power-Quality of the Utility Grid. *eTransportation* **2019**, *1*, 100006. [CrossRef]
185. Kongjeen, Y.; Bhumkittipich, K. Impact of Plug-in Electric Vehicles Integrated into Power Distribution System Based on Voltage-Dependent Power Flow Analysis. *Energies* **2018**, *11*, 1571. [CrossRef]
186. Sbordone, D.; Bertini, I.; Di Pietra, B.; Falvo, M.C.; Genovese, A.; Martirano, L. EV Fast Charging Stations and Energy Storage Technologies: A Real Implementation in the Smart Micro Grid Paradigm. *Electr. Power Syst. Res.* **2015**, *120*, 96–108. [CrossRef]
187. IEC 62196-1:2022 | Plugs, Socket-Outlets, Vehicle Connectors and Vehicle Inlets—Conductive Charging of Electric Vehicles—Part 1: General Requirements. Available online: https://webstore.iec.ch/publication/59922 (accessed on 13 July 2022).
188. J1772_201710: SAE Electric Vehicle and Plug in Hybrid Electric Vehicle Conductive Charge Coupler—SAE International. Available online: https://www.sae.org/standards/content/j1772_201710/ (accessed on 13 July 2022).
189. Mohamed, A.A.S.; Wood, E.; Meintz, A. In-Route Inductive versus Stationary Conductive Charging for Shared Automated Electric Vehicles: A University Shuttle Service. *Appl. Energy* **2021**, *282*, 116132. [CrossRef]
190. Mohamed, A.A.S.; Shaier, A.A.; Metwally, H.; Selem, S.I. A Comprehensive Overview of Inductive Pad in Electric Vehicles Stationary Charging. *Appl. Energy* **2020**, *262*, 114584. [CrossRef]
191. Kim, S.; Tejeda, A.; Covic, G.A.; Boys, J.T. Analysis of Mutually Decoupled Primary Coils for IPT Systems for EV Charging. In Proceedings of the ECCE 2016—IEEE Energy Conversion Congress and Exposition, Milwaukee, WI, USA, 18–22 September 2016; pp. 1–6.
192. Wu, H.H.; Gilchrist, A.; Sealy, K.; Israelsen, P.; Muhs, J. A Review on Inductive Charging for Electric Vehicles. In Proceedings of the IEEE International Electric Machines and Drives Conference, Niagara Falls, ON, Canada, 15–18 May 2011; pp. 143–147.
193. Al-Saadi, M.; Al-Bahrani, L.; Al-Qaisi, M.; Al-Chlaihawi, S.; Crăciunescu, A. Capacitive Power Transfer for Wireless Batteries Charging. *EEA-Electroteh. Electron. Autom.* **2018**, *66*, 40–51.
194. Dai, J.; Ludois, D.C. A Survey of Wireless Power Transfer and a Critical Comparison of Inductive and Capacitive Coupling for Small Gap Applications. *IEEE Trans. Power Electron.* **2015**, *30*, 6017–6029. [CrossRef]
195. Lukic, S.; Pantic, Z. Cutting the Cord: Static and Dynamic Inductive Wireless Charging of Electric Vehicles. *IEEE Electrif. Mag.* **2013**, *1*, 57–64. [CrossRef]
196. Sanguesa, J.A.; Torres-Sanz, V.; Garrido, P.; Martinez, F.J.; Marquez-Barja, J.M. A Review on Electric Vehicles: Technologies and Challenges. *Smart Cities* **2021**, *4*, 372–404. [CrossRef]
197. Is Battery Swapping a Great Alternatives to Charging?—E-Vehicleinfo. Available online: https://e-vehicleinfo.com/is-battery-swapping-a-great-alternatives-to-charging/ (accessed on 13 July 2022).
198. Hannan, M.A.; Lipu, M.S.H.; Ker, P.J.; Begum, R.A.; Agelidis, V.G.; Blaabjerg, F. Power Electronics Contribution to Renewable Energy Conversion Addressing Emission Reduction: Applications, Issues, and Recommendations. *Appl. Energy* **2019**, *251*, 113404. [CrossRef]
199. Enang, W.; Bannister, C. Modelling and Control of Hybrid Electric Vehicles (A Comprehensive Review). *Renew. Sustain. Energy Rev.* **2017**, *74*, 1210–1239. [CrossRef]
200. Norouzi, A.; Heidarifar, H.; Shahbakhti, M.; Koch, C.R.; Borhan, H. Model Predictive Control of Internal Combustion Engines: A Review and Future Directions. *Energies* **2021**, *14*, 6251. [CrossRef]
201. Lin, C.C.; Kang, J.M.; Grizzle, J.W.; Peng, H. Energy Management Strategy for a Parallel Hybrid Electric Truck. In Proceedings of the American Control Conference, Arlington, VA, USA, 25–27 June 2001; pp. 2878–2883.
202. Zeynali, S.; Rostami, N.; Ahmadian, A.; Elkamel, A. Two-Stage Stochastic Home Energy Management Strategy Considering Electric Vehicle and Battery Energy Storage System: An ANN-Based Scenario Generation Methodology. *Sustain. Energy Technol. Assess.* **2020**, *39*, 100722. [CrossRef]
203. Alsharif, A.; Tan, C.W.; Ayop, R.; Lau, K.Y.; Dobi, A.M.D. A Rule-Based Power Management Strategy for Vehicle-to-Grid System Using Antlion Sizing Optimization. *J. Energy Storage* **2021**, *41*, 102913. [CrossRef]
204. Tran, D.D.; Vafaeipour, M.; El Baghdadi, M.; Barrero, R.; Van Mierlo, J.; Hegazy, O. Thorough State-of-the-Art Analysis of Electric and Hybrid Vehicle Powertrains: Topologies and Integrated Energy Management Strategies. *Renew. Sustain. Energy Rev.* **2020**, *119*, 109596. [CrossRef]
205. Hofman, T.; Steinbuch, M.; Van Druten, R.; Serrarens, A. Rule-Based Energy Management Strategies for Hybrid Vehicles. *Int. J. Electr. Hybrid Veh.* **2007**, *1*, 71–94. [CrossRef]

206. Mohd Sabri, M.F.; Danapalasingam, K.A.; Rahmat, M.F. Improved Fuel Economy of Through-the-Road Hybrid Electric Vehicle with Fuzzy Logic-Based Energy Management Strategy. *Int. J. Fuzzy Syst.* **2018**, *20*, 2677–2692. [CrossRef]
207. Nageshrao, S.P.; Jacob, J.; Wilkins, S. Charging Cost Optimization for EV Buses Using Neural Network Based Energy Predictor. *IFAC-PapersOnLine* **2017**, *50*, 5947–5952. [CrossRef]
208. Delprat, S.; Guerra, T.M.; Paganelli, G.; Lauber, J.; Delhom, M. Control Strategy Optimization for an Hybrid Parallel Powertrain. *Proc. Am. Control Conf.* **2001**, *2*, 1315–1320. [CrossRef]
209. Hou, C.; Ouyang, M.; Xu, L.; Wang, H. Approximate Pontryagin's Minimum Principle Applied to the Energy Management of Plug-in Hybrid Electric Vehicles. *Appl. Energy* **2014**, *115*, 174–189. [CrossRef]
210. Kirkpatrick, S.; Gelatt, C.D.; Vecchi, M.P. Optimization by Simulated Annealing. *Science* **1983**, *220*, 671–680. [CrossRef]
211. Chen, Z.; Mi, C.C.; Xia, B.; You, C. Energy Management of Power-Split Plug-in Hybrid Electric Vehicles Based on Simulated Annealing and Pontryagin's Minimum Principle. *J. Power Sources* **2014**, *272*, 160–168. [CrossRef]
212. Trovão, J.P.F.; Member, S.; Santos, V.D.N.; Pereirinha, P.G.; Jorge, H.M.; Antunes, C.H. A Simulated Annealing Approach for Optimal Power Source Management in a Small EV. *IEEE Trans. Sustain. Energy* **2013**, *4*, 867–876. [CrossRef]
213. Song, K.; Wang, X.; Li, F.; Sorrentino, M.; Zheng, B. Pontryagin's Minimum Principle-Based Real-Time Energy Management Strategy for Fuel Cell Hybrid Electric Vehicle Considering Both Fuel Economy and Power Source Durability. *Energy* **2020**, *205*, 118064. [CrossRef]
214. Wang, A.; Yang, W. Design of Energy Management Strategy in Hybrid Electric Vehicles by Evolutionary Fuzzy System Part II: Tuning Fuzzy Controller by Genetic Algorithms. *Proc. World Congr. Intell. Control Autom.* **2006**, *2*, 8329–8333. [CrossRef]
215. Altundogan, T.G.; Yildiz, A.; Karakose, E. Genetic Algorithm Approach Based on Graph Theory for Location Optimization of Electric Vehicle Charging Stations. In Proceedings of the 2021 Innovations in Intelligent Systems and Applications Conference (ASYU), Elazig, Turkey, 6–8 October 2021; pp. 1–5. [CrossRef]
216. Li, H.C.; Lu, C.C.; Eccarius, T.; Hsieh, M.Y. Genetic Algorithm with an Event-Based Simulator for Solving the Fleet Allocation Problem in an Electric Vehicle Sharing System. *Asian Transp. Stud.* **2022**, *8*, 100060. [CrossRef]
217. Hegazy, O.; Van Mierlo, J. Particle Swarm Optimization for Optimal Powertrain Component Sizing and Design of Fuel Cell Hybrid Electric Vehicle. In Proceedings of the International Conference on Optimisation of Electrical and Electronic Equipment, OPTIM, Brasov, Romania, 20–22 May 2010; pp. 601–609.
218. Zhang, X.; Wang, Z.; Lu, Z. Multi-Objective Load Dispatch for Microgrid with Electric Vehicles Using Modified Gravitational Search and Particle Swarm Optimization Algorithm. *Appl. Energy* **2022**, *306*, 118018. [CrossRef]
219. Yin, W.J.; Ming, Z.F. Electric Vehicle Charging and Discharging Scheduling Strategy Based on Local Search and Competitive Learning Particle Swarm Optimization Algorithm. *J. Energy Storage* **2021**, *42*, 102966. [CrossRef]
220. Wang, N.; Li, B.; Duan, Y.; Jia, S. A Multi-Energy Scheduling Strategy for Orderly Charging and Discharging of Electric Vehicles Based on Multi-Objective Particle Swarm Optimization. *Sustain. Energy Technol. Assess.* **2021**, *44*, 101037. [CrossRef]
221. Sadeghi, D.; Hesami Naghshbandy, A.; Bahramara, S. Optimal Sizing of Hybrid Renewable Energy Systems in Presence of Electric Vehicles Using Multi-Objective Particle Swarm Optimization. *Energy* **2020**, *209*, 118471. [CrossRef]
222. Kleimaier, A.; Schroeder, D. Optimization Strategy for Design and Control of a Hybrid Vehicle. In Proceedings of the 6th International Workshop on Advanced Motion Control. Proceedings (Cat. No.00TH8494), Nagoya, Japan, 30 March–1 April 2000; pp. 459–464. [CrossRef]
223. Un-Noor, F.; Padmanaban, S.; Mihet-Popa, L.; Mollah, M.N.; Hossain, E. A Comprehensive Study of Key Electric Vehicle (EV) Components, Technologies, Challenges, Impacts, and Future Direction of Development. *Energies* **2017**, *10*, 1217. [CrossRef]
224. Hussain, S.M.S.; Aftab, M.A.; Ali, I.; Ustun, T.S. IEC 61850 Based Energy Management System Using Plug-in Electric Vehicles and Distributed Generators during Emergencies. *Int. J. Electr. Power Energy Syst.* **2020**, *119*, 105873. [CrossRef]
225. Wang, H.; Xu, D.; Jia, G.; Mao, Z.; Gong, Y.; He, B.; Wang, R.; Fan, H.J. Integration of Flexibility, Cyclability and High-Capacity into One Electrode for Sodium-Ion Hybrid Capacitors with Low Self-Discharge Rate. *Energy Storage Mater.* **2020**, *25*, 114–123. [CrossRef]
226. Deb, N.; Singh, R.; Bai, H. Transformative Role of Silicon Carbide Power Electronics in Providing Low-Cost Extremely Fast Charging of Electric Vehicles. In Proceedings of the 2021 IEEE 4th International Conference on DC Microgrids, Arlington, VA, USA, 18–21 July 2021; pp. 1–6.
227. Aftab, M.A.; Hussain, S.M.S.; Ali, I.; Ustun, T.S. IEC 61850-Based Communication Layer Modeling for Electric Vehicles: Electric Vehicle Charging and Discharging Processes Based on the International Electrotechnical Commission 61850 Standard and Its Extensions. *IEEE Ind. Electron. Mag.* **2020**, *14*, 4–14. [CrossRef]
228. Lin, C.H.; Amir, M.; Tariq, M.; Shahvez, M.; Alamri, B.; Alahmadi, A.; Siddiqui, M.; Beig, A.R. Comprehensive Analysis of IPT v/s CPT for Wireless EV Charging and Effect of Capacitor Plate Shape and Foreign Particle on CPT. *Processes* **2021**, *9*, 1619. [CrossRef]
229. Luo, Z.; Nie, S.; Pathmanathan, M.; Lehn, P.W. Exciter-Quadrature-Repeater Transmitter for Wireless Electric Vehicle Charging with High Lateral Misalignment Tolerance and Low EMF Emission. *IEEE Trans. Transp. Electrif.* **2021**, *7*, 2156–2167. [CrossRef]
230. Bracco, S.; Delfino, F.; Pampararo, F.; Robba, M.; Rossi, M. A Dynamic Optimization-Based Architecture for Polygeneration Microgrids with Tri-Generation, Renewables, Storage Systems and Electrical Vehicles. *Energy Convers. Manag.* **2015**, *96*, 511–520. [CrossRef]
231. Abdolrasol, M.G.M.; Hannan, M.A.; Hussain, S.M.S.; Ustun, T.S.; Sarker, M.R.; Ker, P.J. Energy Management Scheduling for Microgrids in the Virtual Power Plant System Using Artificial Neural Networks. *Energies* **2021**, *14*, 6507. [CrossRef]

232. Okorie, U.; Ukaegbu, J.; Ajuwon, S.; Noah, O.; Oluwo, A.A.; Ogedengbe, E.O.B. Towards an Optimal Wind Energy Harvest for Ev Charging Station: Design of a Data Acquisition System and Aerodynamic Modeling of Hvawt Blade. In Proceedings of the AIAA Propulsion and Energy 2020 Forum, Online, 24–28 August 2020; pp. 1–13. [CrossRef]
233. Englberger, S.; Abo Gamra, K.; Tepe, B.; Schreiber, M.; Jossen, A.; Hesse, H. Electric Vehicle Multi-Use: Optimizing Multiple Value Streams Using Mobile Storage Systems in a Vehicle-to-Grid Context. *Appl. Energy* **2021**, *304*, 117862. [CrossRef]
234. Goel, S.; Sharma, R.; Rathore, A.K. A Review on Barrier and Challenges of Electric Vehicle in India and Vehicle to Grid Optimisation. *Transp. Eng.* **2021**, *4*, 100057. [CrossRef]
235. Gaines, L.L.; Dunn, J.B. *Lithium-Ion Battery Environmental Impacts*; Elsevier: Amsterdam, The Netherlands, 2014; ISBN 9780444595133.
236. Bai, Y.; Muralidharan, N.; Sun, Y.K.; Passerini, S.; Whittingham, M.S.; Belharouak, I. Energy and Environmental Aspects in Recycling Lithium-Ion Batteries: Concept of Battery Identity Global Passport. *Mater. Today* **2020**, *41*, 304–315. [CrossRef]
237. Wang, Y.; Wang, L.; Li, M.; Chen, Z. A Review of Key Issues for Control and Management in Battery and Ultra-Capacitor Hybrid Energy Storage Systems. *eTransportation* **2020**, *4*, 100064. [CrossRef]
238. Second Global Sustainable Transport Conference | United Nations. Available online: https://www.un.org/en/conferences/transport2021 (accessed on 13 July 2022).
239. Nations, U. Sustainable Transport, Sustainable Development. In *Interagency Report for Second Global Sustainable Transport Conference*; United Nations: San Francisco, CA, USA, 2021.
240. Global EV Outlook 2020—Analysis—IEA. Available online: https://www.iea.org/reports/global-ev-outlook-2020 (accessed on 13 July 2022).
241. Chen, X.; Wang, J.; Griffo, A.; Chen, L. Evaluation of Waste Heat Recovery of Electrical Powertrain with Electro-Thermally Coupled Models for Electric Vehicle Applications. *Chin. J. Electr. Eng.* **2021**, *7*, 88–99. [CrossRef]
242. Christensen, C.; Salmon, J. EV Adoption Influence on Air Quality and Associated Infrastructure Costs. *World Electr. Veh. J.* **2021**, *12*, 207. [CrossRef]
243. Horton, D.E.; Schnell, J.L.; Peters, D.R.; Wong, D.C.; Lu, X.; Gao, H.; Zhang, H.; Kinney, P.L. Effect of Adoption of Electric Vehicles on Public Health and Air Pollution in China: A Modelling Study. *Lancet Planet. Health* **2021**, *5*, S8. [CrossRef]
244. Previati, G.; Mastinu, G.; Gobbi, M. Thermal Management of Electrified Vehicles—A Review. *Energies* **2022**, *15*, 1326. [CrossRef]
245. Ibrahim, B.S.K.K.; Aziah, N.M.A.; Ahmad, S.; Akmeliawati, R.; Nizam, H.M.I.; Muthalif, A.G.A.; Toha, S.F.; Hassan, M.K. Fuzzy-Based Temperature and Humidity Control for HVAC of Electric Vehicle. *Procedia Eng.* **2012**, *41*, 904–910. [CrossRef]
246. Dreißigacker, V. Thermal Battery for Electric Vehicles: High-Temperature Heating System for Solid Media Based Thermal Energy Storages. *Appl. Energy* **2021**, *11*, 10500. [CrossRef]
247. Onat, N.C.; Kucukvar, M.; Aboushaqrah, N.N.M.; Jabbar, R. How Sustainable Is Electric Mobility? A Comprehensive Sustainability Assessment Approach for the Case of Qatar. *Appl. Energy* **2019**, *250*, 461–477. [CrossRef]
248. Olatomiwa, L.; Blanchard, R.; Mekhilef, S.; Akinyele, D. Hybrid Renewable Energy Supply for Rural Healthcare Facilities: An Approach to Quality Healthcare Delivery. *Sustain. Energy Technol. Assess.* **2018**, *30*, 121–138. [CrossRef]
249. Adair-Rohani, H.; Zukor, K.; Bonjour, S.; Wilburn, S.; Kuesel, A.C.; Hebert, R.; Fletcher, E.R. Limited Electricity Access in Health Facilities of Sub-Saharan Africa: A Systematic Review of Data on Electricity Access, Sources, and Reliability. *Glob. Health Sci. Pract.* **2013**, *1*, 249–261. [CrossRef]
250. Kang, J.; Kan, C.; Lin, Z. Are Electric Vehicles Reshaping the City? An Investigation of the Clustering of Electric Vehicle Owners' Dwellings and Their Interaction with Urban Spaces. *Int. J. Geo-Inf.* **2021**, *10*, 320. [CrossRef]
251. Shaukat, N.; Khan, B.; Ali, S.M.; Mehmood, C.A.; Khan, J.; Farid, U.; Majid, M.; Anwar, S.M.; Jawad, M.; Ullah, Z. A Survey on Electric Vehicle Transportation within Smart Grid System. *Renew. Sustain. Energy Rev.* **2018**, *81*, 1329–1349. [CrossRef]
252. Andreas, P.; Bergaentzlé, C.; Græsted, I.; Scheller, F. From Passive to Active: Flexibility from Electric Vehicles in the Context of Transmission System Development. *Appl. Energy* **2020**, *277*, 115526. [CrossRef]
253. Dehghani-Sanij, A.R.; Tharumalingam, E.; Dusseault, M.B.; Fraser, R. Study of Energy Storage Systems and Environmental Challenges of Batteries. *Renew. Sustain. Energy Rev.* **2019**, *104*, 192–208. [CrossRef]
254. Chellaswamy, C.; Balaji, L.; Kaliraja, T. Renewable Energy Based Automatic Recharging Mechanism for Full Electric Vehicle. *Eng. Sci. Technol. Int. J.* **2020**, *23*, 555–564. [CrossRef]
255. Valladolid, J.D.; Patino, D.; Gruosso, G.; Correa-Florez, C.A.; Vuelvas, J.; Espinoza, F. A Novel Energy-Efficiency Optimization Approach Based on Driving Patterns Styles and Experimental Tests for Electric Vehicles. *Electronics* **2021**, *10*, 1199. [CrossRef]
256. Ibrahim, A.; Jiang, F. The Electric Vehicle Energy Management: An Overview of the Energy System and Related Modeling and Simulation. *Renew. Sustain. Energy Rev.* **2021**, *144*, 111049. [CrossRef]
257. Cao, Y.; Huang, L.; Li, Y.; Jermsittiparsert, K.; Ahmadi-nezamabad, H.; Nojavan, S. Electrical Power and Energy Systems Optimal Scheduling of Electric Vehicles Aggregator under Market Price Uncertainty Using Robust Optimization Technique. *Electr. Power Energy Syst.* **2020**, *117*, 105628. [CrossRef]
258. Atawi, I.E.; Hendawi, E.; Zaid, S.A. Analysis and Design of a Standalone Electric Vehicle Charging Station Supplied by Photovoltaic Energy. *Processes* **2021**, *9*, 1246. [CrossRef]
259. Bamisile, O.; Babatunde, A.; Adun, H.; Yimen, N.; Mukhtar, M.; Huang, Q.; Hu, W. Electrification and Renewable Energy Nexus in Developing Countries ; an Overarching Analysis of Hydrogen Production and Electric Vehicles Integrality in Renewable Energy Penetration. *Energy Convers. Manag.* **2021**, *236*, 114023. [CrossRef]

260. Li, C.; Shan, Y.; Zhang, L.; Zhang, L.; Fu, R. Techno-Economic Evaluation of Electric Vehicle Charging Stations Based on Hybrid Renewable Energy in China. *Energy Strateg. Rev.* **2022**, *41*, 100850. [CrossRef]
261. Mateo, C.; Reneses, J.; Rodriguez-Calvo, A.; Frias, P.; Sanchez, A. Cost–Benefit Analysis of Battery Storage in Medium-Voltage Distribution Networks. *IET Gener. Transm. Distrib.* **2016**, *10*, 815–821. [CrossRef]
262. Costa, E.; Wells, P.; Wang, L.; Costa, G. The Electric Vehicle and Renewable Energy: Changes in Boundary Conditions That Enhance Business Model Innovations. *J. Clean. Prod.* **2022**, *333*, 130034. [CrossRef]
263. Andrew, Y.; Ng, A.W.; Yu, Z.; Huang, J.; Meng, K.; Dong, Z.Y. A Review of Evolutionary Policy Incentives for Sustainable Development of Electric Vehicles in China: Strategic Implications. *Energy Policy* **2021**, *148*, 111983. [CrossRef]
264. Victor, D.G.; Geels, F.W.; Sharpe, S. *Accelerating the Low Carbon Transition: The Case for Stronger, More Targeted and Coordinated International Action*; U.N. Climate Negotiations, COP 25: Madrid, Spain, 2019.
265. Biden Starts $3 Billion Plan to Boost Battery Production for EVs . Available online: https://www.cnbc.com/2022/05/02/biden-starts-3-billion-plan-to-boost-battery-production-for-evs.html (accessed on 21 July 2022).
266. Lebeau, K.; Van Mierlo, J.; Lebeau, P.; Mairesse, O.; Macharis, C. The Market Potential for Plug-in Hybrid and Battery Electric Vehicles in Flanders: A Choice-Based Conjoint Analysis. *Transp. Res. PART D* **2020**, *17*, 592–597. [CrossRef]
267. Malmgren, I. Quantifying the Societal Benefits of Electric Vehicles. *World Electr. Veh. J.* **2016**, *8*, 996–1007. [CrossRef]
268. Ram, M.; Osorio-Aravena, J.C.; Aghahosseini, A.; Bogdanov, D.; Breyer, C. Job Creation during a Climate Compliant Global Energy Transition across the Power, Heat, Transport, and Desalination Sectors by 2050. *Energy* **2022**, *238*, 121690. [CrossRef]
269. Oliveira, C.; Coelho, D.; Pereira Da Silva, P.; Antunes, C.H. How Many Jobs Can the RES-E Sectors Generate in the Portuguese Context? *Renew. Sustain. Energy Rev.* **2013**, *21*, 444–455. [CrossRef]
270. Mazzeo, D. Nocturnal Electric Vehicle Charging Interacting with a Residential Photovoltaic-Battery System: A 3E (Energy, Economic and Environmental) Analysis. *Energy* **2019**, *168*, 310–331. [CrossRef]
271. Brönner, M.; Hagenauer, M.S.; Lienkamp, M. Sustainability—Recommendations for an Electric Vehicle Manufacturing in Sub-Saharan Africa. *Procedia CIRP* **2019**, *81*, 1148–1153. [CrossRef]
272. Charles, R.G.; Davies, M.L.; Douglas, P.; Hallin, I.L.; Mabbett, I. Sustainable Energy Storage for Solar Home Systems in Rural Sub-Saharan Africa—A Comparative Examination of Lifecycle Aspects of Battery Technologies for Circular Economy, with Emphasis on the South African Context. *Energy* **2019**, *166*, 1207–1215. [CrossRef]
273. Pardo-Bosch, F.; Pujadas, P.; Morton, C.; Cervera, C. Sustainable Deployment of an Electric Vehicle Public Charging Infrastructure Network from a City Business Model Perspective. *Sustain. Cities Soc.* **2021**, *71*, 102957. [CrossRef]
274. Lopez-Behar, D.; Tran, M.; Mayaud, J.R.; Froese, T.; Herrera, O.E.; Merida, W. Putting Electric Vehicles on the Map: A Policy Agenda for Residential Charging Infrastructure in Canada. *Energy Res. Soc. Sci.* **2019**, *50*, 29–37. [CrossRef]
275. Yang, X.; Zhang, Y. A Comprehensive Review on Electric Vehicles Integrated in Virtual Power Plants. *Sustain. Energy Technol. Assess.* **2021**, *48*, 101678. [CrossRef]
276. Luo, L.; Wu, Z.; Gu, W.; Huang, H.; Gao, S.; Han, J. Coordinated Allocation of Distributed Generation Resources and Electric Vehicle Charging Stations in Distribution Systems with Vehicle-to-Grid Interaction. *Energy* **2020**, *192*, 116631. [CrossRef]
277. Singh, B.; Dubey, P.K. Distributed Power Generation Planning for Distribution Networks Using Electric Vehicles: Systematic Attention to Challenges and Opportunities. *J. Energy Storage* **2022**, *48*, 104030. [CrossRef]
278. Jiang, J.; Yu, L.; Zhang, X.; Ding, X.; Wu, C. EV-Based Reconfigurable Smart Grid Management Using Support Vector Regression Learning Technique Machine Learning. *Sustain. Cities Soc.* **2022**, *76*, 103477. [CrossRef]
279. Jiao, F.; Zou, Y.; Zhang, X.; Zhang, B. Online Optimal Dispatch Based on Combined Robust and Stochastic Model Predictive Control for a Microgrid Including EV Charging Station. *Energy* **2022**, *247*, 123220. [CrossRef]
280. Pellow, M.A.; Ambrose, H.; Mulvaney, D.; Betita, R.; Shaw, S. Research Gaps in Environmental Life Cycle Assessments of Lithium Ion Batteries for Grid-Scale Stationary Energy Storage Systems: End-of-Life Options and Other Issues. *Sustain. Mater. Technol.* **2020**, *23*, e00120. [CrossRef]
281. Ciez, R.E.; Whitacre, J.F. Examining Different Recycling Processes for Lithium-Ion Batteries. *Nat. Sustain.* **2019**, *2*, 148–156. [CrossRef]
282. Dewulf, J.; Van der Vorst, G.; Denturck, K.; Van Langenhove, H.; Ghyoot, W.; Tytgat, J.; Vandeputte, K. Recycling Rechargeable Lithium Ion Batteries: Critical Analysis of Natural Resource Savings. *Resour. Conserv. Recycl.* **2010**, *54*, 229–234. [CrossRef]
283. Sofana, R.; Venugopal, P.; Ravi, V.; Alhelou, H.H.; Al-hinai, A.; Siano, P. Analysis of Electric Vehicles with an Economic Perspective for the Future Electric Market. *Futur. Internet* **2022**, *14*, 172.
284. Chen, J.; Todd, J.; Clogston, F. *Creating the Clean Energy Economy: Analysis of Three Clean Energy Industries*; International Economic Development Council: Washington, DC, USA, 2013.
285. Roland-holst, D. *Plug-in Electric Vehicle Deployment in California: An Economic Assessment*; University of California: Berkeley, CA, USA, 2012.
286. Cortright, J. *New York City's Green Dividend*; Senior Policy Advisor, CEOs for Cities. 2010. Available online: https://citeseerx.ist.psu.edu/viewdoc/download?doi=10.1.1.405.3057&rep=rep1&type=pdf (accessed on 18 August 2022).
287. Wesseh, P.K.; Benjamin, N.I.; Lin, B. The Coordination of Pumped Hydro Storage, Electric Vehicles, and Climate Policy in Imperfect Electricity Markets: Insights from China. *Renew. Sustain. Energy Rev.* **2022**, *160*, 112275. [CrossRef]
288. Hansen, K.; Mathiesen, B.V.; Skov, I.R. Full Energy System Transition towards 100% Renewable Energy in Germany in 2050. *Renew. Sustain. Energy Rev.* **2019**, *102*, 1–13. [CrossRef]

289. Zappa, W.; Junginger, M.; Van Den Broek, M.; Cover, C.L. Is a 100% Renewable European Power System Feasible by 2050? *Appl. Energy* **2019**, *233–234*, 1027–1050. [CrossRef]
290. Walmsley, M.R.W.; Walmsley, T.G.; Atkins, M.J.; Kamp, P.J.J.; Neale, J.R. Minimising Carbon Emissions and Energy Expended for Electricity Generation in New Zealand through to 2050. *Appl. Energy* **2014**, *135*, 656–665. [CrossRef]
291. Choma, E.F.; Evans, J.S.; Hammitt, J.K.; Gómez-ibáñez, J.A.; Spengler, J.D. Assessing the Health Impacts of Electric Vehicles through Air Pollution in the United States. *Environ. Int.* **2020**, *144*, 106015. [CrossRef] [PubMed]
292. Haller, M.; Ludig, S.; Bauer, N. Bridging the Scales: A Conceptual Model for Coordinated Expansion of Renewable Power Generation, Transmission and Storage. *Renew. Sustain. Energy Rev.* **2012**, *16*, 2687–2695. [CrossRef]
293. Shu, X.; Guo, Y.; Yang, W.; Wei, K.; Zhu, G. Life-Cycle Assessment of the Environmental Impact of the Batteries Used in Pure Electric Passenger Cars. *Energy Rep.* **2021**, *7*, 2302–2315. [CrossRef]
294. Milev, G.; Hastings, A.; Al-Habaibeh, A. The Environmental and Financial Implications of Expanding the Use of Electric Cars—A Case Study of Scotland. *Energy Built Environ.* **2021**, *2*, 204–213. [CrossRef]
295. He, H.; Tian, S.; Tarroja, B.; Ogunseitan, O.A.; Samuelsen, S.; Schoenung, J.M. Flow Battery Production: Materials Selection and Environmental Impact. *J. Clean. Prod.* **2020**, *269*, 121740. [CrossRef]
296. McManus, M.C. Environmental Consequences of the Use of Batteries in Low Carbon Systems: The Impact of Battery Production. *Appl. Energy* **2012**, *93*, 288–295. [CrossRef]
297. Holland, S.P.; Mansur, E.T.; Muller, N.; Yates, A.J. Decompositions and Policy Consequences of an Extraordinary Decline in Air Pollution from Electricity Generation. *Am. Econ. J. Econ. Policy* **2020**, *12*, 244–274. [CrossRef]

Review

Review on Battery Packing Design Strategies for Superior Thermal Management in Electric Vehicles

Robby Dwianto Widyantara [1,2], Siti Zulaikah [3], Firman Bagja Juangsa [1], Bentang Arief Budiman [1,3,*] and Muhammad Aziz [2,*]

1 Faculty of Mechanical and Aerospace Engineering, Institut Teknologi Bandung, Bandung 40132, Indonesia
2 Institute of Industrial Science, The University of Tokyo, Tokyo 153-8505, Japan
3 National Center for Sustainable Transportation Technology, Institut Teknologi Bandung, Bandung 40132, Indonesia
* Correspondence: bentang@itb.ac.id (B.A.B.); maziz@iis.u-tokyo.ac.jp (M.A.)

Abstract: In the last decades of electric vehicle (EV) development, battery thermal management has become one of the remaining issues that must be appropriately handled to ensure robust EV design. Starting from researching safer and more durable battery cells that can resist thermal exposure, battery packing design has also become important to avoid thermal events causing an explosion or at least to prevent fatal loss if the explosion occurs. An optimal battery packing design can maintain the battery cell temperature at the most favorable range, i.e., 25–40 °C, with a temperature difference in each battery cell of 5 °C at the maximum, which is considered the best working temperature. The design must also consider environmental temperature and humidity effects. Many design strategies have been reported, including novel battery pack constructions, a better selection of coolant materials, and a robust battery management system. However, those endeavors are faced with the main challenges in terms of design constraints that must be fulfilled, such as material and manufacturing costs, limited available battery space and weight, and low energy consumption requirements. This work reviewed and analyzed the recent progress and current state-of-the-art in designing battery packs for superior thermal management. The narration focused on significant findings that have solved the battery thermal management design problem as well as the remaining issues and opportunities to obtain more reliable and enduring batteries for EVs. Furthermore, some recommendations for future research topics supporting the advancement of battery thermal management design were also discussed.

Keywords: battery pack; design strategies; thermal management; lithium-ion battery; electric vehicles

1. Introduction

Lithium-ion batteries (LIBs), as energy storage devices, are considered to have a significant role in determining the performance of electric vehicles (EVs). LIBs depend vastly on temperature for facilitating an optimum performance and lifetime, with the commonly suggested operating temperature being in the range from 25 to 40 °C [1,2]. Operation at lower or higher temperatures than the range can adversely affect or degrade the performance and lifetime of LIBs. Lower-temperature operation with a high current rate leads to lithium plating on the anode, capacity loss, impedance rises, ionic conductivity decreases, and internal short circuits due to metallic lithium dendrites [3,4]. Meanwhile, higher-temperature operation could reduce the active materials and increase the internal resistance, even for overly high temperatures, potentially causing thermal runaways that might prompt fires and explosions [5].

LIBs are packaged in relatively small sizes, called battery cells, to allow effective manufacturing; therefore, LIBs need to be electrically arranged in series and parallel connections (called battery modules) to provide sufficient voltage and capacity in powering

EVs [6]. During EV operation, batteries can experience varying temperatures that lead to unbalanced performances. Each cell generates heat from internal resistance (Joule heating) and redox chemical reactions [7]. In the discharging process, the battery experiences an exothermic chemical reaction, while during the charging process the battery experiences an endothermic chemical reaction [8]. However, as the heat generated from Joule heating is relatively larger than that of the endothermic reactions during charging cycling, the battery cell can still generate heat, with the amount of generated heat depending on the state of charge (SoC) and power flow of the battery cell [9]. By this condition, each battery cell in the battery module might carry a different performance and SoC, leading to different temperatures. Several studies have suggested that the different temperatures between battery cells must not exceed 5 °C [1,2]. Accordingly, battery thermal management systems (BTMS) are demanded to ensure the operating temperatures are within an optimal range and are well distributed across all the battery cells [10–12].

A BTMS, as a part of battery packing, has a role in maintaining the thermal condition of batteries under optimal conditions by performing heat transfer from the battery to the outside when the battery is prone to overheat, and occasionally vice versa when the battery is in a low-temperature application. Several prior studies have reported various developed BTMS designs and strategies, which can be categorized based on the working principle and the coolant material used to transport the heat. Under the working principle, the heat can either be transported by direct contact between the coolant and battery cell or by indirect contact through a pipe as a heat transfer medium (see Figure 1a) [13,14]. The significant factors influencing the thermal management performance of a direct contact BTMS are the battery pack geometry [15], the space between batteries [16], and the battery layout configuration [17]. When using a heat transfer medium, the significant factors that must be considered include the contact area [18] and pipe dimensions, such as height and thickness [13]. Both types also share some of the same significant factors, such as the coolant material [19], velocity [20], and temperature [21]. In the meantime, the coolant materials include air-cooled [21–23], liquid-cooled [24,25], and phase-change materials (PCMs) [26,27] (see Figure 1b). Each of these BTMS types have particular advantages and disadvantages. An air-cooled BTMS has a simple construction that makes it low-cost, but it still suffers from the low heat capacity of the air as the heat transfer medium.

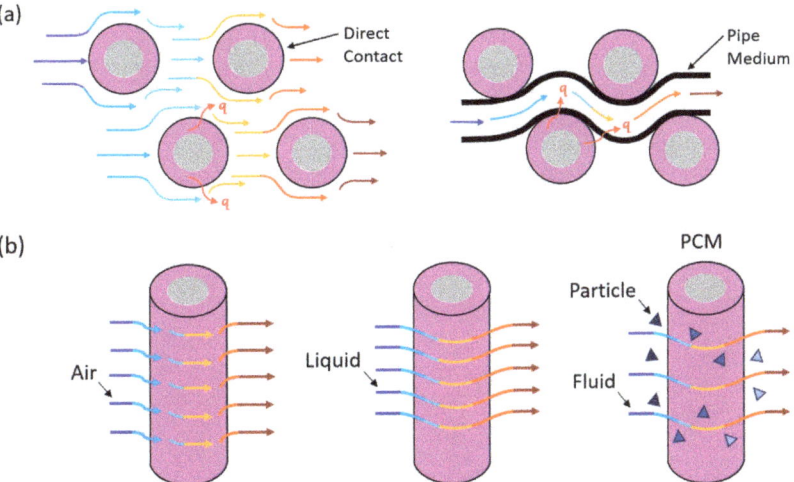

Figure 1. (a) Working principle of thermal transport from the battery to the environment and vice versa and (b) typical coolant materials for battery thermal management. The battery illustration only shows cylindrical cell types, but it also applies to prismatic and pouch cells.

On the other hand, a liquid-cooled BTMS utilizes liquid with a higher heat capacity than air, but practical EV applications require a high conductivity pipe, which leads to a complicated construction [21]. The pipe also possesses leakage risks and potentially prompts battery cell corrosion and short circuit. Subsequently, solid–liquid-phase-transitioning PCMs, mostly based on paraffin, have been developed to increase heat capacity and are widely suggested for EV battery cooling systems [28]. A PCM-based BTMS has a high efficiency and stable performance under extreme conditions due to its ability to store heat during the phase change process; however, the material has a low heat conductivity and needs to be regenerated after being completely melted [29]. Based on the working principle and the coolant materials used, more advanced battery packing design strategies have been presently proposed, such as air-cooled batteries with liquid cooling [30], liquid cooling with a heat pipe [31], and PCMs with a heat pipe [32]. For low-temperature operation, a preheating mechanism is utilized to initially condition the batteries under a working temperature. Preheating can be categorized into external preheating using air [33], liquid [34], or PCMs [35] and internal preheating using excitation by direct current [36], alternating current [37], and pulse preheating [38]. These strategies are applicable to different battery geometries, such as cylindrical, prismatic, and pouch, even though each type has its own thermal characteristics. Concretely, pouch cells tend to have maximum temperature near the terminal tabs due to the high concentration of local current densities. Cylindrical cells can have higher core temperatures than surface temperatures due to adiabatic conditions and higher thermal resistance at the cell core. In terms of thermal management, cylindrical cells have the advantage of a high heat transfer area, which can improve heat transfer by the BTMS. Prismatic cells also have their maximum temperature at their core and different temperatures across their thickness. However, in prismatic cells, the temperature rise might be mitigated by its large thermal mass. An additional fin also can be added in the case of prismatic cells to improve heat transfer by the BTMS [39,40]. BTMSs are expected to ensure that the battery temperature can be managed at optimal operating temperatures with a minimal temperature difference for each battery cell. In achieving the main objective of providing the optimal temperature and performance of LIBs, some challenges must be tackled in designing a battery pack. The advancement of LIB technology using different kinds of battery constituent materials that keep bringing higher energy density batteries must also consider thermal properties as an essential battery issue. The limited space of EV compartments for the battery and the BTMS is also the reason for the urgency of LIB packing technology advancement. The climates and seasons that vary in different places of the world and at different times of the year must also be considered as one of the battery packing design requirements. Furthermore, the safety of the battery, drivers, environment, and the BTMS itself must also become the main points of view in the design. This review paper provides a comprehensive discussion of the challenges, the recent advancement of technology, and the opportunities for research in the design and development of BTMS for EVs.

2. Challenges in Managing the Thermal Aspect of Batteries

2.1. Novel Battery Materials for Higher Energy and Power Density Demand

From 2008 up to now, LIB technology has been improved by an increase in the volumetric energy density to more than five times, with nearly a 90% cost reduction in the battery pack level. This substantial progress was achieved due to research and development in active materials, electrode processing, and cell manufacturing [41]. Early LIB commercialization was conducted by Sony in 1990 for electronic devices, in which a $LiCoO_2$ (LCO) cathode and a petroleum coke anode developed by the Goodenough and Asahi Kasei Corporation, respectively, were combined to create a fully rechargeable battery with an energy density of 80 Wh/kg and a volumetric energy density of 200 Wh/L [42,43]. The creation of LIBs with a graphite anode, proposed by Sanyo Electric, together with ethylene carbonate (EC) as the co-solvent in the liquid electrolyte based on Dahn's work, could increase the voltage and volumetric energy to 4.2 V and 400 Wh/L, respectively. For its increased oxidation stability, Guyomard and Tarascon published a novel electrolyte formulation, namely $LiPF_6$ in EC/DMC, in 1993 [44]. This electrolyte is still widely used today and allows LCO-based LIBs to have three times the energy density (250 Wh/kg and 600 Wh/L) of Sony's first-generation batteries [44]. Such an immense improvement is expected to be progressively performed for considerable opportunities in improving LIB technology [45]. Despite this, at present, the technology still cannot beat internal combustion engine (ICE) vehicles in terms of energy density.

Other cathode materials besides LCO have also been introduced for commercial EVs to ensure robustness and reliability during operation. They are $LiFePO_4$ (LFP), $LiNi_xMn_yCo_{1-x-y}O_2$ (NMC), $LiNi_xCo_yAl_{1-x-y}O_2$ (NCA) [46], $LiMn_2O_4$ (LMO) and $LiN_{0.5}Mn_{1.5}O_4$ (LNMO) [47]. Each cathode has superiorities over others. For example, LFP is considered the safest cathode but has the lowest energy density. LFP is usually used for low-cost EVs, light EVs, and electric motorcycles. LFP does not require a complex cooling system because it might not explode if a thermal event occurs [48]. In contrast, NCA has the highest energy density, but its explosive content requires a special battery packing design. Due to its characteristics, NCA batteries are used for luxurious cars requiring high energy and power densities. The safety issues are solved by implementing a more advanced BTMS, which might be expensive and complicated but reliable in their design. Furthermore, depending on its nickel content, NMC has moderate energy density and safety [49]. NMC-based batteries are suitable for heavy vehicles, such as electric buses and trucks, which require high density and moderate risk due to their size and the related implications when an accident happens. Spinel-type cathode materials, such as LMO and LNMO, were introduced as cobalt-free cathodes to counter the scarcity and toxicity of cobalt [50,51]. LMO cathodes are cheap and environmentally friendly but have a low energy density, making them less preferable for practical applications [52]. Nickel-substituted spinel cathodes, LMNO, known for their high voltage stability, have a high power-density and fast Li^+ diffusion [47,53]. On the other hand, the impurity of $Li_xNi_{1-x}O$ in LNMO drops the electrical performance, which makes the material difficult to prepare [54].

Significant work has been conducted by Murashko et al., in which they investigated the heat properties of LIB materials based on LFP, NMC, and NCA cathodes with graphite anodes. The specific heat capacity and thermal conductivity as a function of the SoC of the batteries were found to be different [55]. NMC and LFP have a higher specific heat capacity and a lower thermal conductivity, so they are more challenging to heat up and cool down, but this gives them more thermal stability. NCA, in comparison, has a lower specific heat capacity and a higher thermal conductivity. This means it is less thermally stable and is easy to heat up, but it is also easier to cool down. Such behavior needs to be accommodated by the BTMS to keep the LIBs in the optimal temperature range and with a minimum temperature difference.

In general, the higher energy density of LIBs results in a longer usage time, leading to more heat generation. The cooling load requirements for BTMSs are increasing along with the development of batteries with a high energy density. Different LIB constituent materials result in different heat generation rates and thermal properties. The main reason that NMC and NCA are more active than LFP is the adoption of nickel as the reactive material [49]. It might ensure high energy and power densities, but at the same time, it can easily generate exothermic reactions [56]. Golubkov et al. showed that NCA produced more flammable CO, CO_2, and H_2 gases than LFP during a temperature ramp test, indicating that NCA was more reactive than LFP [57]. The potential for flammable gas production is bigger when the SoC of the battery is full or overcharged. Table 1 shows a characteristic comparison of cathode materials generally used for EVs.

Figure 2 illustrates the heat flux curves during the charging and discharging processes of a common battery cell. During the charging process, the heat flux increases until a steady state value is obtained. The value depends on the charging rate. A steady-state condition is achieved when there is a heat flow balance among the environmental conditions, Joule heating, and endothermic chemical reactions during the charging process [58]. In contrast, rather than reaching a certain steady state value, the discharging process tends to generate heat in a transient manner [58]. The discharging process results in Joule heating and exothermic chemical reactions, which cause the heat flux to increase proportionally. Identical to the charging process, this heat flux also depends on the discharging rate. In the relationship of thermal management design, the charging process might implement a high charging rate as long as the steady-state heat flux does not cause a thermal event. For the discharging process, the heat flux can be limited by ensuring that the battery does not discharge with a high C rate in the long term. This condition can be controlled by the battery management system.

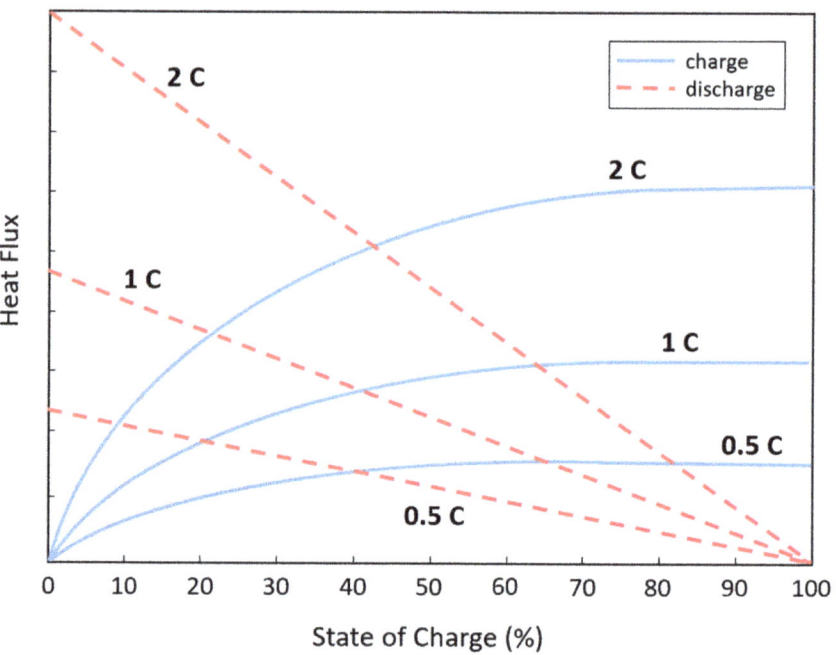

Figure 2. Typical heat generation curves of battery cells during the charging and discharging process.

Table 1. Characteristics of cathodes for lithium-ion batteries.

Cathode Material	Specific Capacity (mAh/g)	Average Potential (V, vs. LiO/Li$^+$)	Max Discharge Current (A)	Acceptable Temperature (°C)	Refs.
LiCoO$_2$	140	3.0	8	0 to 50	[59,60]
LiFePO$_4$	95	3.5	70	−20 to 65	[61–63]
LiNi$_x$Co$_y$Al$_{1-x-y}$O$_2$	199	3.75	6	0 to 45	[64,65]
LiNi$_{0.8}$Mn$_{0.1}$Co$_{0.1}$O$_2$	180	3.8	20	−5 to 50	[46,66]

2.2. Limited Vehicle Compartment Space and Safety

BTMSs of EVs need to have robust cooling and heating performances. This can be realized by increasing the system size to maximize the heat transfer area and the cooling medium. However, when designing the battery packing of EVs, we have only a limited space for the vehicles to work with ergonomically; hence, the design needs to be simultaneously as compact as possible and still sufficiently reliable. The limited space in EVs also makes the battery packs susceptible to heat accumulation, especially during fast charging and discharging [67]. Weng et al. showed that in a limited space with a limited cooling medium rate, increasing the heat transfer area does not always give a better cooling efficiency [68]. Indeed, a compact battery pack design is also required to ensure the high energy density of EVs and to compete with other vehicle types, such as ICE and hybrid vehicles.

Figure 3a shows the total carried energy versus the weight of commercial vehicles. ICE and hybrid vehicles are still leading in terms of energy storage. However, it is worth noticing that the efficiency of ICEs might be much lower (around 20%) compared with electric motors (about 80%) [69]. Furthermore, the efficiency of hybrid vehicles can be calculated based on whether the energy is stored in the fuel tank (20%) or the batteries (80%). Those efficiencies also depend on the power train configuration used for the vehicles [70]. A more complicated power train system requires more components, consequently reducing power train efficiency. Figure 3b shows the effective energy used for vehicles considering the energy efficiency of power trains. EVs still have lower energy storage, even though this is after considering the drive train efficiency. This problem is related to the battery's stored energy density, which is highly determined by active material characteristics and the heavy cooling system carried by the vehicles. Therefore, adding more batteries to EVs might not be beneficial, since this can increase the total weight of the vehicle. EVs have an advantage due to the use of electric motors having a constant torque in a wide range of rotational speeds, eliminating the transmission and gearbox. However, this advantage must be compensated by a heavy battery pack due to the low stored energy density. To compete with ICE vehicles, EVs must have at least a two times higher stored energy density than the current technology battery, as shown in Figure 3b.

Three common types of battery cells are available in the market that can accommodate the limited space for battery packing, namely cylindrical cells, prismatic cells, and pouch cells [71]. A cylindrical cell uses steel to wrap the battery with various diameters and lengths, which ensures full protection from mechanical loading [72]. The size is considerably small to ensure the battery cell has good thermal stability. The small size also helps the cell to be arranged within the available space. Furthermore, the small size also makes the cell adaptive to various battery modules or pack designs. However, these advantages must be paid with their low effective volumetric and gravimetric energy densities. In contrast, pouch batteries use aluminum polymer foils to wrap the cell [73]. This makes the cell have an effective high gravimetric energy density, but at the same time the strength and rigidity of the battery become questionable [74]. Such battery cells also require additional packs when they are used for EVs, reducing their densities. To improve the density, a prismatic cell is introduced with a relatively bigger size than other cells. This might be promising for EVs requiring a high energy capacity. However, their bigger size causes the cell to be not favorable for thermal management. The quality of each cell might also be different since it is more difficult to control the quality during the manufacturing process of these battery cells.

Besides the battery cell design, battery modules and packing are optimally designed to meet the available space and to provide good thermal management. The battery cells must unavoidably be arranged in series and parallel connections. A series connection is created to fulfill the required voltage, whereas a parallel connection is created to fulfill the required energy capacity [75]. In a series connection, the nonuniform charging/discharging process of each cell is unavoidable. Consequently, each cell has a different SoC, which implies different temperatures and different levels of Joule heating [76]. This phenomenon makes designing BTMS more complicated, which must be handled by optimized design strategies. It is worth noting that in designing battery packing, thermal management is only one of the main issues that must be handled. Other issues, such as vibration and crash safety, must also be considered during the design process [77].

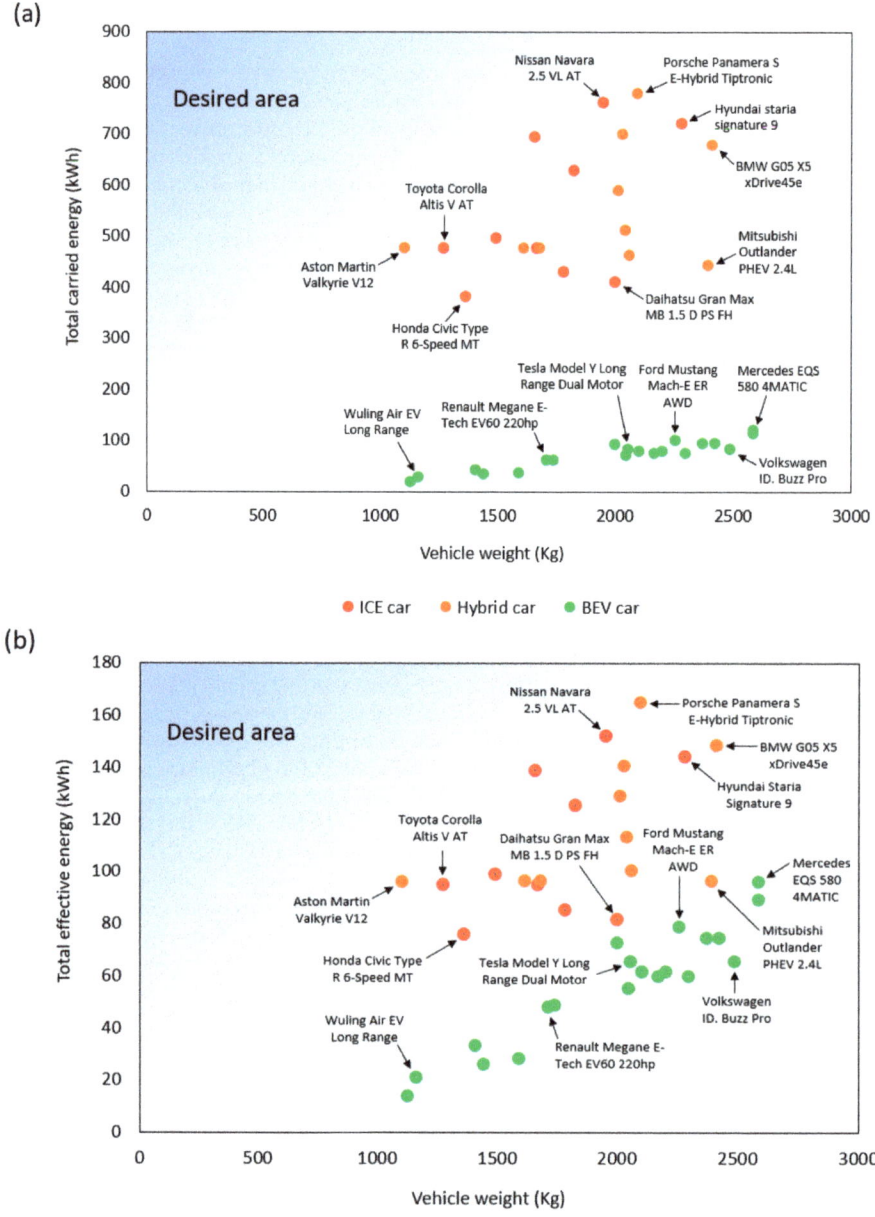

Figure 3. (**a**) Total carried energy by vehicles versus their weights and (**b**) total effective energy of vehicles versus their weights. The effective energy is calculated by considering total power train efficiencies of 20% for ICE vehicles and 80% for EVs.

2.3. Climate and Seasonal/Environmental Temperature

For obtaining the best performance and lifetime, LIBs should generally be operated at a temperature between 25 and 40 °C with a maximum difference between the batteries of 5 °C [1,2]. The optimum operating temperature might depend on the battery's active materials [78]. The battery's cooling and heating requirements should be considered depending on environmental conditions such as the climate, season, and driving circumstances. To improve the performance and maintain the battery's ideal state of health, the temperature across the battery pack should be kept inside the optimal range and as uniform as possible. As a result, a sophisticated BTMS with accurate temperature management capabilities in locations with varying temperatures throughout the year is greatly desired for EVs [79]. Since the heat transfer rate and specific heat of air significantly decrease during a hot summer, an air-cooled BTMS might not be suitable for year-round usage. An air-cooling system cannot provide the necessary cooling load to the battery pack [80]. A hot environmental temperature can trigger the battery to accelerate the redox chemical reactions, which directly causes an abundance of heat generation (thermal event). In the long run, the phenomenon can cause thermal runaway in the battery.

For subtropical climates, EVs require a heating system to ensure the battery does not freeze. A low temperature makes the battery lose its performance due to the increased liquid electrolyte viscosity, causing high internal resistance [81,82]. This makes the chemical reactions in the battery slow. As a result, the current output can be low, causing the electric motor of the EV drive train to not have the power to rotate. At an environmental temperature of 0 °C, the discharging capacity can drop by more than 20%, and the dropped capacity proportionally increases as the temperature decreases [83]. Even though a low temperature is usually less hazardous for a battery than a high temperature, the malfunctioning of the battery makes the EV unable to operate normally, which is less favorable than ICE vehicles. Figure 4 shows the effects of environmental temperature on battery performance for several battery types [84,85]. Operating the battery outside the optimal temperature causes low performance for both charging and discharging conditions.

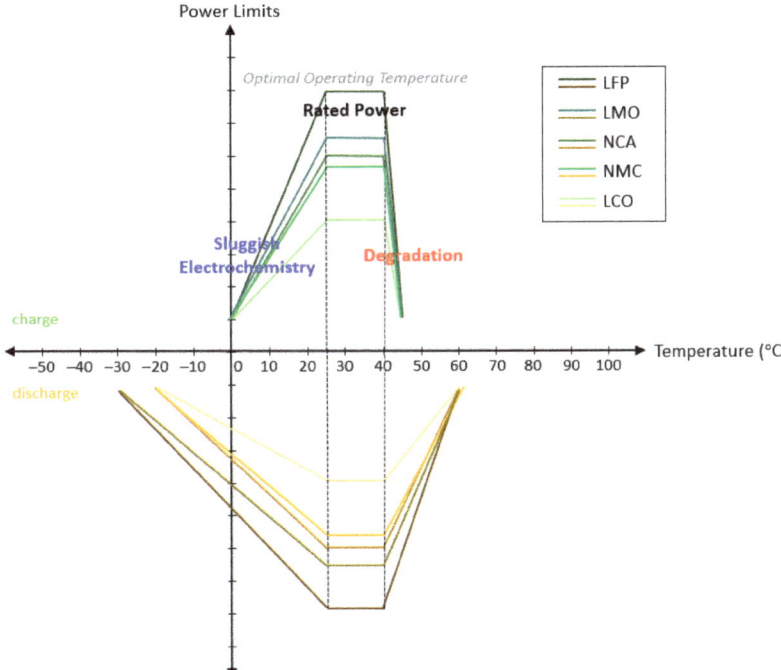

Figure 4. Typical LIBs performance during charging–discharging process for different temperature conditions.

3. Recent Advancements in Research and Development of LIBs

As a summary of the discussion in the previous section, three fundamental problems must be solved in designing battery packing with superior thermal management in EVs.

These problems have become design constraints that might be difficult to solve simultaneously. For example, considering the current technology of the constituent material of batteries, high-energy-capacity cathodes (NMC or NCA) release a lot of heat during the charging and discharging processes, increasing the thermal runaway risk [86,87]. Extreme climate or season conditions also require robust and complicated battery packing designs that might increase the vehicle weight and consume a lot of energy [79,88]. Optimizing a battery design that accommodates all constraints is usually suggested to solve such problems. In this section, we explain the significant findings from the recent decade that can solve these problems. Some ideas are still under ongoing research, which might become promising game changers in battery packing design and technology, even though they still need to be further proven before their implementation.

The thermal problems of LIBs, which are high heat generation and varying environmental temperatures, are caused by the narrow working temperature of LIBs. Recent advancements in battery technology give potential solutions to these problems, for instance by implementing a functional co-solvent for the electrolytes—or utilizing solid-state batteries. Functional co-solvents and electrolyte additives have been vastly studied and developed, with significant progress being achieved in recent decades such as in obtaining a longer cycle life, allowing a fast charging process, hindering electrolyte decomposition at the electrodes, enhancing the energy density, enabling a wider temperature range of operation, or producing lower-cost LIBs [89–95]. Lu et al. [96] proposed 2,2,2-trifluoroethyl N-caproate (TFENH) as a co-solvent for LIB electrolytes. The TFENH co-solvent improved the low- and high-temperature performance of a $LiCoO_2$/graphite battery by keeping the volume ratio of TFENH within 17 to 25 vol%. An X-ray photoelectron spectroscopy (XPS) test showed that a film of $CH_3(CH_2)_4COOLi$ was formed, which decomposed from the TFENH. The film reduced the other reduction reaction products from the carbonate solvent and lithium salt, which in turn improved the low-temperature performance and cycling stability of the LIB. Another co-solvent application was proposed by Ouyang et al. [97], which used a combination of fluoroethylene carbonate (FEC) and dimethyl carbonate (DMC) at a ratio of 3:7, named FD37. The co-solvent was used for an NMC LIB and showed a significant improvement in cycling features due to the formation of a cathode–electrode interface. The interface inhibited the electrolyte decomposition and further improved the stability of the electrode/electrolyte interface. The co-solvent also improved the thermal stability of the LIB, which was shown in high-temperature cycling and storage tests. The LIBs with an FD37 co-solvent showed a better capacity retention and Coulombic efficiency than the traditional LIB. However, employing this strategy in the LIBs with liquid-based electrolytes does not mean fully omitting the natural characteristics of liquid electrolytes that tend to have a consequence in easier fire ignition when it reacts with particular chemicals, even if they are non-flammable [98]. For small-scale applications, such as power sources for portable devices, attempting this strategy could be favorable and promising, while the drawback case might be conveniently handled or prevented. On the contrary, aside from the enhancement of the electrochemical performances required for large-scale demands, especially in EV applications, this technology needs more concern regarding safety, owing to the severe impacts, such as catastrophic fires or explosions, which will result in the case where the battery gets into trouble.

Therefore, in recent LIB development, it is believed that solid-state battery (SSB) technology could replace the LIB technology based on liquid electrolytes currently commercialized in the market, considering the exponential progress of SSB research pursued in the last decades, which is much improved compared to the initially revealed idea many years past [99,100]. Furthermore, a functional additive that previously accounted for significant advantages in LIBs could also be employed in solid electrolytes, allowing SSBs to become more promising candidates for next-generation power sources, especially for EV applications [101]. The use of materials such as oxide inorganic solid-electrolyte NASICON-type LATP ($Li_{1+x}Al_xTi_{2-x}(PO_4)_3$), NASICON-type LGPS ($Li_{10}GeP_2S_{12}$), NASICON-type LAGP ($Li_{1+x}Al_xGe_{2-x}(PO_4)_3$) or garnet-type LLZO ($La_3Li_7O_{12}Zr_2$) instead of liquid electrolytes with a separator enables the significant enhancement of a battery's capacity by up to four times due to generally of having a wide electrochemical stability window [102]. The main capacity of LIBs is in the range of 150–200 mAh [103]. Liu et al. recently constructed a solid-state Li–air battery with lithium foil anode, LAGP and a solid electrolyte, and an LAGP–nanoparticle composite/single-walled carbon nanotubes (SWCNTs) as the air electrode [104]. A remarkably high capacity of 2800 mAh/g was shown in the first cycle despite the good cycling performances being limited to 1000 mAh/g, and the electrochemical process may vastly differ between an air and a pure oxygen atmosphere. Tao et al. reported that the use of LLZO-nanoparticle-filled poly(ethylene oxide) electrolytes could generate favorable capacities of >900, 1210, and 1556 mAh/g at successive temperatures of 37, 50, and 70 °C, respectively, even though at the first charge/discharge cycle they ran into an irreversible electrochemical reaction, which lead to capacity decay [105]. Accordingly, the use of solid electrolytes also allows the battery to have a wider operational temperature owing to the high thermal resistance, and this substantially increases the battery safety significantly, even in harsh environments [105–109]. The current method proposed to extend the temperature range is thus available by coupling the types of solid electrolyte material and implementing such filler/substitute material. These superiority aspects lead SSBs to be the primary candidates for solving the low-density and safety concerns that mostly arise in conventional batteries. Furthermore, the manufacturing process of SSBs can be considerably easier when producing a solid electrolyte using additive manufacturing, spark plasma sintering, or conventional sintering [106]. The installation of solid electrolyte in SSBs could also prevent leakage that commonly arises from the liquid electrolyte, which can cause a short circuit and thermal runaway [107].

The promising nature of SSBs does not mean they can be implemented immediately. In fact, one remaining fundamental problem must be solved, i.e., mechanical damages that always appear after several charging–discharging cycles [108]. These damages occur due to the solid electrolytes that always bear the pressure loading from the expansion–shrinkage of electrodes, which is different to liquid electrolytes that can redistribute the pressure around the cell pack. The most common solutions to avoid these damages involve using zero-strain cathodes such as LTO or limiting the SoC in the battery application to ensure the solid electrolyte is not subjected to overpressure. The high internal resistance of SSBs also occurs due to the low quality of the manufacturing process. Toyota has spent research funding to solve the SSB problem so that they can be implemented in their cars [110]. Bolloré Group launched the BlueIndy electric-car-sharing program to demonstrate an EV prototype installed SSB of 30 kWh [111]. Other established companies such as Volkswagen, BMW, Daimler, and Hyundai also keep paying attention to the research and development of SSB technology [112]. The immediate solution to overcome the high heat generation of high-energy-density batteries and their environmental problems is through advanced BTMSs, which have better thermal management performances. The advancement of BTMS research has resulted in hybrid BTMSs, which combine one type of BTMS with another to utilize the advantages of both systems while overcoming the weaknesses of each system. Yang et al. [30] combined mini-channel liquid and air cooling to improve the BTMS cooling performance for cylindrical batteries. The maximum temperature and temperature distribution were reduced to 2.22 and 2.04 K, respectively. Another hybrid BTMS combining liquid cooling with a heat pipe was proposed by Jang et al. for a prismatic battery [31]. The liquid cooling utilized a mini-channel heat sink placed on top of the battery, while a heat pipe was placed on the front side of each battery. The heat pipe transferred the heat generated by the batteries up to the heat sink area to be further transferred by the liquid coolant out of the battery pack. The proposed BTMS successfully reduced the maximum temperature of the battery up to 9.4 °C. A BTMS with a combination of a liquid PCM and a heat pipe for a pouch battery was proposed by Zhou et al. [32]. The PCM took the heat from the battery through convection, and then the heat pipe transferred the heat out of the battery pack by air. The system lowered the temperature difference between the batteries by up to 67% compared to a forced air-cooled BTMS. A comparison of several of the recent

BTMSs is presented in Table 2. The progress of BTMS strategies in providing the optimal temperature conditions of batteries with a maximum temperature of 25 to 40 °C and a maximum temperature difference of 5 °C, which LIBs require, is summarized in Figure 5.

Table 2. BTMS comparison.

BTMS	Maximum Temperature (°C)	Maximum Temperature Difference (°C)	Refs.
Air-cooled (AC)	34	4.4	[22]
Liquid-cooled (LC)	37.36	1.96	[24]
Heat pipe (HP)	31	4.6	[20]
PCM	44.5	0.42	[26]
Air–liquid-cooled (AC/LC)	29.61	2.09	[30]
Liquid-cooled heat pipe (LC/HP)	24.6	4.27	[31]
PCM heat pipe (PCM/HP)	48	2	[32]

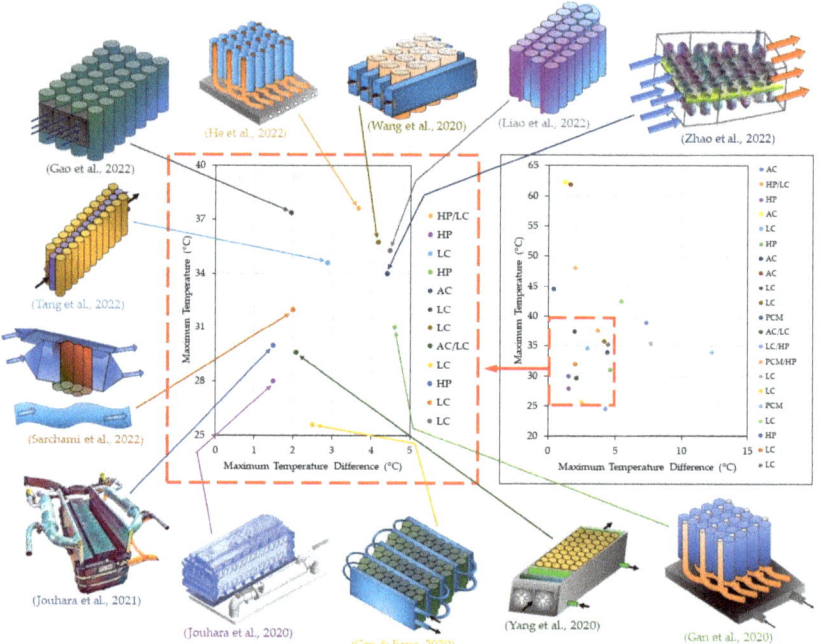

Figure 5. The progress and achievement of BTMS strategies in providing the optimal temperature conditions for LIBs in the form of AC [7], HP/LC [13], HP [14], AC [15], LC [18], HP [20], AC [22], AC [23], LC [24], LC [25], PCM [26], AC/LC [30], LC/HP [31], PCM/HP [32], LC [67], LC [80], PCM [113], LC [114], HP [115], LC [116], and LC [117].

The mentioned research [30–32] proved that hybrid BTMSs could improve the cooling performance of ordinary BTMSs. However, the drawback of hybrid BTMSs is that they require a higher energy consumption and a bigger compartment space. Therefore, due to their compactness (handling the limited compartment space) and minimum power consumption, without sacrificing safety, direct-cooling BTMSs have been introduced in the last decades [80]. Their cooling design is superior to air- and liquid-cooling designs. In addition, they can be integrated with the air conditioning system of an EV to achieve a compact design without sacrificing the cooling performance [113,118,119]. The basic idea of direct cooling BTMSs is that the battery requires an operational temperature similar to the comfort temperature of the passenger. Thus, the cooling system of the battery and cabin can be integrated. A direct cooling system solves the basic problem of air-cooling system, which is usually unreliable without implementing a complicated structure, such as liquid cooling. Together with an advanced control system, direct cooling BTMSs could be promising for solving the thermal management of EV batteries. At this time, the Nissan Leaf and Toyota Prius use a direct cooling system and show favorable performance.

A battery preheating mechanism was a method introduced and proposed by researchers for EV applications in cold environmental temperatures, which can be further categorized into external and internal preheating. External heating uses a heat source outside the battery to heat the battery by transferring the heat through a medium, such as air, liquid, or PCM. Air preheating uses power from the battery to heat a resistance heater and then transfers the heat to the battery via air convection produced by a fan [33]. In contrast, Wang et al. [34] proposed an immersive preheating system by immersing prismatic batteries in a flowing heat transfer liquid. The system used an external heat exchanger to heat the liquid for heating the battery. Ruan et al. [36] proposed an internal preheating method by direct current discharge due to its simplicity and high heat generation. The research successfully optimized the heating time to 103 s and reduced the capacity degradation to 1.4%. Zhang et al. [37] proposed a preheating mechanism using a sinusoidal alternating current, causing the battery to be only heated by the irreversible heat of Joule heating, since the reversible heat of the electrochemical reaction was canceled out after one period. The study found that the heating process can take less than 15 min with no capacity loss after repeated preheating, which was slower than DC preheating but had less capacity degradation. Moreover, a pulse internal preheating system heats the battery by pulse excitation, which Wu et al. [38] successfully optimized by varying the amplitude and frequency. The results showed that the heating time was 308 s with a capacity degradation of 0.035% after 30 cycles.

Advanced control systems always become the main solution to control complicated BTMSs, considering that heat generation in the battery caused by internal resistance (Joule heating) and redox chemical reactions cannot be avoided. The same thing also goes for the non-uniform temperature of each cell, which is also unable to be averted since the battery must be arranged in both series and parallel connections. Hence, a control system was implemented to ensure a constant and uniform battery temperature, which can be approached by two methods. The first approach is to control the SoC and voltage level of each cell using a battery management system (BMS) so that the battery can generate heat uniformly.

In addition, a BMS can also restrict the excessive current flow during the charging–discharging process, which commonly raises the heat generation drastically. The second approach is installing a temperature sensor, heat flow sensor, fuse, and gas sensor for identifying a heating spike. The data obtained from the sensor can then be analyzed to determine the necessary action of a cooling system set. The development of machine learning in this situation can plentifully help researchers to find robust control systems that are urgently needed, especially for extreme climates and weather. Figure 6 maps the efforts conducted to solve the thermal management issues, and it can be seen that many solutions have been proposed to tackle the thermal management issues. However, the facts show that the researchers are only capable of dealing with it partially. This implies that thermal management issues are still the biggest concern, which is directly related to the main safety concerns of EVs, aside from the driver's range anxiety.

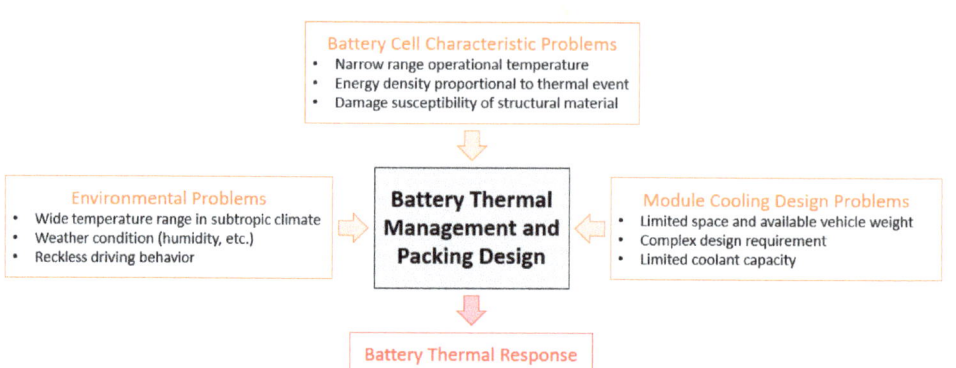

Figure 6. The remaining problems of battery thermal management and packing design that must be solved in future research.

4. Future Perspectives of BTMS Design

In recent decades, a great deal of research has been devoted to preventing thermal events and thermal runaways in batteries, which practically bears out the fundamental problem of EVs. Various proposals of solutions to ensure a constant, steady, and uniform operational temperature were introduced and substantiated to resolve these problems, regardless of the limited and controlled conditions encountered, such as in lab-scale or prototype models. Consistent progress drawn from these conditions is expected to lead to further research opportunities until a superior battery thermal management is realized for general application.

To systematically discuss and analyze the recent progress and future prospects of thermal management in the battery packing of EVs, we divided the solutions into two main categories: practical and fundamental solutions. The difference between both solutions is in how they define the problem and propose solutions in terms of complying with the thermal management requirements of battery packing design. In practical solutions, the researchers strive to develop advanced module cooling strategies to ensure the battery cells are at operational temperature and has a low temperature difference among its battery cells. Significant efforts have been deployed to address this concern. In contrast, fundamental solutions focus on improving the battery cell to resist environmental conditions and thermal self-generation (see Figure 6).

Under practical solutions, the ongoing research on battery management technology is presumably capable of reaching maturity, which can be implemented in future EV design. Some promising technologies have been developed, including the implementation of machine learning for advanced control systems [114,120], high thermal diffusivity fluids by nanofluids for liquid coolants [121], and reconstructions of battery packing to obtain the temperature distribution among the cell levels in a compact design [115]. Tang et al. implemented machine learning to analyze the performance of liquid-cooled batteries [120]. Implementing machine learning can predict cooling parameters that can significantly improve the liquid-cooling performance without expensive and time-consuming experiments. Another significant piece of work has been conducted by Khan et al. who implemented and developed machine learning to optimize the design of liquid-cooled thermal management in a battery [114]. These studies proved that machine learning could effectively increase the liquid-cooling performance in the design process and operation.

Nanofluids are another promising method for further improving the cooling capacity by altering the thermal transport properties of the cooling liquid. Nano-sized solid materials, mainly metal oxides and carbon nanotubes (CNT), with superior thermal transport properties can be mixed into the liquid to improve its thermal transport properties. Dielectric liquids (aliphatic liquids, silicone liquids, fluorocarbons) and non-dielectric liquids (water, ethylene glycol, oils) have been used as base fluids to combine with nanoparticles to improve heat transmission. The fluid's poor thermal conductivity determines the heat transfer rate, a problem for well-known heat transfer fluids including water, glycol, and oil. Mixing nanoparticles into these fluids is one of the most acceptable ways to boost their thermal conductivity. An essential step in nanofluid research is preparing the fluid to produce a stable fluid that will not agglomerate at high temperatures or after a specific time. Nanoparticle agglomeration is a fundamental issue with all nanopowder technologies that needs to be addressed in order to produce effective nanoparticle suspensions. The creation and suspension of non-agglomerated or uniformly distributed nanoparticles are necessary to significantly improve the heat transfer properties of nanofluids.

A number of experimental and numerical studies have been reported on the thermal conductivity of nanofluids as a function of temperature, volume fraction, nanoparticle properties, size, and shape. An experimental study with an improved cooling efficiency has also been reported for the water cooling of a battery pack by adding alumina nanoparticles to water as a base fluid [116]. With a relatively low volume fraction of aluminum oxide at 2 vf%, an increase in the heat transfer coefficient was observed, resulting in an enhancement of the maximum temperature and temperature difference by 7.76% and 117.668 and 0.26 K, respectively [116]. Aluminum oxide with a much higher thermal conductivity could increase the overall thermal conductivity and density of nanofluids, improving the heat transfer rate of the cooling liquid [122]. A numerical simulation of a liquid-cooling BTMS for 31 lithium-ion batteries reported a maximum temperature difference decreases of 12.6% by employing Cu–water-based nanofluids [117]. The simulation results also showed a proportional effect of the cooling performance as the volume fraction increased to 5%.

However, the addition of nanoparticles in the cooling liquid increases the density of the liquid, affecting the pressure drop along the liquid flow. A higher pressure drop requires more powerful and expensive pumping equipment, reducing the overall battery performance. At a low Reynolds number of 920, the pressure drop increased significantly to over 60% at a 2 vf% nanoparticle addition [116]. A lesser effect of nanoparticles on the pressure drop was shown with a high Reynolds number of 1840, generating only 8.33% of additional pressure drop with 2 vf% of nanoparticles [120]. In summary, nanoparticle addition in cooling liquids in the form of nanofluids may boost the cooling efficiency, which is a promising practical solution for BTMSs. The addition of nanoparticles is, however, limited by the dispersion conditions in the liquid and the operational flow characteristics, which determine the effects of nanoparticle addition on the pressure drop.

The reconstruction of more robust battery packing is also one of the practical solutions to handle battery packing design problems. Arora et al. show that for commercial cars, relative battery cell movement and displacement are commonly used as the failure criteria of the packing. The source of movement or displacement can come from a thermal event, vehicle vibration, or impact loading when an accident happens [77]. Most cars position battery packs integrated into the chassis to ensure they can have sufficient structural rigidity and be protected from front, back, and side impacts. Another common method is rearranging the battery cells to achieve the most efficient space usage without sacrificing thermal management. Many studies have proposed cell arrangements for cylindrical battery cell types, including inline or staggered formations, a number of series and parallel connections, and customized cell sizes.

Furthermore, the arrangement can only be focused on the cell size and geometry for other types of battery cells since they can be more flexibly arranged. An innovative battery pack is demonstrated by Tesla, in which their battery packs in their cars use a small wire fuse to protect the battery cells from thermal events [123]. The Nissan Leaf considers using direct cooling rather than liquid cooling to manage the battery temperature. It is worth noting that even though the practical solutions might work for actual applications considering the currently available battery technology, they still require complicated design and manufacturing processes, which might not be economically beneficial for commercial cars. Each implemented battery pack technology must be tested and proven safe, following standards, to have a high reliability and, simultaneously, not significantly increase the manufacturing costs to compete with ICE cars. A permanent solution that solves the fundamental problems of battery management systems has to be found as soon as possible.

To overcome the fundamental problems, the characteristics of non-flammable designs, -wide-range operational temperatures, and non-poisonous materials must be unveiled in the future by conducting research and development. The material couple of anode–electrolyte–cathode also ought to have low Joule heating and low heat generation considering the internal resistance and redox chemical reactions. SSBs have been admitted as a permanent solution to tackle the low mechanical stiffness and battery strength induced by liquid-based electrolyte use. In addition, SSBs can produce wider operational temperatures. As studied by Ogawa et al. [124] solid-state thin film lithium batteries can be operated at a low temperature of –40 °C with a high temperature of 170 °C, and, correspondingly, a recent study from Wang et al. [125] showed that lithium-metal-based SSBs can be operated at the temperature range from −73 to 120 °C. Using solid materials to replace flammable organic liquid-based electrolytes would also allow the battery to resist thermal events [126]. Furthermore, the possible much lower freezing point could make SSBs operate at a lower temperature, making them more promising and capable of running in subtropical climates [127].

However, as discussed in the previous section, the mechanical damages of SSBs must be initially solved, such as by implementing functionally graded material (FGM), which is currently predicted as one solution that can solve this problem by eliminating the thermal management component in the battery packing [128].

SSBs with future ideal materials make it possible to integrate battery cells into the EV frame or body [129,130]. This idea can be a game changer in which the battery will also become a load-bearing structure. This would be possible if a novel multi-functional material that not only stores energy but also acts as a load-bearing structure could be found. At this stage, the voltage and capacity of the battery can be achieved by one cell only, in which the design of the series and parallel connections of the battery is not required. This proposed research topic can possibly and interestingly be further explored, but it would still require much groundbreaking research to be conducted before implementing this technology. In other words, all this SSB research is still on the laboratory scale and probably needs time to be ready and to be implemented in EVs.

5. Conclusions and Recommendation

This paper reviewed the fundamental problems of BTMSs in EVs. The design strategies of battery packing to solve these problems were also explained comprehensively. It was shown that many research publications have tried to analyze and summarize the strategies for solving these problems. Finally, complicated battery packing, including advanced control for thermal management, was introduced to ensure that the battery cell temperatures are kept at optimal operating temperatures. This is a practical solution due to the current state of battery technology.

Thermal management systems might not be required in the future after novel material constituents of batteries are found. The new materials should possess the properties of being non-flammable, having a wide-range operational temperature, being non-poisonous, and generating low Joule heating and redox chemical heating. In fact, the current battery operating temperature is identical to environmental temperatures, which might not require additional battery temperature control as long as novel material constituents are found. SSB types of battery are believed to be the next generation of battery technology for EVs. Furthermore, materials that can be used for the EV frame are also desired, which means that they must be able to bear external loading.

Author Contributions: Conceptualization, B.A.B. and M.A.; methodology, F.B.J., B.A.B. and M.A.; validation, F.B.J. and M.A.; formal analysis, R.D.W. and S.Z.; investigation, R.D.W.; resources, B.A.B.; data curation, S.Z.; writing—original draft preparation, R.D.W., S.Z. and B.A.B.; writing—review and editing, F.B.J. and M.A.; visualization, S.Z.; supervision, B.A.B. and M.A.; project administration, B.A.B.; funding acquisition, B.A.B. All authors have read and agreed to the published version of the manuscript.

Funding: This work was funded by the Institute of Science and Technology Development under the Excellent Research Program for Center/Research Center with contract no. 650B/IT1.A/SK-KP/2022 and the Indonesia Endowment Fund for Education (LPDP) under the Research and Innovation Program (RISPRO) with contract no. PRJ-85/LPDP/2020.

Institutional Review Board Statement: Not applicable.

Informed Consent Statement: Not applicable.

Data Availability Statement: Not applicable.

Conflicts of Interest: The authors declare no conflict of interest. The funders had no role in the design of the study; in the collection, analyses, or interpretation of data; in the writing of the manuscript; or in the decision to publish the results.

References

1. Malik, M.; Dincer, I.; Rosen, M.; Fowler, M. Experimental Investigation of a New Passive Thermal Management System for a Li-Ion Battery Pack Using Phase Change Composite Material. *Electrochim. Acta* **2017**, *257*, 345–355. [CrossRef]
2. Pesaran, A.A. Battery Thermal Models for Hybrid Vehicle Simulations. *J. Power Sources* **2002**, *110*, 377–382. [CrossRef]
3. Petzl, M.; Kasper, M.; Danzer, M.A. Lithium Plating in a Commercial Lithium-Ion Battery—A Low-Temperature Aging Study. *J. Power Sources* **2015**, *275*, 799–807. [CrossRef]
4. Tippmann, S.; Walper, D.; Balboa, L.; Spier, B.; Bessler, W.G. Low-Temperature Charging of Lithium-Ion Cells Part I: Electrochemical Modeling and Experimental Investigation of Degradation Behavior. *J. Power Sources* **2014**, *252*, 305–316. [CrossRef]
5. Ma, S.; Jiang, M.; Tao, P.; Song, C.; Wu, J.; Wang, J.; Deng, T.; Shang, W. Temperature Effect and Thermal Impact in Lithium-Ion Batteries: A Review. *Prog. Nat. Sci. Mater. Int.* **2018**, *28*, 653–666. [CrossRef]
6. Panchal, S.; Mathew, M.; Dincer, I.; Agelin-Chaab, M.; Fraser, R.; Fowler, M. Thermal and Electrical Performance Assessments of Lithium-Ion Battery Modules for an Electric Vehicle under Actual Drive Cycles. *Electric. Power Syst. Res.* **2018**, *163*, 18–27. [CrossRef]
7. Jiaqiang, E.; Yue, M.; Chen, J.; Zhu, H.; Deng, Y.; Zhu, Y.; Zhang, F.; Wen, M.; Zhang, B.; Kang, S. Effects of the Different Air Cooling Strategies on Cooling Performance of a Lithium-Ion Battery Module with Baffle. *Appl. Therm. Eng.* **2018**, *144*, 231–241. [CrossRef]
8. Onda, K.; Ohshima, T.; Nakayama, M.; Fukuda, K.; Araki, T. Thermal Behavior of Small Lithium-Ion Battery during Rapid Charge and Discharge Cycles. *J. Power Sources* **2006**, *158*, 535–542. [CrossRef]
9. al Hallaj, S.; Prakash, J.; Selman, J.R. Characterization of Commercial Li-Ion Batteries Using Electrochemical-Calorimetric Measurements. *J. Power Sources* **2000**, *87*, 186–194. [CrossRef]
10. Wu, W.; Wang, S.; Wu, W.; Chen, K.; Hong, S.; Lai, Y. A Critical Review of Battery Thermal Performance and Liquid Based Battery Thermal Management. *Energy Convers. Manag.* **2019**, *182*, 262–281. [CrossRef]
11. Liu, H.; Wei, Z.; He, W.; Zhao, J. Thermal Issues about Li-Ion Batteries and Recent Progress in Battery Thermal Management Systems: A Review. *Energy Convers. Manag.* **2017**, *150*, 304–330. [CrossRef]
12. Jaguemont, J.; Boulon, L.; Dubé, Y. A Comprehensive Review of Lithium-Ion Batteries Used in Hybrid and Electric Vehicles at Cold Temperatures. *Appl. Energy* **2016**, *164*, 99–114. [CrossRef]
13. He, L.; Tang, X.; Luo, Q.; Liao, Y.; Luo, X.; Liu, J.; Ma, L.; Dong, D.; Gan, Y.; Li, Y. Structure Optimization of a Heat Pipe-Cooling Battery Thermal Management System Based on Fuzzy Grey Relational Analysis. *Int. J. Heat Mass Transf.* **2022**, *182*. [CrossRef]
14. Jouhara, H.; Serey, N.; Khordehgah, N.; Bennett, R.; Almahmoud, S.; Lester, S.P. Investigation, Development and Experimental Analyses of a Heat Pipe Based Battery Thermal Management System. *Int. J.* **2020**, *1*, 100004. [CrossRef]
15. Zhang, J.; Wu, X.; Chen, K.; Zhou, D.; Song, M. Experimental and Numerical Studies on an Efficient Transient Heat Transfer Model for Air-Cooled Battery Thermal Management Systems. *J. Power Sources* **2021**, *490*, 229539. [CrossRef]
16. Zhao, J.; Rao, Z.; Huo, Y.; Liu, X.; Li, Y. Thermal Management of Cylindrical Power Battery Module for Extending the Life of New Energy Electric Vehicles. *Appl. Therm. Eng.* **2015**, *85*, 33–43. [CrossRef]
17. Kang, D.; Lee, P.-Y.; Yoo, K.; Kim, J. Internal Thermal Network Model-Based Inner Temperature Distribution of High-Power Lithium-Ion Battery Packs with Different Shapes for Thermal Management. *J. Energy Storage* **2020**, *27*, 101017. [CrossRef]
18. Tang, Z.; Zhao, Z.; Yin, C.; Cheng, J. Orthogonal Optimization of a Liquid Cooling Structure with Straight Microtubes and Variable Heat Conduction Blocks for Battery Module. *J. Energy Eng.* **2022**, *148*, 04022017. [CrossRef]
19. Chen, J.; Kang, S.; Jiaqiang, E.; Huang, Z.; Wei, K.; Zhang, B.; Zhu, H.; Deng, Y.; Zhang, F.; Liao, G. Effects of Different Phase Change Material Thermal Management Strategies on the Cooling Performance of the Power Lithium Ion Batteries: A Review. *J. Power Sources* **2019**, *442*, 227228. [CrossRef]
20. Gan, Y.; He, L.; Liang, J.; Tan, M.; Xiong, T.; Li, Y. A Numerical Study on the Performance of a Thermal Management System for a Battery Pack with Cylindrical Cells Based on Heat Pipes. *Appl. Therm. Eng.* **2020**, *179*, 115740. [CrossRef]
21. Widyantara, R.D.; Naufal, M.A.; Sambegoro, P.L.; Nurprasetio, I.P.; Triawan, F.; Djamari, D.W.; Nandiyanto, A.B.D.; Budiman, B.A.; Aziz, M. Low-Cost Air-Cooling System Optimization on Battery Pack of Electric Vehicle. *Energies* **2021**, *14*, 7954. [CrossRef]

22. Zhao, G.; Wang, X.; Negnevitsky, M.; Zhang, H.; Li, C. Performance Improvement of a Novel Trapezoid Air-Cooling Battery Thermal Management System for Electric Vehicles. *Sustainability* **2022**, *14*, 4975. [CrossRef]
23. Chen, K.; Chen, Y.; She, Y.; Song, M.; Wang, S.; Chen, L. Construction of Effective Symmetrical Air-Cooled System for Battery Thermal Management. *Appl. Therm. Eng.* **2020**, *166*, 114679. [CrossRef]
24. Gao, R.; Fan, Z.; Liu, S. A Gradient Channel-Based Novel Design of Liquid-Cooled Battery Thermal Management System for Thermal Uniformity Improvement. *J. Energy Storage* **2022**, *48*, 104014. [CrossRef]
25. Wang, H.; Tao, T.; Xu, J.; Mei, X.; Liu, X.; Gou, P. Cooling Capacity of a Novel Modular Liquid-Cooled Battery Thermal Management System for Cylindrical Lithium Ion Batteries. *Appl. Therm. Eng.* **2020**, *178*, 115591. [CrossRef]
26. Huang, R.; Li, Z.; Hong, W.; Wu, Q.; Yu, X. Experimental and Numerical Study of PCM Thermophysical Parameters on Lithium-Ion Battery Thermal Management. *Energy Rep.* **2020**, *6*, 8–19. [CrossRef]
27. Heyhat, M.M.; Mousavi, S.; Siavashi, M. Battery Thermal Management with Thermal Energy Storage Composites of PCM, Metal Foam, Fin and Nanoparticle. *J. Energy Storage* **2020**, *28*, 101235. [CrossRef]
28. Subramanian, M.; Hoang, A.T.; Kalidasan, B.; Nižetić, S.; Solomon, J.M.; Balasubramanian, D.; Subramaniyan, C.; Thenmozhi, G.; Metghalchi, H.; Nguyen, X.P. A Technical Review on Composite Phase Change Material Based Secondary Assisted Battery Thermal Management System for Electric Vehicles. *J. Clean. Prod.* **2021**, *322*, 129079. [CrossRef]
29. Lazrak, A.; Fourmigué, J.F.; Robin, J.F. An Innovative Practical Battery Thermal Management System Based on Phase Change Materials: Numerical and Experimental Investigations. *Appl. Therm. Eng.* **2018**, *128*, 20–32. [CrossRef]
30. Yang, W.; Zhou, F.; Zhou, H.; Wang, Q.; Kong, J. Thermal Performance of Cylindrical Lithium-Ion Battery Thermal Management System Integrated with Mini-Channel Liquid Cooling and Air Cooling. *Appl. Therm. Eng.* **2020**, *175*, 15331. [CrossRef]
31. Jang, D.S.; Yun, S.; Hong, S.H.; Cho, W.; Kim, Y. Performance Characteristics of a Novel Heat Pipe-Assisted Liquid Cooling System for the Thermal Management of Lithium-Ion Batteries. *Energy Convers. Manag.* **2022**, *251*, 115001. [CrossRef]
32. Zhou, H.; Dai, C.; Liu, Y.; Fu, X.; Du, Y. Experimental investigation of battery thermal management and safety with heat pipe and immersion phase change liquid. *J. Power Sources* **2020**, *473*, 228545. [CrossRef]
33. Ji, Y.; Wang, C.Y. Heating Strategies for Li-Ion Batteries Operated from Subzero Temperatures. *Electrochim. Acta* **2013**, *107*, 664–674. [CrossRef]
34. Wang, Y.; Rao, Z.; Liu, S.; Li, X.; Li, H.; Xiong, R. Evaluating the Performance of Liquid Immersing Preheating System for Lithium-Ion Battery Pack. *Appl. Therm. Eng.* **2021**, *190*, 116811. [CrossRef]
35. He, F.; Li, X.; Zhang, G.; Zhong, G.; He, J. Experimental Investigation of Thermal Management System for Lithium Ion Batteries Module with Coupling Effect by Heat Sheets and Phase Change Materials. *Int. J. Energy Res.* **2018**, *42*, 3279–3288. [CrossRef]
36. Ruan, H.; Jiang, J.; Sun, B.; Su, X.; He, X.; Zhao, K. An Optimal Internal-Heating Strategy for Lithium-Ion Batteries at Low Temperature Considering Both Heating Time and Lifetime Reduction. *Appl. Energy* **2019**, *256*, 113797. [CrossRef]
37. Zhang, J.; Ge, H.; Li, Z.; Ding, Z. Internal Heating of Lithium-Ion Batteries Using Alternating Current Based on the Heat Generation Model in Frequency Domain. *J. Power Sources* **2015**, *273*, 1030–1037. [CrossRef]
38. Wu, X.; Cui, Z.; Chen, E.; Du, J. Capacity Degradation Minimization Oriented Optimization for the Pulse Preheating of Lithium-Ion Batteries under Low Temperature. *J. Energy Storage* **2020**, *31*, 101746. [CrossRef]
39. Yeow, K.; Teng, H.; Thelliez, M.; Tan, E. 2012 SIMULIA Community Conference 3D Thermal Analysis of Li-Ion Battery Cells with Various Geometries and Cooling Conditions Using Abaqus. In Proceedings of the SIMULIA Community Conference, Providence, RI, USA, 15–17 May 2012.
40. Verma, A.; Prajapati, A.; Rakshit, D. A Comparative Study on Prismatic and Cylindrical Lithium-Ion Batteries Based on Their Performance in High Ambient Environment. *J. Inst. Eng. Ser. C* **2022**, *103*, 149–166. [CrossRef]
41. Li, J.; Fleetwood, J.; Hawley, W.B.; Kays, W. From Materials to Cell: State-of-the-Art and Prospective Technologies for Lithium-Ion Battery Electrode Processing. *Chem. Rev.* **2022**, *122*, 903–956. [CrossRef]
42. Goodenough, J.B. How We Made the Li-Ion Rechargeable Battery. *Nat. Electron.* **2018**, *1*, 204. [CrossRef]
43. Nishi, Y. Lithium Ion Secondary Batteries; Past 10 Years and the Future. *J. Power Sources* **2001**, *100*, 101–106. [CrossRef]
44. Xie, J.; Lu, Y.C. A Retrospective on Lithium-Ion Batteries. *Nat. Commun.* **2020**, *11*, 2499. [CrossRef]
45. Grey, C.P.; Hall, D.S. Prospects for Lithium-Ion Batteries and beyond—A 2030 Vision. *Nat. Commun.* **2020**, *11*, 6279. [CrossRef]
46. Preger, Y.; Barkholtz, H.M.; Fresquez, A.; Campbell, D.L.; Juba, B.W.; Romàn-Kustas, J.; Ferreira, S.R.; Chalamala, B. Degradation of Commercial Lithium-Ion Cells as a Function of Chemistry and Cycling Conditions. *J. Electrochem. Soc.* **2020**, *167*, 120532. [CrossRef]
47. Huang, Y.; Dong, Y.; Li, S.; Lee, J.; Wang, C.; Zhu, Z.; Xue, W.; Li, Y.; Li, J. Lithium Manganese Spinel Cathodes for Lithium-Ion Batteries. *Adv. Energy Mater.* **2021**, *11*, 2000997. [CrossRef]
48. Li, W.; Wang, H.; Zhang, Y.; Ouyang, M. Flammability Characteristics of the Battery Vent Gas: A Case of NCA and LFP Lithium-Ion Batteries during External Heating Abuse. *J. Energy Storage* **2019**, *24*, 100775. [CrossRef]
49. Bak, S.M.; Hu, E.; Zhou, Y.; Yu, X.; Senanayake, S.D.; Cho, S.J.; Kim, K.B.; Chung, K.Y.; Yang, X.Q.; Nam, K.W. Structural Changes and Thermal Stability of Charged LiNixMnyCozO2 Cathode Materials Studied by Combined in Situ Time-Resolved XRD and Mass Spectroscopy. *ACS Appl. Mater. Interfaces* **2014**, *6*, 22594–22601. [CrossRef]
50. Alves Dias, P.; Blagoeva, D.; Pavel, C.; Arvanitidis, N. *Cobalt: Demand-Supply Balances in the Transition to Electric Mobility*; Publications Office of the European Union: Luxembourg, 2018.

51. Flexer, V.; Baspineiro, C.F.; Galli, C.I. Lithium Recovery from Brines: A Vital Raw Material for Green Energies with a Potential Environmental Impact in Its Mining and Processing. *Sci. Total Environ.* **2018**, *639*, 1188–1204. [CrossRef]
52. Iskandar Radzi, Z.; Helmy Arifin, K.; Zieauddin Kufian, M.; Balakrishnan, V.; Rohani Sheikh Raihan, S.; Abd Rahim, N.; Subramaniam, R. Review of Spinel $LiMn_2O_4$ Cathode Materials under High Cut-off Voltage in Lithium-Ion Batteries: Challenges and Strategies. *J. Electroanal. Chem.* **2022**, *920*, 116623. [CrossRef]
53. Manthiram, A.; Chemelewski, K.; Lee, E.-S. A Perspective on the High-Voltage $LiMn_{1.5}Ni_{0.5}O_4$ Spinel Cathode for Lithium-Ion Batteries. *Energy Environ. Sci.* **2014**, *7*, 1339. [CrossRef]
54. Zhong, Q.; Bonakdarpour, A.; Zhang, M.; Gao, Y.; Dahn, J.R. Synthesis and Electrochemistry of $LiNi_x Mn_{2-x} O_4$. *J. Electrochem. Soc.* **1997**, *144*, 205–213. [CrossRef]
55. Murashko, K.; Li, D.; Danilov, D.L.; Notten, P.H.L.; Pyrhönen, J.; Jokiniemi, J. Applicability of Heat Generation Data in Determining the Degradation Mechanisms of Cylindrical Li-Ion Batteries. *J. Electrochem. Soc.* **2021**, *168*, 010511. [CrossRef]
56. Ma, L.; Nie, M.; Xia, J.; Dahn, J.R. A Systematic Study on the Reactivity of Different Grades of Charged $Li[Ni_xMn_yCo_z]O_2$ with Electrolyte at Elevated Temperatures Using Accelerating Rate Calorimetry. *J. Power Sources* **2016**, *327*, 145–150. [CrossRef]
57. Golubkov, A.W.; Scheikl, S.; Planteu, R.; Voitic, G.; Wiltsche, H.; Stangl, C.; Fauler, G.; Thaler, A.; Hacker, V. Thermal Runaway of Commercial 18650 Li-Ion Batteries with LFP and NCA Cathodes—Impact of State of Charge and Overcharge. *RSC Adv.* **2015**, *5*, 57171–57186. [CrossRef]
58. Liu, G.; Ouyang, M.; Lu, L.; Li, J.; Han, X. Analysis of the Heat Generation of Lithium-Ion Battery during Charging and Discharging Considering Different Influencing Factors. *J. Therm. Anal. Calorim.* **2014**, *116*, 1001–1010. [CrossRef]
59. Tan, K.S.; Reddy, M.V.; Rao, G.V.S.; Chowdari, B.V.R. High-Performance $LiCoO_2$ by Molten Salt ($LiNO_3$:LiCl) Synthesis for Li-Ion Batteries. *J. Power Sources* **2005**, *147*, 241–248. [CrossRef]
60. Matasso, A.; Wong, D.; Wetz, D.; Liu, F. Effects of High-Rate Cycling on the Bulk Internal Pressure Rise and Capacity Degradation of Commercial $LiCoO\ 2$ Cells. *J. Electrochem. Soc.* **2015**, *162*, A885–A891. [CrossRef]
61. Prosini, P.P.; Zane, D.; Pasquali, M. Improved Electrochemical Performance of a $LiFePO_4$-Based Composite Cathode. *Electrochim. Acta* **2001**, *46*, 3517–3523. [CrossRef]
62. Forman, J.C.; Moura, S.J.; Stein, J.L.; Fathy, H.K. Genetic Identification and Fisher Identifiability Analysis of the Doyle-Fuller-Newman Model from Experimental Cycling of a $LiFePO_4$ Cell. *J. Power Sources* **2012**, *210*, 263–275. [CrossRef]
63. Kurpiel, W.; Polnik, B.; Orzech, Ł.; Lesiak, K.; Miedziński, B.; Habrych, M.; Debita, G.; Zamłyńska, M.; Falkowski-gilski, P. Influence of Operation Conditions on Temperature Hazard of Lithium-Iron-Phosphate ($LiFePO_4$) Cells. *Energies* **2021**, *14*, 6728. [CrossRef]
64. Wagner, N.P.; Asheim, K.; Vullum-Bruer, F.; Svensson, A.M. Performance and Failure Analysis of Full Cell Lithium Ion Battery with $LiNi_{0.8}Co_{0.15}Al_{0.05}O_2$ and Silicon Electrodes. *J. Power Sources* **2019**, *437*, 226884. [CrossRef]
65. Barkholtz, H.M.; Fresquez, A.; Chalamala, B.R.; Ferreira, S.R. A Database for Comparative Electrochemical Performance of Commercial 18650-Format Lithium-Ion Cells. *J. Electrochem. Soc.* **2017**, *164*, A2697–A2706. [CrossRef]
66. Li, J.; Downie, L.E.; Ma, L.; Qiu, W.; Dahn, J.R. Study of the Failure Mechanisms of $LiNi_{0.8}Mn_{0.1}Co_{0.1}O_2$ Cathode Material for Lithium Ion Batteries. *J. Electrochem. Soc.* **2015**, *162*, A1401–A1408. [CrossRef]
67. Li, W.; Peng, X.; Xiao, M.; Garg, A.; Gao, L. Multi-Objective Design Optimization for Mini-Channel Cooling Battery Thermal Management System in an Electric Vehicle. *Int. J. Energy Res.* **2019**, *43*, 3668–3680. [CrossRef]
68. Weng, J.; Ouyang, D.; Yang, X.; Chen, M.; Zhang, G.; Wang, J. Optimization of the Internal Fin in a Phase-Change-Material Module for Battery Thermal Management. *Appl. Therm. Eng.* **2020**, *167*, 114698. [CrossRef]
69. Travesset-Baro, O.; Rosas-Casals, M.; Jover, E. Transport Energy Consumption in Mountainous Roads. A Comparative Case Study for Internal Combustion Engines and Electric Vehicles in Andorra. *Transp. Res. D Transp. Environ.* **2015**, *34*, 16–26. [CrossRef]
70. Wahid, M.R.; Budiman, B.A.; Joelianto, E.; Aziz, M. A Review on Drive Train Technologies for Passenger Electric Vehicles. *Energies* **2021**, *14*, 6742. [CrossRef]
71. Halimah, P.N.; Rahardian, S.; Budiman, B.A. Battery Cells for Electric Vehicles. *Int. J. Sustain. Transp. Technol.* **2019**, *2*, 54–57. [CrossRef]
72. Wierzbicki, T.; Sahraei, E. Homogenized Mechanical Properties for the Jellyroll of Cylindrical Lithium-Ion Cells. *J. Power Sources* **2013**, *241*, 467–476. [CrossRef]
73. Zeng, F.; Chen, J.; Yang, F.; Kang, J.; Cao, Y.; Xiang, M. Effects of Polypropylene Orientation on Mechanical and Heat Seal Properties of Polymer-Aluminum-Polymer Composite Films for Pouch Lithium-Ion Batteries. *Materials* **2018**, *11*, 144. [CrossRef] [PubMed]
74. Budiman, B.A.; Rahardian, S.; Saputro, A.; Hidayat, A.; Pulung Nurprasetio, I.; Sambegoro, P. Structural Integrity of Lithium-Ion Pouch Battery Subjected to Three-Point Bending. *Eng. Fail. Anal.* **2022**, *138*, 106307. [CrossRef]
75. Cordoba-Arenas, A.; Onori, S.; Rizzoni, G. A Control-Oriented Lithium-Ion Battery Pack Model for Plug-in Hybrid Electric Vehicle Cycle-Life Studies and System Design with Consideration of Health Management. *J. Power Sources* **2015**, *279*, 791–808. [CrossRef]
76. Wang, B.; Ji, C.; Wang, S.; Sun, J.; Pan, S.; Wang, D.; Liang, C. Study of Non-Uniform Temperature and Discharging Distribution for Lithium-Ion Battery Modules in Series and Parallel Connection. *Appl. Therm. Eng.* **2020**, *168*, 114831. [CrossRef]
77. Arora, S.; Shen, W.; Kapoor, A. Review of Mechanical Design and Strategic Placement Technique of a Robust Battery Pack for Electric Vehicles. *Renew. Sus. Energy Rev.* **2016**, *60*, 1319–1331. [CrossRef]

78. Omar, N.; van den Bossche, P.; Mulder, G.; Daowd, M.; Timmermans, J.M.; van Mierlo, J.; Pauwels, S. Assessment of Performance of Lithium Iron Phosphate Oxide, Nickel Manganese Cobalt Oxide and Nickel Cobalt Aluminum Oxide Based Cells for Using in Plug-in Battery Electric Vehicle Applications. In Proceedings of the 2011 IEEE Vehicle Power and Propulsion Conference, Chicago, IL, USA, 6–9 September 2011. [CrossRef]
79. Guo, J.; Jiang, F. A Novel Electric Vehicle Thermal Management System Based on Cooling and Heating of Batteries by Refrigerant. *Energy Convers. Manag.* **2021**, *237*, 114145. [CrossRef]
80. Cen, J.; Jiang, F. Li-Ion Power Battery Temperature Control by a Battery Thermal Management and Vehicle Cabin Air Conditioning Integrated System. *Energy Sus.Dev.* **2020**, *57*, 141–148. [CrossRef]
81. Fan, R.; Zhang, C.; Wang, Y.; Ji, C.; Meng, Z.; Xu, L.; Ou, Y.; Chin, C.S. Numerical Study on the Effects of Battery Heating in Cold Climate. *J. Energy Storage* **2019**, *26*, 100969. [CrossRef]
82. Zhang, S.S.; Xu, K.; Jow, T.R. The Low Temperature Performance of Li-Ion Batteries. *J. Power Sources* **2003**, *115*, 137–140. [CrossRef]
83. Lei, Z.; Zhang, Y.; Lei, X. Improving Temperature Uniformity of a Lithium-Ion Battery by Intermittent Heating Method in Cold Climate. *Int. J. Heat Mass Transf.* **2018**, *121*, 275–281. [CrossRef]
84. Hebert, A.; McCalla, E. The Role of Metal Substitutions in the Development of Li Batteries, Part I: Cathodes. *Mater. Adv.* **2021**, *2*, 3474–3518. [CrossRef]
85. Lighting Global. *Lithium-Ion Battery Overview*; International Finance Group: Washington, DC, USA, 2012.
86. Li, W.; Erickson, E.M.; Manthiram, A. High-Nickel Layered Oxide Cathodes for Lithium-Based Automotive Batteries. *Nat. Energy* **2020**, *5*, 26–34. [CrossRef]
87. Zhang, S.; Ma, J.; Hu, Z.; Cui, G.; Chen, L. Identifying and Addressing Critical Challenges of High-Voltage Layered Ternary Oxide Cathode Materials. *Chem. Mater.* **2019**, *31*, 6033–6065. [CrossRef]
88. Zou, H.; Wang, W.; Zhang, G.; Qin, F.; Tian, C.; Yan, Y. Experimental Investigation on an Integrated Thermal Management System with Heat Pipe Heat Exchanger for Electric Vehicle. *Energy Convers. Manag.* **2016**, *118*, 88–95. [CrossRef]
89. Zhang, C.-M.; Li, F.; Zhu, X.-Q.; Yu, J.-G. Triallyl Isocyanurate as an Efficient Electrolyte Additive for Layered Oxide Cathode Material-Based Lithium-Ion Batteries with Improved Stability under High-Voltage. *Molecules* **2022**, *27*, 3107. [CrossRef]
90. Park, S.; Jeong, S.Y.; Lee, T.K.; Park, M.W.; Lim, H.Y.; Sung, J.; Cho, J.; Kwak, S.K.; Hong, S.Y.; Choi, N.-S. Replacing Conventional Battery Electrolyte Additives with Dioxolone Derivatives for High-Energy-Density Lithium-Ion Batteries. *Nat. Commun.* **2021**, *12*, 838. [CrossRef]
91. Xu, G.; Huang, S.; Cui, Z.; Du, X.; Wang, X.; Lu, D.; Shangguan, X.; Ma, J.; Han, P.; Zhou, X.; et al. Functional Additives Assisted Ester-Carbonate Electrolyte Enables Wide Temperature Operation of a High-Voltage (5 V-Class) Li-Ion Battery. *J. Power Sources* **2019**, *416*, 29–36. [CrossRef]
92. Chen, M.; Liu, Z.; Zhao, X.; Li, K.; Wang, K.; Liu, Z.; Xia, L.; Yuan, J.; Zhao, R. Fluorinated Co-Solvent Electrolytes for High-Voltage Ni-Rich LiNi$_{0.8}$Co$_{0.1}$Mn$_{0.1}$O$_2$ (NCM811) Positive Electrodes. *Front. Energy Res.* **2022**, *10*, 973336. [CrossRef]
93. Qin, Y.; Ren, Y.; Wang, Q.; Li, Y.; Liu, J.; Liu, Y.; Guo, B.; Wang, D. Simplifying the Electrolyte Systems with the Functional Co-solvent. *ACS Appl. Mater. Interfaces* **2019**, *11*, 27854–27861. [CrossRef]
94. Chatterjee, K.; Pathak, A.D.; Lakma, A.; Sharma, C.S.; Sahu, K.K.; Singh, A.K. Synthesis, Characterization and Application of a Non-Flammable Dicationic Ionic Liquid in Lithium-Ion Battery as Electrolyte Additive. *Sci. Rep.* **2020**, *10*, 9606. [CrossRef]
95. Liu, Q.Q.; Petibon, R.; Du, C.Y.; Dahn, J.R. Effects of Electrolyte Additives and Solvents on Unwanted Lithium Plating in Lithium-Ion Cells. *J. Electrochem. Soc.* **2017**, *164*, A1173–A1183. [CrossRef]
96. Lu, W.; Xie, K.; Chen, Z.; Xiong, S.; Pan, Y.; Zheng, C. A New Co-Solvent for Wide Temperature Lithium Ion Battery Electrolytes: 2,2,2-Trifluoroethyl n-Caproate. *J. Power Sources* **2015**, *274*, 676–684. [CrossRef]
97. Ouyang, D.; Wang, K.; Yang, Y.; Wang, Z. Fluoroethylene Carbonate as Co-Solvent for Li(Ni$_{0.8}$Mn$_{0.1}$Co$_{0.1}$)O$_2$ Lithium-Ion Cells with Enhanced High-Voltage and Safety Performance. *J. Power Sources* **2022**, *542*, 231780. [CrossRef]
98. Gond, R.; van Ekeren, W.; Mogensen, R.; Naylor, A.J.; Younesi, R. Non-Flammable Liquid Electrolytes for Safe Batteries. *Mater. Horiz.* **2021**, *8*, 2913–2928. [CrossRef] [PubMed]
99. Knodler, R. Thermal Properties of Sodium-Sulphur Cells. *J. Appl. Electrochem.* **1984**, *14*, 39–46. [CrossRef]
100. Manthiram, A.; Yu, X.; Wang, S. Lithium Battery Chemistries Enabled by Solid-State Electrolytes. *Nat. Rev. Mater.* **2017**, *2*, 16103. [CrossRef]
101. Chen, H.; Zheng, M.; Qian, S.; Ling, H.Y.; Wu, Z.; Liu, X.; Yan, C.; Zhang, S. Functional Additives for Solid Polymer Electrolytes in Flexible and High-energy-density Solid-state Lithium-ion Batteries. *Carbon Energy* **2021**, *3*, 929–956. [CrossRef]
102. Xia, S.; Wu, X.; Zhang, Z.; Cui, Y.; Liu, W. Practical Challenges and Future Perspectives of All-Solid-State Lithium-Metal Batteries. *Chem* **2019**, *5*, 753–785. [CrossRef]
103. Kanno, R. Secondary Batteries—Lithium Rechargeable Systems | Electrolytes: Solid Sulfide. In *Encyclopedia of Electrochemical Power Sources*; Elsevier: Amsterdam, The Netherlands, 2009; pp. 129–137. [CrossRef]
104. Liu, Y.; Li, B.; Kitaura, H.; Zhang, X.; Han, M.; He, P.; Zhou, H. Fabrication and Performance of All-Solid-State Li–Air Battery with SWCNTs/LAGP Cathode. *ACS Appl. Mater. Interfaces* **2015**, *7*, 17307–17310. [CrossRef]
105. Yao, P.; Yu, H.; Ding, Z.; Liu, Y.; Lu, J.; Lavorgna, M.; Wu, J.; Liu, X. Review on Polymer-Based Composite Electrolytes for Lithium Batteries. *Front. Chem.* **2019**, *7*, 522. [CrossRef]

106. Liu, B.; Zhang, L.; Xu, S.; McOwen, D.W.; Gong, Y.; Yang, C.; Pastel, G.R.; Xie, H.; Fu, K.; Dai, J.; et al. 3D Lithium Metal Anodes Hosted in Asymmetric Garnet Frameworks toward High Energy Density Batteries. *Energy Storage Mater.* **2018**, *14*, 376–382. [CrossRef]
107. Singh, V.K.; Faisal, M.; Khan, J. Analytical Study and Comparison of Solid and Liquid Batteries for Electric Vehicles and Thermal Management Simulation. *United Int. J. Res.Technol. (UIJRT)* **2019**, *1*, 27–33.
108. Budiman, B.A.; Saputro, A.; Rahardian, S.; Aziz, M.; Sambegoro, P.; Nurprasetio, I.P. Mechanical Damages in Solid Electrolyte Battery Due to Electrode Volume Changes. *J. Energy Storage* **2022**, *52*, 104810. [CrossRef]
109. Chang, Z.; Yang, H.; Zhu, X.; He, P.; Zhou, H. A Stable Quasi-Solid Electrolyte Improves the Safe Operation of Highly Efficient Lithium-Metal Pouch Cells in Harsh Environments. *Nat. Commun.* **2022**, *13*, 1510. [CrossRef]
110. Robinson, A.L.; Janek, J. Solid-State Batteries Enter EV Fray. *MRS Bull* **2014**, *39*, 1046–1047. [CrossRef]
111. Motavalli, J. Technology: A Solid Future. *Nature* **2015**, *526*, S96–S97. [CrossRef]
112. Bindra, A. Electric Vehicle Batteries Eye Solid-State Technology: Prototypes Promise Lower Cost, Faster Charging, and Greater Safety. *IEEE Power Electr.Mag.* **2020**, *7*, 16–19. [CrossRef]
113. Al-Zareer, M.; Dincer, I.; Rosen, M.A. Novel Thermal Management System Using Boiling Cooling for High-Powered Lithium-Ion Battery Packs for Hybrid Electric Vehicles. *J. Power Sources* **2017**, *363*, 291–303. [CrossRef]
114. Khan, S.A.; Eze, C.; Dong, K.; Shahid, A.R.; Patil, M.S.; Ahmad, S.; Hussain, I.; Zhao, J. Design of a New Optimized U-Shaped Lightweight Liquid-Cooled Battery Thermal Management System for Electric Vehicles: A Machine Learning Approach. *Int. Commun. Heat Mass Transf.* **2022**, *136*, 106209. [CrossRef]
115. Jouhara, H.; Delpech, B.; Bennett, R.; Chauhan, A.; Khordehgah, N.; Serey, N.; Lester, S.P. Heat Pipe Based Battery Thermal Management: Evaluating the Potential of Two Novel Battery Pack Integrations. *Int. J. Therm.* **2021**, *12*, 100115. [CrossRef]
116. Sarchami, A.; Najafi, M.; Imam, A.; Houshfar, E. Experimental Study of Thermal Management System for Cylindrical Li-Ion Battery Pack Based on Nanofluid Cooling and Copper Sheath. *Int. J. Therm. Sci.* **2022**, *171*, 107244. [CrossRef]
117. Liao, G.; Wang, W.; Zhang, F.; E, J.; Chen, J.; Leng, E. Thermal Performance of Lithium-Ion Battery Thermal Management System Based on Nanofluid. *Appl. Therm. Eng.* **2022**, *216*, 118997. [CrossRef]
118. Lu, M.; Zhang, X.; Ji, J.; Xu, X.; Zhang, Y. Research Progress on Power Battery Cooling Technology for Electric Vehicles. *J. Energy Storage* **2020**, *27*, 101155. [CrossRef]
119. Zhang, G.; Qin, F.; Zou, H.; Tian, C. Experimental Study on a Dual- Parallel-Evaporator Heat Pump System for Thermal Management of Electric Vehicles. *Energy Procedia* **2017**, *105*, 2390–2395. [CrossRef]
120. Tang, X.; Guo, Q.; Li, M.; Wei, C.; Pan, Z.; Wang, Y. Performance Analysis on Liquid-Cooled Battery Thermal Management for Electric Vehicles Based on Machine Learning. *J. Power Sources* **2021**, *494*, 229727. [CrossRef]
121. Can, A.; Selimefendigil, F.; Öztop, H.F. A Review on Soft Computing and Nanofluid Applications for Battery Thermal Management. *J. Energy Storage* **2022**, *53*, 105214. [CrossRef]
122. Minea, A.A. A Study on Brinkman Number Variation on Water Based Nanofluid Heat Transfer in Partially Heated Tubes. *Mech. Res. Commun.* **2016**, *73*, 7–11. [CrossRef]
123. Berdichevsky, G.; Kelty, K.; Straubel, J.B.; Toomre, E. *The Tesla Roadster Battery System*; Tesla Motors Inc.: San Carlos, CA, USA, 2006.
124. Ogawa, M.; Yoshida, K.; Harada, K. All-Solid-State Lithium Batteries with Wide Operating Temperature Range. *Environ. Energy Res.* **2012**, *74*, 88–90.
125. Wang, S.; Song, H.; Song, X.; Zhu, T.; Ye, Y.; Chen, J.; Yu, L.; Xu, J.; Chen, K. An Extra-Wide Temperature All-Solid-State Lithium-Metal Battery Operating from −73 °C to 120 °C. *Energy Storage Mater.* **2021**, *39*, 139–145. [CrossRef]
126. Guo, Y.; Wu, S.; He, Y.-B.; Kang, F.; Chen, L.; Li, H.; Yang, Q.-H. Solid-State Lithium Batteries: Safety and Prospects. *EScience* **2022**, *2*, 138–163. [CrossRef]
127. Hughes, R.; Vagg, C. Assessing the Feasibility of a Cold Start Procedure for Solid State Batteries in Automotive Applications. *Batteries* **2022**, *8*, 13. [CrossRef]
128. Li, H.; Wang, H.; Xu, Z.; Wang, K.; Ge, M.; Gan, L.; Zhang, Y.; Tang, Y.; Chen, S. Thermal-Responsive and Fire-Resistant Materials for High-Safety Lithium-Ion Batteries. *Small* **2021**, *17*, 2103679. [CrossRef]
129. Jin, C.; Sun, Y.; Yao, J.; Feng, X.; Lai, X.; Shen, K.; Wang, H.; Rui, X.; Xu, C.; Zheng, Y.; et al. No Thermal Runaway Propagation Optimization Design of Battery Arrangement for Cell-to-Chassis Technology. *ETransportation* **2022**, *14*, 100199. [CrossRef]
130. Roper, S.W.K.; Kim, I.Y. Integrated Topology and Packaging Optimization for Conceptual-Level Electric Vehicle Chassis Design via the Component-Existence Method. *Proc. Inst. Mech.Eng. Part D J. Automob. Eng.* **2022**, 09544070221113895. [CrossRef]

Article

Parametric Evaluation of Thermal Behavior for Different Li-Ion Battery Chemistries

Thomas Imre Cyrille Buidin and Florin Mariasiu *

Automotive Engineering and Transports Department, Technical University of Cluj-Napoca, 400114 Cluj-Napoca, Romania
* Correspondence: florin.mariasiu@auto.utcluj.ro

Abstract: The prediction of thermal behavior is essential for an efficient initial design of thermal management systems which equip energy sources based on electrochemical cells. In this study, the surface temperature of various cylindrical types of Li-ion batteries is monitored at multiple points during discharge. Three different battery chemistries and two sizes (18650 and 21700) are considered in this study, allowing the comparison of the influence these parameters have on the temperature rise considering different discharge rates (1C, 2C and 3C). Based on repeated experimental measurements, a simple equation that describes the thermal behavior of batteries is proposed and further used to create 3D thermal maps for each analyzed battery (generally error is below 1 °C but never exceeds 3 °C). The practical utility of such an equation is that it can drastically reduce the time spent with experimental measurements required to characterize the thermal behavior of cylindrical Li-ion batteries, necessary for the initial design process of energy sources' thermal management system.

Keywords: battery; Li-ion; temperature; thermal map; parametric equation

1. Introduction

In recent years, the need to reduce environmental pollution has been increasingly acknowledged, offering a wider market share to electric vehicles year by year [1]. Lithium-ion (Li-ion) batteries are the preferred energy source for electric vehicles, due to their proven supremacy compared to other chemistries in terms of power and energy density, self-discharge rate and cycle life [2]. However, these properties are strongly affected by temperature [3]. During charging and discharging operations the ongoing electrochemical processes generate a considerable quantity of heat, resulting in the rise of the battery temperature [4]. The heat generation of Li-ion batteries generally varies in time and influenced by the working conditions, such as state of charge, discharge rate and ambient temperature [4,5].

Studies show that operating Li-ion batteries at high temperatures can quicken chemical changes, such as the growth of solid electrolyte interphase in cells, loss of active material or electrolytic corrosion [6–9]. This leads to the reduction in the electrodes' available surface area for electrochemical reactions. Therefore, it is crucial to be able to predict heat generation characteristics and temperature rise for the right battery thermal management system design, for maintaining the performance and safety of the battery cells.

The different processes that take place inside the batteries are strongly connected one to another. The reactions generate heat which affects temperature uniformity inside the cell [10,11]. The temperature gradient consecutively dictates the electrochemical reaction kinetics, the transfer of ionic charge and the crystalline phase equilibria of the electrodes. The dynamics of these phenomena are heavily connected [12]. The interactions between thermal, electrical, and electrochemical phenomena are illustrated in Figure 1. The processes include species diffusion, charge transport, chemical kinetics and thermal transport and are directed by physical laws including several transport properties, such as thermal conductivity, mass diffusivity or reaction rates [12].

Citation: Buidin, T.I.C.; Mariasiu, F. Parametric Evaluation of Thermal Behavior for Different Li-Ion Battery Chemistries. *Batteries* **2022**, *8*, 291. https://doi.org/10.3390/batteries8120291

Academic Editor: King Jet Tseng

Received: 17 November 2022
Accepted: 14 December 2022
Published: 17 December 2022

Publisher's Note: MDPI stays neutral with regard to jurisdictional claims in published maps and institutional affiliations.

Copyright: © 2022 by the authors. Licensee MDPI, Basel, Switzerland. This article is an open access article distributed under the terms and conditions of the Creative Commons Attribution (CC BY) license (https://creativecommons.org/licenses/by/4.0/).

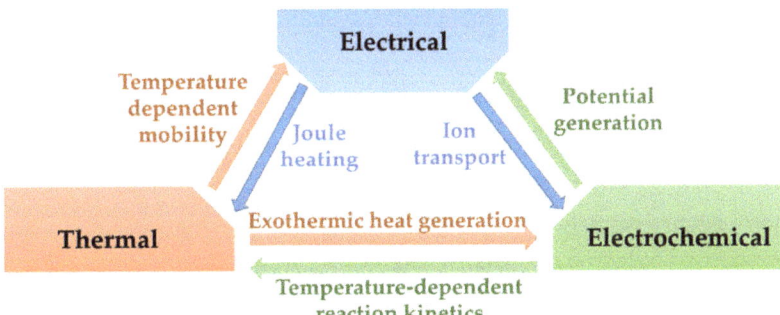

Figure 1. Schematic illustrating processes inside Li-ion batteries.

The elementary components of Li-ion batteries are the two electrodes, anode and cathode, and the electrolyte which allows the transfer of lithium ions between the electrodes. The negative electrode (anode) is usually made from carbon (graphite) or lithium titanate ($Li_4Ti_5O_{12}$), while other materials such as Li metals or Li(Si) alloys are considered. The electrolyte is typically an organic liquid containing a variety of organic carbonates and a salt, such as lithium hexafluorophosphate (LiPF6) [13]. Within the family of Li-ion batteries, there are several positive electrode (cathode) materials, which define the battery chemistry, for example: $LiCoO_2$ (LCO), $LiFePO_4$ (LFP), $Li(Ni_{1-x-y}Mn_xCo_y)O_2$ (NMC), $Li(Ni_{1-x-y}Co_xAl_y)O_2$ (NCA) and $LiMn_2O_4$ (LMO) [14]. LCO batteries have high energy density but are very reactive and therefore have weak thermal stability. LFP batteries have high power density and thermal stability and are inexpensive. NMC batteries have high specific energy and exceptional thermal properties, obtained by the right proportions of nickel and manganese to enhance each other's strengths. Therefore, they are in high demand for their use in electric vehicle batteries. NCA batteries also offer high specific power and specific energy and a long lifespan, but are not as safe as the other chemistries and are more expensive.

The heat generation and temperature rise of Li-ion batteries have been investigated by a large number of papers [4,5,10,15–19]. Generally, there are two main causes attributed to battery heat generation, namely the over-potential and the entropic heat flow [20,21]. The former is due to ohmic losses, charge transfer at the interfaces and mass transfer constraints, as well as the current flowing from one electrode to the other, originating in an irreversible heat flow. Meanwhile, the latter is due to the reactions at the anode and cathode, respectively, during charge and discharge [17]. Experimental measurements and mathematical models have shown that the thermal influence of the two electrodes can be different, with a more concentrated heat generation at the positive terminal [22,23]. Different values at the two terminals also exist for parameters such as lithium diffusion coefficient, reaction rate and entropy change, influencing the characterization of thermal behavior at the electrodes [24].

The prediction of Li-ion batteries' thermal behavior can be conducted in several ways. Equivalent circuit models are used due to their simplicity and suitable performance, describing the state of charge, current and heat generation [25]. Multiphysics modeling can be used to analyze the influence of the battery's active components on the contribution of reversible and irreversible heat generation [13]. Another numerical prediction method of the battery core temperature is the use of coupled linear differential equations, based on measured ambient and battery surface temperatures [26]. Drake et al. [5] monitored the variation of internal and external temperatures of cylindrical Li-ion batteries, as well as the heat flux on their outer surface, to determine the heat generation rate. Well-established experimental methods are the accelerating rate calorimetry (ARC) and isothermal heat conduction calorimetry (IHC) [27]. Under adiabatic conditions (ARC method), heat generation can be calculated based on the specific heat capacity and temperature rise of the battery [28].

To minimize heat exchange with the surroundings, Arora et al. [18] placed pouch battery cells in slots inside HDPE slabs. A novel isothermal calorimetric method is proposed by Hu et al. [29], which allows the simultaneous measurement of internal resistance and entropy coefficient by applying a sinusoidal current and analyzing the heat generation responses in the frequency domain. An additional IHC calorimeter was designed by Yin et al. [30] with highly dynamic properties, obtained from the use of thermoelectric devices. The proposed calorimeter allows the monitoring of variations in heat generation rate during dynamic charge/discharge processes, due to the reduced thermal inertia. A similar concept was proposed by Diaz et al. [31], where the entire experimental apparatus, placed in an environmental chamber, was set to be maintained at constant temperature by the means of thermoelectric elements. The rate of heat extraction must be equal to the rate of heat generation, resulting in 4D maps of heat generation, as a function of frequency, current, state of charge and temperature.

Most of the studies limit temperature and heat flux measurement to the exterior surface of the batteries, although temperature distribution inside the cell can be experimentally investigated using micro thermocouples [11]. Such measurements show that due to the poor radial thermal conductivity of the cylindrical batteries, significant temperature gradients can form within the cells, especially at high discharge rates. This non-uniform temperature distribution amplifies the non-uniform distribution of current density, caused by the dependencies between the state of charge (SoC), current and temperature, which in extreme conditions can cause a short circuit or overcharge [32,33]. Therefore, measuring the internal temperature provides more information about the state of health of the battery, while external measurements can significantly underestimate the maximum temperature.

Consequently, the prediction of battery heat generation and temperature rise for a given battery shape, capacity and chemistry must be addressed before designing the battery pack and its thermal management system [34]. Regarding temperature indication methods, key characteristics are accuracy, resolution, and measurement range. Thermistors are solid semiconductor devices, which show a rapid change in their electrical resistance with temperature [35]. Resistance Temperature Detectors are equipment consisting of metallic conductors and presenting an increase in electrical resistance with temperature [36]. Thermocouples are devices functioning on the principle of the Seebeck effect, which consists of the formation of an electromotive force by exposing two distinct conductors to a temperature difference [37]. They are a preferred solution due to their low cost, robustness, size and temperature range [38]. Even though a higher accuracy can be obtained, it normally lies within 1 or 2 °C and is thus ordinary. Amid the numerous accessible scientific papers that conduct experimental measurements regarding battery temperature, commercial thermocouples are regularly adopted. A justification might be that devices are generally equipped with thermocouple input channels. The greater part of investigations uses the common K-type or T-type thermocouples to measure the battery temperature [39–48].

The authors analyzed bibliographic sources related to this subject and failed to identify a mathematical relationship of the thermal behavior that considers the chemical particularities of cylindrical Li-ion batteries. For this purpose, in this paper, cylindrical Li-ion batteries of different sizes and chemistries are discharged at several rates while measuring the surface temperature in three points using K-type thermocouples. The objective is to create a 3D thermal map for each of the tested batteries, in which temperature rise is described as a function of the state of charge and discharge rate and finally a mathematical equation is proposed that characterizes the thermal behavior of the batteries. Fitting equations describing temperature rise or heat generation appear in numerous papers, usually in the form of simple polynomial equations of the second or third degree. Although these have excellent accuracy, their implementation requires the definition of all the individual unknown parameters present in polynomial equations (up to nine in the case of a third-degree two-variable polynomial equation), which is a time-consuming process. The novelty of this research lies in the general character of the proposed fitting equation, which includes some of the main parameters of Li-ion batteries, such as nominal voltage and capacity, reducing the number of unknown parameters to just one. The function of the battery's chemistry, this parameter can be quickly determined even by a single measurement, reducing considerably the time necessary for creating the sought thermal maps.

2. Materials and Methods

Three types of cylindrical Li-ion batteries from different manufacturers were used during the measurements, their parameters are listed in Table 1. Cylindrical Li-ion cells have a significantly higher energy density than prismatic or pouch cells [49]. Even though they cannot be packed as optimally as the other two cell shapes, cylindrical cells maintain a slight advantage even at a system (battery pack) level. Temperature measurements were made using K-type thermocouples with PTFE insulation and 0.2 mm twisted pair conductor fixed to the battery surface using thermal insulation tape. To capture possible differences in heat generation and therefore temperature at the battery terminals, three thermocouples were used for the tests, in the proximity of the cathode, in the proximity of the anode and at the middle of the battery, as shown in Figure 2. Nickel sheets were welded to the cell terminals using spot welding.

Table 1. Battery cell parameters.

Parameters	KeepPower	Panasonic	Samsung
Chemistry	LCO	NCA	NMC
Size	18650	18650	21700
Anode active material	Graphite	Graphite	Graphite
Cathode active material	$LiMn_2O_4$	$LiNi_{0.8}Co_{0.15}Al_{0.05}O_2$	$Li(NiMnCo)O_2$
Anode electrode thickness	126 μm	126 μm	126 μm
Cathode electrode thickness	125 μm	125 μm	125 μm
Anode current collector foil thicknesses (copper)	10 μm	10 μm	10 μm
Cathode current collector foil thicknesses (aluminum)	20 μm	20 μm	20 μm
Electrolyte	Lithium Hexafluorophosphate (LiPF6)		
Separator (polypropylene (PP))	20 μm	20 μm	22 μm
Nominal voltage [V]	3.7	3.6	3.6
Nominal capacity [Ah]	2.6	3.1	4
Maximum discharge current [A]	15	10	45

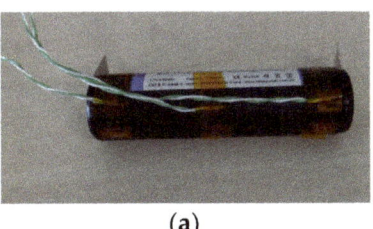
(a)

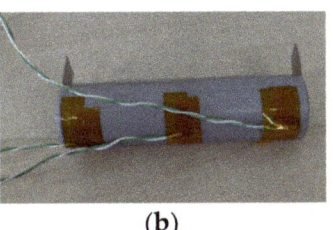
(b)

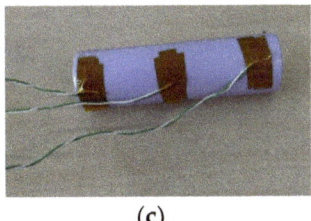
(c)

Figure 2. Positioning of thermocouples on the battery surface: (**a**) LCO; (**b**) NCA; (**c**) NMC.

The thermocouples were connected to a Pico TC-08 data logger, for which the temperature accuracy is the sum of ±0.2% of reading and ±0.5 °C. Temperatures were recorded from the date logger at an interval of one second on the personal computer. The constant current discharge of the batteries was conducted using an East Tester ETS5410 programmable electronic load, with a current accuracy of ±(0.05% + 0.05% FS). Additionally, a thermal camera with a sensor resolution of 76,800 pixels and a thermal sensitivity of 70 mK was used as a second temperature monitoring tool for validation. The whole experimental setup is shown in Figure 3.

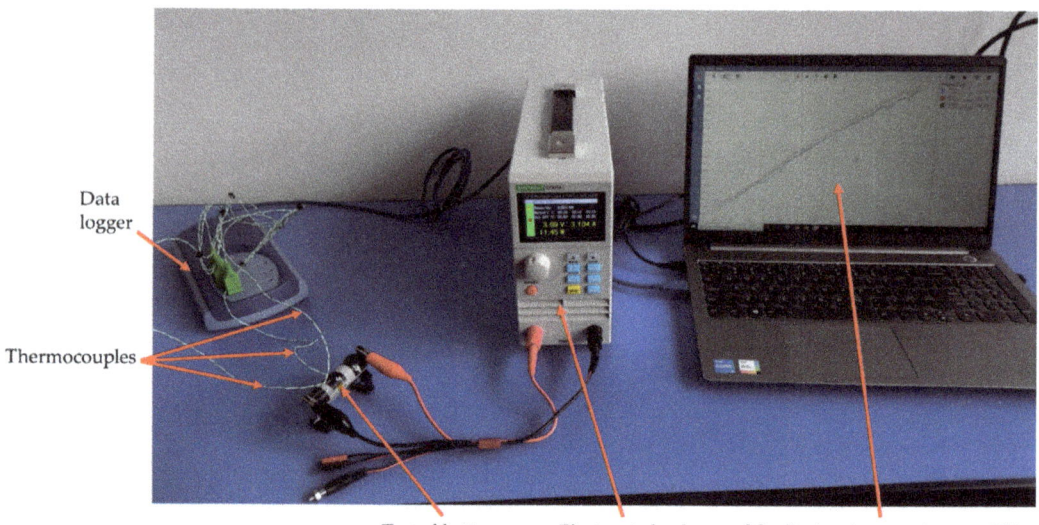

Figure 3. Experimental setup.

Experimental Procedure

In this experimental study, for each battery type, three individual cells from the same batch were selected. The nine selected cells were numbered as follows: #1, 4, 7–NCA (Panasonic, Beijing, China); #2, 5, 8–NMC (Samsung); #3, 6, 9–LCO (KeepPower). Every cell was successively discharged three times at discharge rates of 1C, 2C and 3C. For each discharge process, the average temperature rise for all three thermocouples was determined. In the case where at a single thermocouple a difference larger than 5% of the average temperature rise between measurements was detected, a fourth measurement was conducted at the respective discharge rate. The four measurements were compared and only the three with the smallest added differences were considered in the following. Completing the above-mentioned experimental procedure required a total of approximately 60 h of experimental measurements.

The hereby selected experimental data collections were then plotted on a single diagram for each battery type, resulting in a total of nine diagrams (3 battery types × 3 discharge rates). Each diagram illustrates a total of 27 curves, representing the measured temperature rise on all three locations on the battery surface. For every location, a 3rd-order polynomial equation was generated to describe the thermal behavior at the positive and negative terminal and in the middle of the cell, respectively. Additionally, another equation was generated to describe the trendline of all measured data.

The nine diagrams with their respective equations are illustrated in Figure 4, while the equations' coefficients are represented in Table 2.

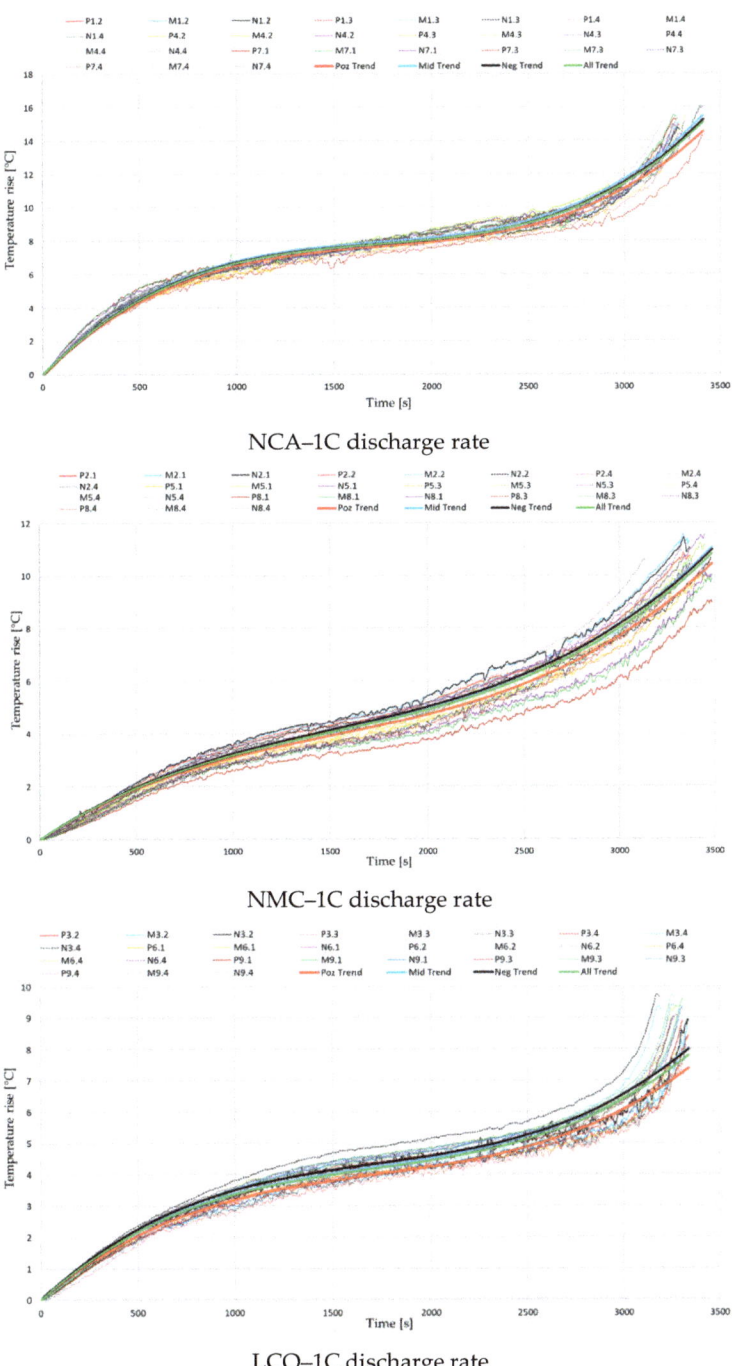

NCA–1C discharge rate

NMC–1C discharge rate

LCO–1C discharge rate

Figure 4. *Cont.*

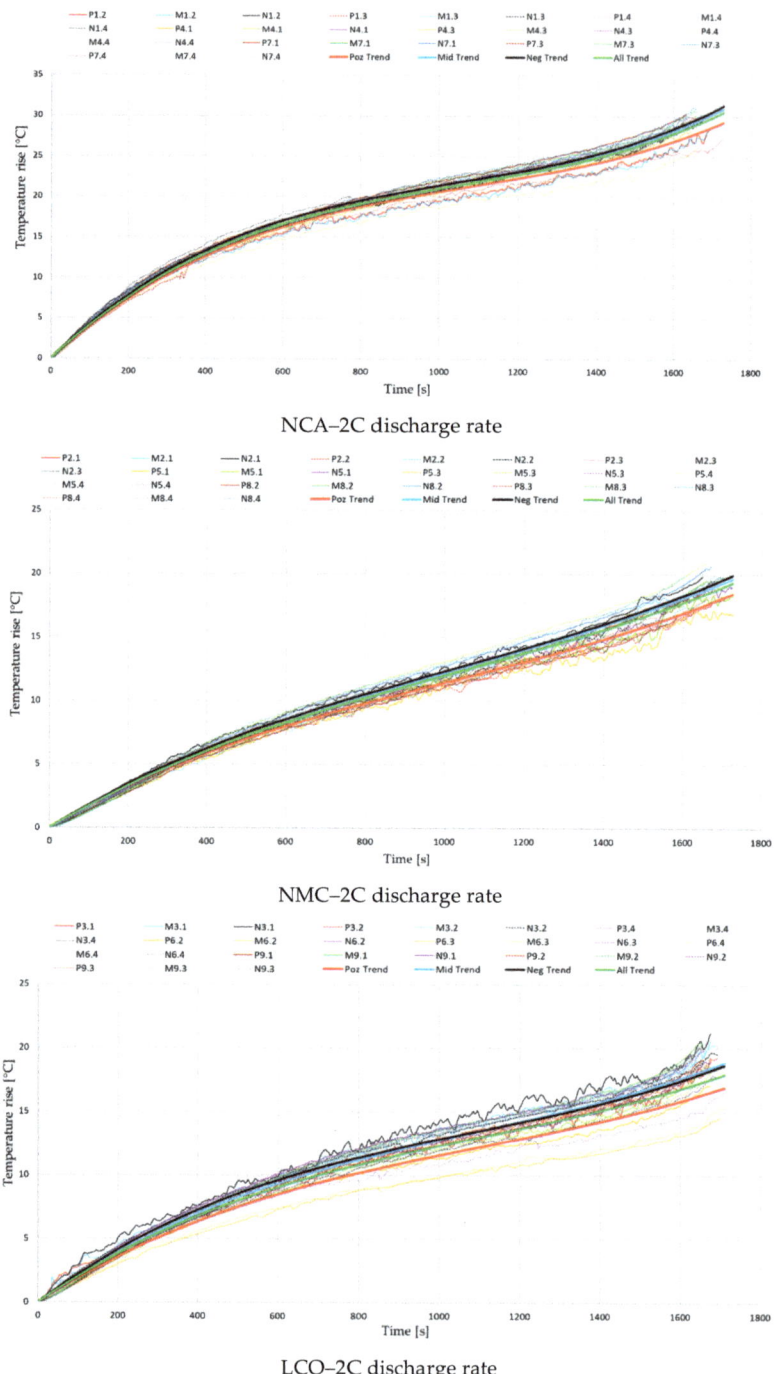

Figure 4. *Cont.*

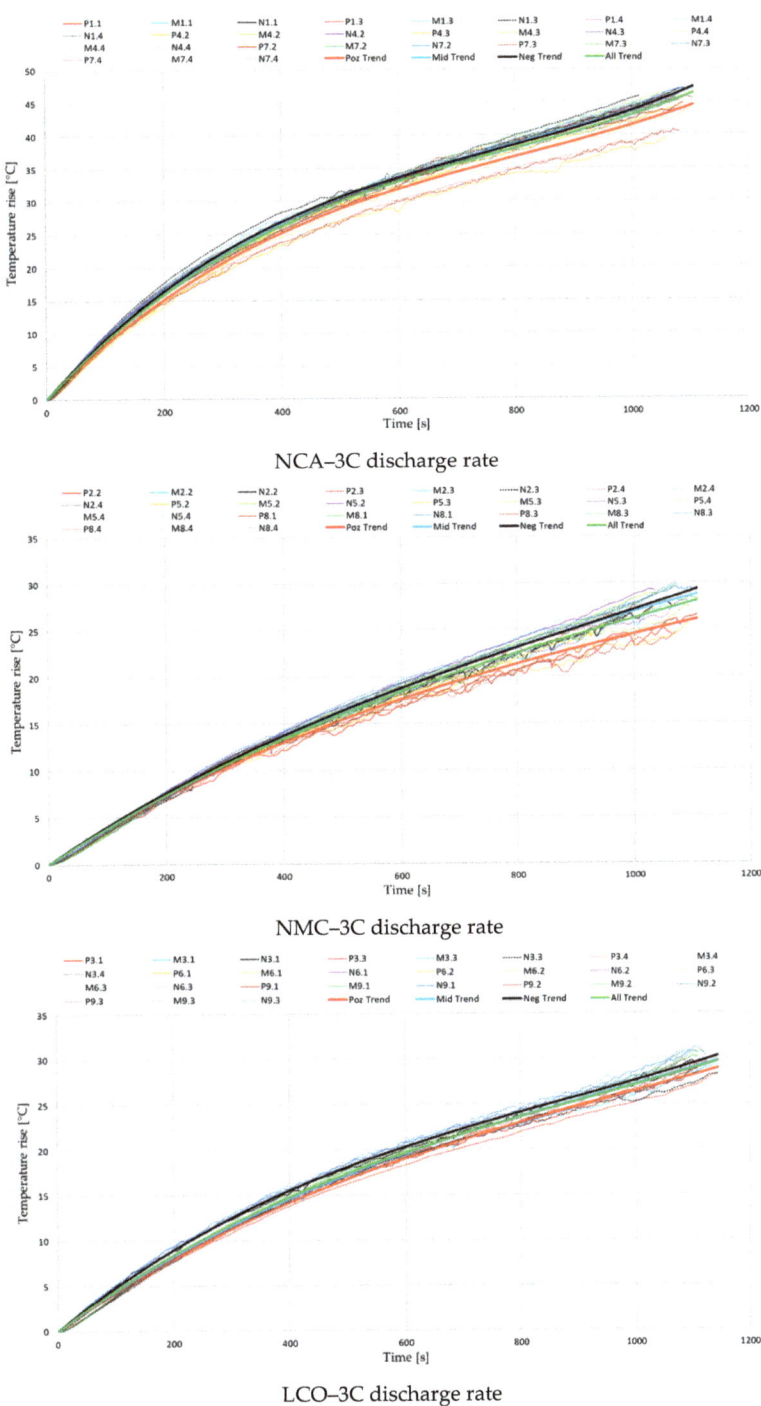

Figure 4. Temperature variations during the discharge process, for different cell chemistries.

Table 2. Third-order polynomial equations' coefficients.

Discharge Rate	Battery's Chemistry	Location of Measurement	Polynomial Equations' Coefficients			R-Square
			a	b	c	
1C	NCA	Positive terminal	1.14×10^{-9}	5.95×10^{-6}	1.13×10^{-2}	0.9714
		Middle	1.25×10^{-9}	6.46×10^{-6}	1.20×10^{-2}	0.9828
		Negative terminal	1.23×10^{-9}	6.37×10^{-6}	1.18×10^{-2}	0.9854
		All data	1.21×10^{-9}	6.26×10^{-6}	1.17×10^{-2}	0.9774
	NMC	Positive terminal	4.60×10^{-10}	2.10×10^{-6}	4.73×10^{-3}	0.9626
		Middle	4.68×10^{-10}	2.11×10^{-6}	4.84×10^{-3}	0.9686
		Negative terminal	4.62×10^{-10}	2.11×10^{-6}	4.88×10^{-3}	0.97
		All data	4.63×10^{-10}	2.11×10^{-6}	4.82×10^{-3}	0.9644
	LCO	Positive terminal	4.90×10^{-10}	2.54×10^{-6}	5.24×10^{-3}	0.9763
		Middle	5.37×10^{-10}	2.75×10^{-6}	5.61×10^{-3}	0.9625
		Negative terminal	5.50×10^{-10}	2.86×10^{-6}	5.80×10^{-3}	0.959
		All data	5.26×10^{-10}	2.72×10^{-6}	5.55×10^{-3}	0.9581
2C	NCA	Positive terminal	9.86×10^{-9}	3.19×10^{-5}	4.26×10^{-2}	0.9786
		Middle	1.10×10^{-8}	3.42×10^{-5}	4.42×10^{-2}	0.9915
		Negative terminal	1.15×10^{-8}	3.60×10^{-5}	4.58×10^{-2}	0.996
		All data	1.08×10^{-8}	3.40×10^{-5}	4.42×10^{-2}	0.9863
	NMC	Positive terminal	3.17×10^{-9}	9.68×10^{-6}	1.80×10^{-2}	0.9901
		Middle	2.68×10^{-9}	8.44×10^{-6}	1.79×10^{-2}	0.9864
		Negative terminal	3.23×10^{-9}	9.90×10^{-6}	1.90×10^{-2}	0.9926
		All data	3.02×10^{-9}	9.33×10^{-6}	1.83×10^{-2}	0.9841
	LCO	Positive terminal	3.68×10^{-9}	1.23×10^{-5}	2.02×10^{-2}	0.934
		Middle	4.08×10^{-9}	1.35×10^{-5}	2.22×10^{-2}	0.98
		Negative terminal	4.73×10^{-9}	1.55×10^{-5}	2.36×10^{-2}	0.9638
		All data	3.91×10^{-9}	1.31×10^{-5}	2.15×10^{-2}	0.9481
3C	NCA	Positive terminal	3.43×10^{-8}	8.40×10^{-5}	9.11×10^{-2}	0.9838
		Middle	3.80×10^{-8}	9.14×10^{-5}	9.71×10^{-2}	0.9977
		Negative terminal	4.27×10^{-8}	9.91×10^{-5}	0.1	0.9971
		All data	3.83×10^{-8}	9.15×10^{-5}	9.61×10^{-2}	0.9887
	NMC	Positive terminal	6.03×10^{-9}	2.17×10^{-5}	4.02×10^{-2}	0.9885
		Middle	3.68×10^{-9}	1.66×10^{-5}	3.98×10^{-2}	0.9938
		Negative terminal	8.40×10^{-9}	2.39×10^{-5}	4.26×10^{-2}	0.9948
		All data	6.04×10^{-9}	2.07×10^{-5}	4.08×10^{-2}	0.9858
	LCO	Positive terminal	9.28×10^{-9}	2.80×10^{-5}	4.51×10^{-2}	0.9959
		Middle	9.47×10^{-9}	2.86×10^{-5}	4.63×10^{-2}	0.9953
		Negative terminal	1.46×10^{-8}	3.91×10^{-5}	5.20×10^{-2}	0.9924
		All data	1.11×10^{-8}	3.19×10^{-5}	4.78×10^{-2}	0.9911

The given coefficients are from a general 3rd-order polynomial equation of the following form:

$$\Delta T = a \cdot t^3 + b \cdot t^2 + c \cdot t \qquad (1)$$

where ΔT is the temperature rise in °C and t is the discharge time in seconds.

3. Results and Discussions

It can be observed that especially at the lowest discharge rate of 1C, near the start and end of the discharge process, a steeper temperature rise appears. When applying the discharge current, a sudden drop in battery voltage occurs, increasing the irreversible part of the generated heat. Similarly, near the end of discharge, the battery voltage drops at a faster rate, increasing again the irreversible component. The initial stage of discharge is also the moment that presents the highest differences in local current density inside the cell, while near the terminal stage of the process the gradient of electrolyte Li$^+$ concentration is at its highest [50].

In Figure 5, pictures taken with the thermal camera at the end of the 2C discharge process are illustrated, which seem to be in agreement with the temperature rise values presented in Figure 4. One can also notice the different color shades of the wires in the presented three pictures, which can be due to the variable color scales present in the color bars, but mostly to differences in current values corresponding to the discharge rate for the presented battery capacities. The differences in wire temperature for the same C-rates also influenced the results presented in Figure 4 and the comparative analysis between battery chemistries.

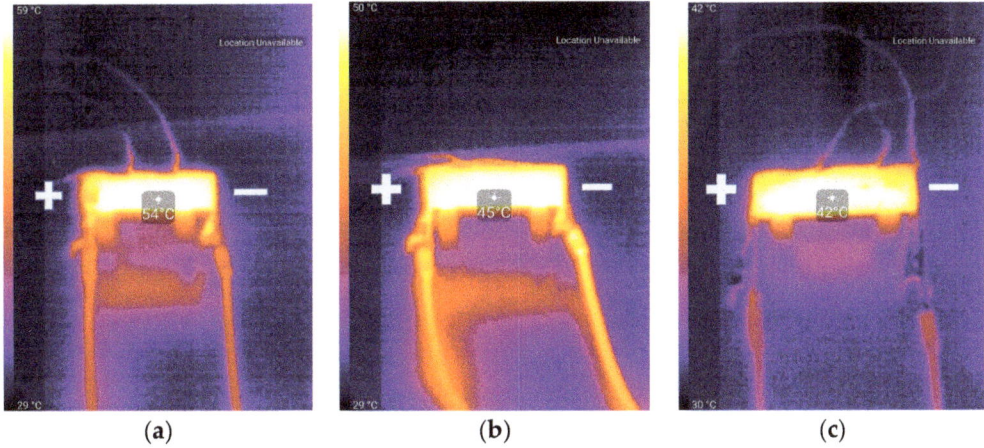

Figure 5. Thermal camera pictures at end of 2C discharge process: (**a**) NCA; (**b**) NMC; (**c**) LCO.

Contrarily to what was expected, the results from Figure 4 indicate that for every case the lowest temperatures were recorded at the positive terminal. While the differences between the negative terminal and the middle of the battery were negligible, those to the positive terminal were up to even 3 °C at the highest analyzed discharge rate. This disparity could be attributed to the limits of measuring temperature on the battery surface or the moderate accuracy of the thermocouples and other equipment. However, given the high number of tests completed and the repeatability of the performed experimental procedure, the authors consider that this ever-present trend's origins should be further investigated.

Regarding the comparison of the different battery chemistries and capacities, the conducted experiments show that the temperature rise for the analyzed NCA battery is far higher. Temperatures at the end of the discharge process were regularly 10 °C above the ones measured at the other two chemistries, but in some cases exceeded even 20 °C. Based on the ambient conditions, one can see that the end surface temperature for the analyzed NCA batteries can easily exceed 70 °C when discharged at 3C, a case in which a thermal management system is necessary. This finding can be aligned with other investigations where the thermal stability of different cathode chemistries was compared. Barkholtz et al. [51] stated that NCA and LCO cathodes are metastable, with NCA batteries showing the highest thermal runaway rates, while LFP cathodes are stable. Another study [52] also showed that from the thermal stability and reactivity point of view, when comparing three cathode materials for Li-ion batteries, the electrochemically delithiated NCA was found to be the least stable.

The magnitude of temperature rise was similar for the analyzed LCO and NMC batteries, although the variation with increasing discharge rate was slightly different. If at 1C the average temperature rise for the LCO battery was 3 °C lower than for the NMC, at 2C the results became closely the same, and finally, at 3C it increased above the values recorded for the NMC battery. Additionally, different studies [13,53] have already proved that a higher nominal capacity results in an increase in the irreversible heat generation's contribution, due to the ohmic polarization in the cathode, separator and anode. This information, in correlation with the obtained results, indicates the conclusion that of the analyzed batteries, the NMC chemistry presents the best thermal behavior, and offers the potential to be used in the future in lower-cost and higher-specific-energy batteries for electric vehicles [54]. However, for a stronger validation of this statement, a similar test procedure with a 18650-type NMC battery is suggested for further investigations.

It is observed that no relevant temperature gradient was detected at the various thermocouple measuring locations on the batteries' surface nor on the thermal camera images. Given the prior observation as well as the large number of data and generated coefficients, only the general trendline for all measured data will be considered in the following, with the aim of simplifying the results' interpretation process. Even though only one dataset will be used in the following, indicating the global thermal behavior of batteries, measurements in several locations on the battery surface gave a more accurate calculation of the average temperature rise and offered the possibility to analyze thermal behavior also on a local level. Furthermore, if more relevant temperature differences arose at the different measuring locations, thermal maps and subsequently temperature predictions could be generated for multiple locations on the batteries. The obtained equations (presented in Table 2) were introduced in Matlab software and a thermal map was generated for each considered battery, as shown in Figure 6.

Based on the presented thermal maps an equation is proposed, which can describe relatively simply and with good accuracy the illustrated surfaces with as few coefficients as possible. For this purpose, several battery parameters are integrated into the equation and a single coefficient is left to be determined for every case based on the battery chemistry.

The proposed equation is the following:

$$\Delta T = B\left[n \cdot U\left(D + S^2 - D \cdot S^2\right) - S^3 - atan\left(e^D\right)\right] - 1.5 \cdot Q \cdot S \qquad (2)$$

where ΔT–temperature rise [°C]; n–size coefficient: $n = 1$ for 18650-type cell; $n = 2$ for 21700-type cell; U–battery nominal voltage [V]; D–discharge rate (C-rate) [-]; S – State of Charge (SoC) [-]; Q–battery nominal capacity [Ah]; B–chemistry coefficient to be determined for each battery.

From our study, a general indicative value for the B coefficient should be between 0 and 5. Together with the original polynomial equations from Figure 6, the surfaces obtained from the proposed fitting equation are illustrated in Figure 7, while in Table 3 the values for the B coefficient and main fitting parameters are indicated.

It can be seen that the proposed equation is in good agreement with the experimental results (according to R-square values). The largest differences are at values of the battery state of charge below 0.1 (SoC-10%), where the equation underestimates battery temperatures by up to 3 °C. Other than that, the error generally is below 1 °C but never exceeds 2 °C. Naturally, the proposed fitting equation represents only an approximation of battery temperature rise and cannot be yet applied for precise value estimation. Moreover, all results are limited to cases with a constant discharge rate and therefore cannot be taken for good in the case of dynamic discharge profiles. Additionally, more experiments with other battery chemistries and sizes are needed to truly validate the equation and possibly refine its current form. However, the utility of such an equation is that it can drastically reduce the time spent with experimental measurements required to characterize the thermal behavior of Li-ion batteries. With the mean of only a few measurements, the equation can be used to determine the B coefficient for any battery type and then apply it to other discharge rates, eliminating the necessity of performing all the experimental and data processing work presented in this study.

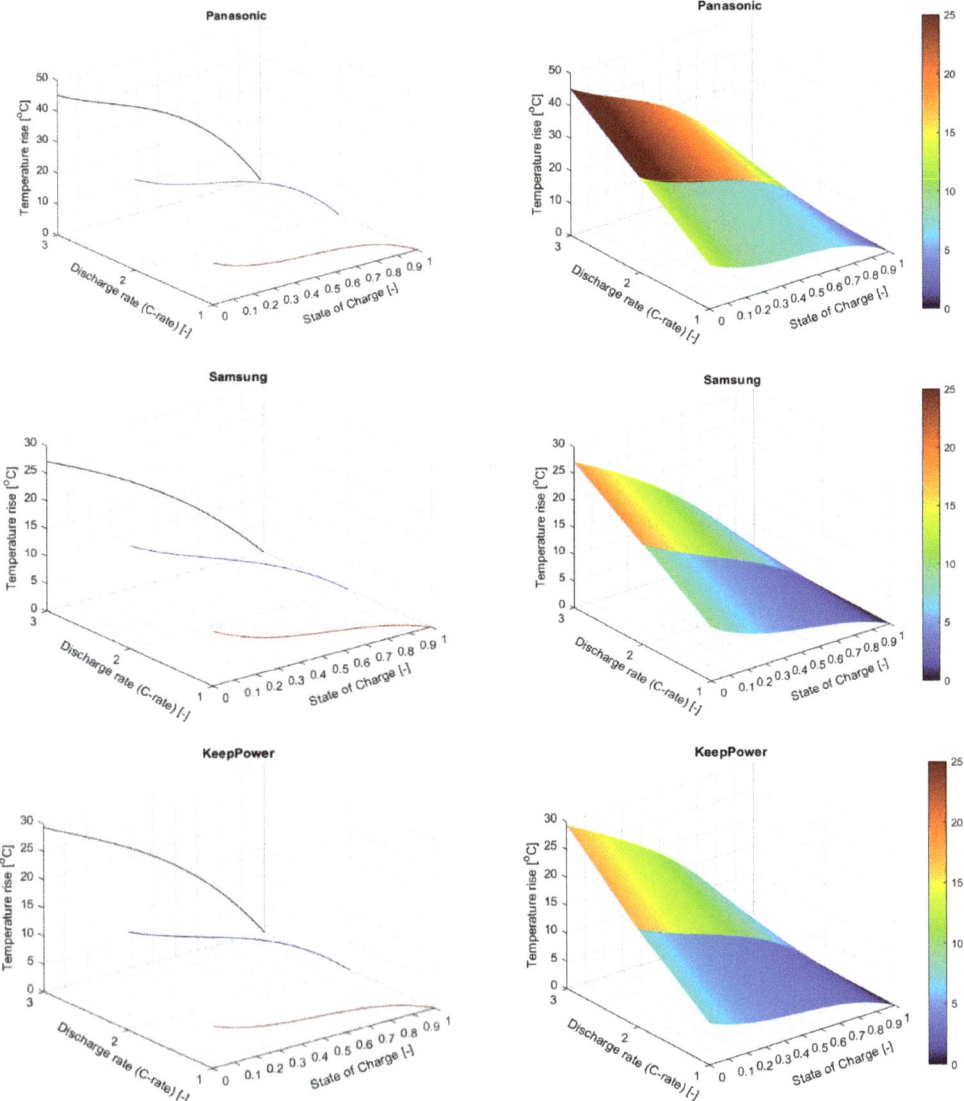

Figure 6. Thermal maps for the studied batteries (Panasonic- NCA, Samsung-NMC, KeepPower-LCO).

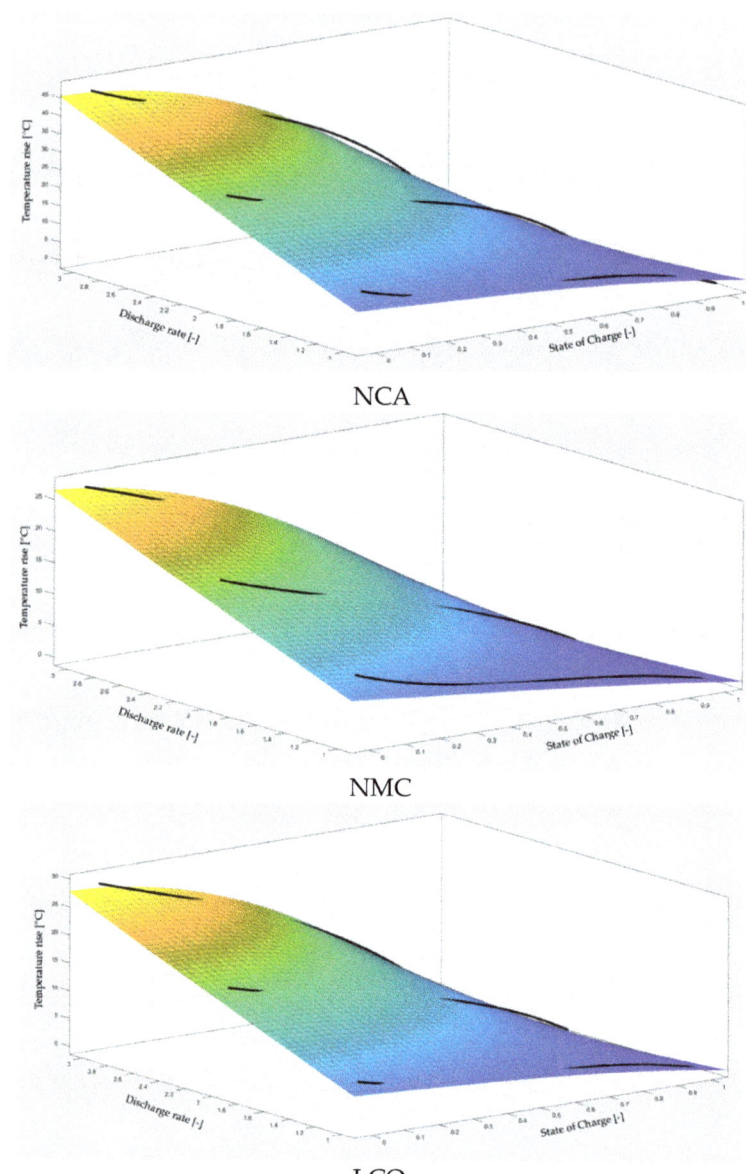

Figure 7. Thermal maps obtained from the proposed equation.

Table 3. Fitting parameters.

Parameters	Panasonic NCA	Samsung NMC	KeepPower LCO
B coefficient	4.697	1.26	2.786
R-square	0.993	0.991	0.993
RMSE	1.04	0.662	0.632

4. Conclusions

The article presents a research topic in accordance with the need to introduce sustainable transport systems, which led to the research and analysis of efficient energy sources for electric vehicles. In the presented study, the temperature rise was experimentally analyzed during discharge mode at 1C, 2C and 3C for three electrochemistry types (NCA, NMC and LCO) used in the construction of cylindrical batteries.

Based on the obtained results, it can be observed that the temperature rise for the analyzed NCA battery was by far the highest, while the magnitude of temperature rises for the analyzed LCO and NMC batteries were similar. Furthermore, through a more accurate analysis of the results, it can be stated that the NMC chemistry presents the best thermal behavior and offers the potential to be used in the future in lower-cost and higher-specific-energy batteries for electric vehicles.

By analyzing the data related to the thermal variation of batteries with different chemistries, it was possible to propose a parametric equation with good statistical correlation (generally error is below 1 °C but never exceeds 3 °C, compared to the experimental data), describing temperature rise as the main function of the state of charge and discharge rate. The simple form of the equation offers the advantage of being easily used in the initial design and construction processes of an energy source consisting of cylindrical batteries.

Future research directions can aim to investigate whether the proposed form of the parametric equation offers a potential application for other chemistries, sizes of cylindrical batteries or even constructive types of batteries (prismatic, pouch).

An important future research orientation is increasing the number of parameters that are used in describing the prediction function. This desiderate can be achieved in several ways. Co-simulations with finite element analysis can be conducted to simultaneously evaluate the thermal, mechanical and chemical performances of the batteries. Such an approach would allow us to monitor the aging mechanisms of the cells, the electrodes' structural deterioration due to mechanical strains and the generation of reversible and irreversible heat [55]. The second possibility is the implementation of machine learning methods, given the large quantity of measured data. Such a future development would enable a precise prediction of battery performance in various working conditions, with the possibility of increasing the dataset's complexity by measuring other input parameters, such as state of health or internal resistance.

Author Contributions: Conceptualization, T.I.C.B. and F.M.; methodology, T.I.C.B. and F.M.; validation, T.I.C.B.; formal analysis, T.I.C.B.; investigation, T.I.C.B.; writing—original draft preparation, T.I.C.B. and F.M.; writing—review and editing, T.I.C.B. and F.M.; supervision, F.M. All authors have read and agreed to the published version of the manuscript.

Funding: This research received no external funding.

Institutional Review Board Statement: Not applicable.

Informed Consent Statement: Not applicable.

Data Availability Statement: Not applicable.

Conflicts of Interest: The authors declare no conflict of interest.

References

1. *New Registrations of Electric Cars, EU-27 2022*. Available online: https://www.eea.europa.eu/data-and-maps/daviz/new-electric-vehicles-in-eu-2#tab-chart_3 (accessed on 2 December 2022).
2. Bukhari, S.M.A.S.; Maqsood, J.; Baig, M.Q.; Ashraf, S.; Khan, T.A. Comparison of Characteristics—Lead Acid, Nickel Based, Lead Crystal and Lithium Based Batteries. In Proceedings of the 2015 17th UKSim-AMSS International Conference on Modelling and Simulation (UKSim), Cambridge, UK, 25–27 March 2016; pp. 444–450. [CrossRef]
3. Spitthoff, L.; Shearing, P.; Burheim, O. Temperature, Ageing and Thermal Management of Lithium-Ion Batteries. *Energies* **2021**, *14*, 1248. [CrossRef]
4. Liu, S.; Zhang, H.; Xu, X. A study on the transient heat generation rate of lithium-ion battery based on full matrix orthogonal experimental design with mixed levels. *J. Energy Storage* **2021**, *36*, 102446. [CrossRef]

5. Drake, S.; Martin, M.; Wetz, D.; Ostanek, J.; Miller, S.; Heinzel, J.; Jain, A. Heat generation rate measurement in a Li-ion cell at large C-rates through temperature and heat flux measurements. *J. Power Sources* **2015**, *285*, 266–273. [CrossRef]
6. Liu, P.; Wang, J.; Hicks-Garner, J.; Sherman, E.; Soukiazian, S.; Verbrugge, M.; Tataria, H.; Musser, J.; Finamore, P. Aging Mechanisms of LiFePO[sub 4] Batteries Deduced by Electrochemical and Structural Analyses. *J. Electrochem. Soc.* **2010**, *157*, A499–A507. [CrossRef]
7. Amine, K.; Liu, J.; Belharouak, I. High-temperature storage and cycling of C-LiFePO$_4$/graphite Li-ion cells. *Electrochem. Commun.* **2005**, *7*, 669–673. [CrossRef]
8. Ramadass, P.; Haran, B.; White, R.; Popov, B.N. Capacity fade of Sony 18650 cells cycled at elevated temperatures: Part I. Cycling performance. *J. Power Sources* **2002**, *112*, 606–613. [CrossRef]
9. Maleki, H.; Deng, G.; Anani, A.; Howard, J. Thermal Stability Studies of Li-Ion Cells and Components. *J. Electrochem. Soc.* **1999**, *146*, 3224–3229. [CrossRef]
10. Drake, S.; Wetz, D.; Ostanek, J.; Miller, S.; Heinzel, J.; Jain, A. Measurement of anisotropic thermophysical properties of cylindrical Li-ion cells. *J. Power Sources* **2014**, *252*, 298–304. [CrossRef]
11. Zhang, G.; Cao, L.; Ge, S.; Wang, C.-Y.; Shaffer, C.E.; Rahn, C.D. In Situ Measurement of Radial Temperature Distributions in Cylindrical Li-Ion Cells. *J. Electrochem. Soc.* **2014**, *161*, A1499–A1507. [CrossRef]
12. Shah, K.; Vishwakarma, V.; Jain, A. Measurement of Multiscale Thermal Transport Phenomena in Li-Ion Cells: A Review. *J. Electrochem. Energy Convers. Storage* **2016**, *13*. [CrossRef]
13. Nazari, A.; Farhad, S. Heat generation in lithium-ion batteries with different nominal capacities and chemistries. *Appl. Therm. Eng.* **2017**, *125*, 1501–1517. [CrossRef]
14. Miao, Y.; Hynan, P.; von Jouanne, A.; Yokochi, A. Current Li-Ion Battery Technologies in Electric Vehicles and Opportunities for Advancements. *Energies* **2019**, *12*, 1074. [CrossRef]
15. Jindal, P.; Katiyar, R.; Bhattacharya, J. Evaluation of accuracy for Bernardi equation in estimating heat generation rate for continuous and pulse-discharge protocols in LFP and NMC based Li-ion batteries. *Appl. Therm. Eng.* **2021**, *201*, 117794. [CrossRef]
16. Xie, Y.; Shi, S.; Tang, J.; Wu, H.; Yu, J. Experimental and analytical study on heat generation characteristics of a lithium-ion power battery. *Int. J. Heat Mass Transf.* **2018**, *122*, 884–894. [CrossRef]
17. Christen, R.; Martin, B.; Rizzo, G. New Experimental Approach for the Determination of the Heat Generation in a Li-Ion Battery Cell. *Energies* **2021**, *14*, 6972. [CrossRef]
18. Arora, S.; Kapoor, A. Experimental Study of Heat Generation Rate during Discharge of LiFePO$_4$ Pouch Cells of Different Nominal Capacities and Thickness. *Batteries* **2019**, *5*, 70. [CrossRef]
19. Huang, Y.; Lu, Y.; Huang, R.; Chen, J.; Chen, F.; Liu, Z.; Yu, X.; Roskilly, A.P. Study on the thermal interaction and heat dissipation of cylindrical Lithium-Ion Battery cells. *Energy Procedia* **2017**, *142*, 4029–4036. [CrossRef]
20. Bandhauer, T.M.; Garimella, S.; Fuller, T.F. A Critical Review of Thermal Issues in Lithium-Ion Batteries. *J. Electrochem. Soc.* **2011**, *158*, R1. [CrossRef]
21. Zhang, J.; Huang, J.; Li, Z.; Wu, B.; Nie, Z.; Sun, Y.; An, F.; Wu, N. Comparison and validation of methods for estimating heat generation rate of large-format lithium-ion batteries. *J. Therm. Anal.* **2014**, *117*, 447–461. [CrossRef]
22. Liu, F.; Lan, F.; Chen, J. Dynamic thermal characteristics of heat pipe via segmented thermal resistance model for electric vehicle battery cooling. *J. Power Sources* **2016**, *321*, 57–70. [CrossRef]
23. Kleiner, J.; Singh, R.; Schmid, M.; Komsiyska, L.; Elger, G.; Endisch, C. Influence of heat pipe assisted terminal cooling on the thermal behavior of a large prismatic lithium-ion cell during fast charging in electric vehicles. *Appl. Therm. Eng.* **2020**, *188*, 116328. [CrossRef]
24. Hosseinzadeh, E.; Marco, J.; Jennings, P. The impact of multi-layered porosity distribution on the performance of a lithium ion battery. *Appl. Math. Model.* **2018**, *61*, 107–123. [CrossRef]
25. Mahfoudi, N.; Boutaous, M.; Xin, S.; Buathier, S. Thermal Analysis of LMO/Graphite Batteries Using Equivalent Circuit Models. *Batteries* **2021**, *7*, 58. [CrossRef]
26. Surya, S.; Marcis, V.; Williamson, S. Core Temperature Estimation for a Lithium Ion 18650 Cell. *Energies* **2020**, *14*, 87. [CrossRef]
27. Thakur, A.K.; Prabakaran, R.; Elkadeem, M.; Sharshir, S.W.; Arıcı, M.; Wang, C.; Zhao, W.; Hwang, J.-Y.; Saidur, R. A state of art review and future viewpoint on advance cooling techniques for Lithium–ion battery system of electric vehicles. *J. Energy Storage* **2020**, *32*, 101771. [CrossRef]
28. Lin, C.; Xu, S.; Liu, J. Measurement of heat generation in a 40 Ah LiFePO$_4$ prismatic battery using accelerating rate calorimetry. *Int. J. Hydrogen Energy* **2018**, *43*, 8375–8384. [CrossRef]
29. Hu, Y.; Choe, S.-Y.; Garrick, T.R. Measurement of heat generation rate and heat sources of pouch type Li-ion cells. *Appl. Therm. Eng.* **2021**, *189*, 116709. [CrossRef]
30. Yin, Y.; Zheng, Z.; Choe, S.-Y. Design of a Calorimeter for Measurement of Heat Generation Rate of Lithium Ion Battery Using Thermoelectric Device. *SAE Int. J. Altern. Powertrains* **2017**, *6*, 252–260. [CrossRef]
31. Diaz, L.B.; Hales, A.; Marzook, M.W.; Patel, Y.; Offer, G. Measuring Irreversible Heat Generation in Lithium-Ion Batteries: An Experimental Methodology. *J. Electrochem. Soc.* **2022**, *169*, 030523. [CrossRef]
32. Chen, S.; Wan, C.; Wang, Y. Thermal analysis of lithium-ion batteries. *J. Power Sources* **2005**, *140*, 111–124. [CrossRef]
33. Chen, S.C.; Wang, Y.-Y.; Wan, C.-C. Thermal Analysis of Spirally Wound Lithium Batteries. *J. Electrochem. Soc.* **2006**, *153*, A637–A648. [CrossRef]

34. Buidin, T.; Mariasiu, F. Battery Thermal Management Systems: Current Status and Design Approach of Cooling Technologies. *Energies* **2021**, *14*, 4879. [CrossRef]
35. Becker, J.A.; Green, C.B.; Pearson, G.L. Properties and Uses of Thermistors—Thermally Sensitive Resistors. *Trans. Am. Inst. Electr. Eng.* **1946**, *65*, 711–725. [CrossRef]
36. Childs, P.R.N.; Greenwood, J.R.; Long, C.A. Review of temperature measurement. *Rev. Sci. Instrum.* **2000**, *71*, 2959–2978. [CrossRef]
37. Raijmakers, L.; Danilov, D.; Eichel, R.-A.; Notten, P. A review on various temperature-indication methods for Li-ion batteries. *Appl. Energy* **2019**, *240*, 918–945. [CrossRef]
38. Duff, M.; Towey, J. Two Ways to Measure Temperature Using Thermocouples Feature Simplicity, Accuracy, and Flexibility. *Analog Dialogue* **2010**, *44*, 1–6.
39. Lv, Y.; Zhou, D.; Yang, X.; Liu, X.; Li, X.; Zhang, G. Experimental investigation on a novel liquid-cooling strategy by coupling with graphene-modified silica gel for the thermal management of cylindrical battery. *Appl. Therm. Eng.* **2019**, *159*, 113885. [CrossRef]
40. Weng, J.; Yang, X.; Zhang, G.; Ouyang, D.; Chen, M.; Wang, J. Optimization of the detailed factors in a phase-change-material module for battery thermal management. *Int. J. Heat Mass Transf.* **2019**, *138*, 126–134. [CrossRef]
41. Yan, J.; Li, K.; Chen, H.; Wang, Q.; Sun, J. Experimental study on the application of phase change material in the dynamic cycling of battery pack system. *Energy Convers. Manag.* **2016**, *128*, 12–19. [CrossRef]
42. Lv, Y.; Situ, W.; Yang, X.; Zhang, G.; Wang, Z. A novel nanosilica-enhanced phase change material with anti-leakage and anti-volume-changes properties for battery thermal management. *Energy Convers. Manag.* **2018**, *163*, 250–259. [CrossRef]
43. Zhang, J.; Li, X.; Zhang, G.; Wang, Y.; Guo, J.; Wang, Y.; Huang, Q.; Xiao, C.; Zhong, Z. Characterization and experimental investigation of aluminum nitride-based composite phase change materials for battery thermal management. *Energy Convers. Manag.* **2019**, *204*, 112319. [CrossRef]
44. Ling, Z.; Wen, X.; Zhang, Z.; Fang, X.; Gao, X. Thermal management performance of phase change materials with different thermal conductivities for Li-ion battery packs operated at low temperatures. *Energy* **2017**, *144*, 977–983. [CrossRef]
45. Zhong, G.; Zhang, G.; Yang, X.; Li, X.; Wang, Z.; Yang, C.; Yang, C.; Gao, G. Researches of composite phase change material cooling/resistance wire preheating coupling system of a designed 18650-type battery module. *Appl. Therm. Eng.* **2017**, *127*, 176–183. [CrossRef]
46. Lyu, Y.; Siddique, A.; Majid, S.; Biglarbegian, M.; Gadsden, S.; Mahmud, S. Electric vehicle battery thermal management system with thermoelectric cooling. *Energy Rep.* **2019**, *5*, 822–827. [CrossRef]
47. Li, X.; Zhong, Z.; Luo, J.; Wang, Z.; Yuan, W.; Zhang, G.; Yang, C.; Yang, C. Experimental Investigation on a Thermoelectric Cooler for Thermal Management of a Lithium-Ion Battery Module. *Int. J. Photoenergy* **2019**, *2019*, 3725364. [CrossRef]
48. Behi, H.; Karimi, D.; Behi, M.; Jaguemont, J.; Ghanbarpour, M.; Behnia, M.; Berecibar, M.; Van Mierlo, J. Thermal management analysis using heat pipe in the high current discharging of lithium-ion battery in electric vehicles. *J. Energy Storage* **2020**, *32*, 101893. [CrossRef]
49. Löbberding, H.; Wessel, S.; Offermanns, C.; Kehrer, M.; Rother, J.; Heimes, H.; Kampker, A. From Cell to Battery System in BEVs: Analysis of System Packing Efficiency and Cell Types. *World Electr. Veh. J.* **2020**, *11*, 77. [CrossRef]
50. Liang, J.; Gan, Y.; Li, Y.; Tan, M.; Wang, J. Thermal and electrochemical performance of a serially connected battery module using a heat pipe-based thermal management system under different coolant temperatures. *Energy* **2019**, *189*, 116233. [CrossRef]
51. Barkholtz, H.M.; Preger, Y.; Ivanov, S.; Langendorf, J.; Torres-Castro, L.; Lamb, J.; Chalamala, B.; Ferreira, S.R. Multi-scale thermal stability study of commercial lithium-ion batteries as a function of cathode chemistry and state-of-charge. *J. Power Sources* **2019**, *435*, 226777. [CrossRef]
52. Huang, Y.; Lin, Y.-C.; Jenkins, D.M.; Chernova, N.A.; Chung, Y.; Radhakrishnan, B.; Chu, I.-H.; Fang, J.; Wang, Q.; Omenya, F.; et al. Thermal Stability and Reactivity of Cathode Materials for Li-Ion Batteries. *ACS Appl. Mater. Interfaces* **2016**, *8*, 7013–7021. [CrossRef]
53. Mevawalla, A.; Panchal, S.; Tran, M.-K.; Fowler, M.; Fraser, R. Mathematical Heat Transfer Modeling and Experimental Validation of Lithium-Ion Battery Considering: Tab and Surface Temperature, Separator, Electrolyte Resistance, Anode-Cathode Irreversible and Reversible Heat. *Batteries* **2020**, *6*, 61. [CrossRef]
54. Houache, M.S.E.; Yim, C.-H.; Karkar, Z.; Abu-Lebdeh, Y. On the Current and Future Outlook of Battery Chemistries for Electric Vehicles—Mini Review. *Batteries* **2022**, *8*, 70. [CrossRef]
55. Mirsalehian, M.; Beykirch, R. Thermal Investigation and Physical Modeling of Lithium-ion Batteries. *ATZ Worldw.* **2020**, *122*, 36–41. [CrossRef]

Article

Topographical Optimization of a Battery Module Case That Equips an Electric Vehicle

Ioan Szabo [1], Liviu I. Scurtu [1], Horia Raboca [2] and Florin Mariasiu [1,*]

[1] Automotive Engineering and Transport Department, Technical University of Cluj-Napoca, 103–105 Muncii Avenue, 400114 Cluj-Napoca, Romania
[2] PACS Faculty, Babes-Bolyai University of Cluj-Napoca, 71 General Traian Mosoiu St., 400347 Cluj-Napoca, Romania
* Correspondence: florin.mariasiu@auto.utcluj.ro

Abstract: The exponential development and successful application of systems-related technologies that can put electric vehicles on a level playing field in direct competition with vehicles powered by internal combustion engines mean that the foreseeable future of the automobile (at least) will be dominated by vehicles that have electric current stored in batteries as a source of energy. The problem at the European level related to the dependence on battery suppliers from Asia directly correlates with the need to use batteries as energy storage media for energy from renewable sources (photovoltaic and wind), and leads to the need for research into the possibilities for their reuse, remanufacturing or recycling (at the end of their life or purpose of use), and reintroduction, either fully or partially, back into the economy. This article presents possibilities for increasing the protection of the integrity of the cells that form a battery in the event of an impact/road accident, by the numerical analysis of a topographically optimized battery module case. The proposed solution/method is innovative and offers a cell protection efficiency of between 16.6–60% (19.7% to 40.7% if the mean values for all three impact velocities are considered). The efficiency of a cell's protection decreases with the increase in impact velocity and provides the premise for a greater part of the saved cells to be reintegrated into other energy storage systems (photovoltaic and/or wind), avoiding future problems relating to environmental pollution.

Keywords: battery; electric vehicle; topographical optimization; mechanical stresses; circular economy

Citation: Szabo, I.; Scurtu, L.I.; Raboca, H.; Mariasiu, F. Topographical Optimization of a Battery Module Case That Equips an Electric Vehicle. *Batteries* **2023**, *9*, 77. https://doi.org/10.3390/batteries9020077

Academic Editors: Federico Baronti and Carlos Ziebert

Received: 18 November 2022
Revised: 9 January 2023
Accepted: 19 January 2023
Published: 23 January 2023

Copyright: © 2023 by the authors. Licensee MDPI, Basel, Switzerland. This article is an open access article distributed under the terms and conditions of the Creative Commons Attribution (CC BY) license (https://creativecommons.org/licenses/by/4.0/).

1. Introduction

The potential of electric vehicles (EVs) as a suitable solution to the massive reduction in greenhouse gases caused by transportation is increasingly evident in the automotive industry, with numerous electric vehicles already on the market. Because, at this time, the field of transport mainly uses means of transport equipped with internal combustion engines, the pollution caused by them makes the field of transport one of the biggest contributors to greenhouse emissions and toxic emissions that affect the global population. The use of internal combustion engines powered by fossil fuels (diesel or petrol) leads to toxic emissions consisting of carbon dioxide (CO_2), carbon monoxide (CO), nitrogen oxides (NOx), hydrocarbons (HC), and suspended particles (PM), volatile organic compounds (VOCs) [1,2], with a direct and negative impact on the environment and human health [3,4]. For example, in 2017, in Europe, road transport was responsible for almost 72% of total greenhouse gas emissions from transport, and of these emissions, 44% were from passenger cars, 9% from light commercial vehicles, and 19% came from heavy-duty vehicles [5].

For these main reasons, the development and use of electric vehicles directly and immediately contribute to the reduction in greenhouse gas pollution, as well as to the reduction in pollution in large urban agglomerations. Another aspect that should not be neglected is the operating/utilization costs of vehicles with conventional combustion engines and vehicles with an electric powertrain. It is estimated that it costs about six times less to operate an electric vehicle than an internal combustion engine vehicle, as the cost difference is primarily due to the low efficiency of internal combustion engines (30–35%) compared to the efficiency of electric motors (90–92%). The flow of cumulative mechanical and thermal losses leads to a decrease in the efficiency of the thermal engine by three times compared to an electric vehicle, and also, an electric vehicle consumes up to 75% of the energy for travel (motion), with the necessary energy stored in the batteries, which is about 6–10 km/kWh [6]. However, in addition to the primary advantages presented, there are two major barriers to massive entry into the automotive market for electric vehicles: the high cost of the energy source (25–45% of the purchase cost of an electric vehicle [6,7]) and the low storage capacity for electrical energy. A high battery energy capacity also brings higher costs, but at the same time increases the weight characteristics, which directly affects the autonomy of electric vehicles. For these reasons, there are currently several trends to increase the energy efficiency of electric vehicles by [7–12]:

- The development of the energy capacity of cells through other technologies;
- The functional optimization of the components of a battery cell;
- High-performance and energy-efficient battery management systems (BMS);
- The reduction of mechanical losses by reducing the number of components in the electric powertrain;
- The development of energy flow management systems at the level of the entire vehicle;
- The protection, safety, and security assurances in the use of energy sources.

Improving the safety of the energy sources used in electric vehicles requires major and constant attention from manufacturers [13,14]. Currently, there is much research that is focused on the safe integration of energy sources in electric vehicles to solve problems that can result in various safety incidents (which can lead to ignition and fire hazards), which are a pressing priority already in the design phase [13,15–20]. It is considered that the stresses that may appear in the exploitation of EV energy sources (Figure 1) are multiple from the point of view of physical phenomena.

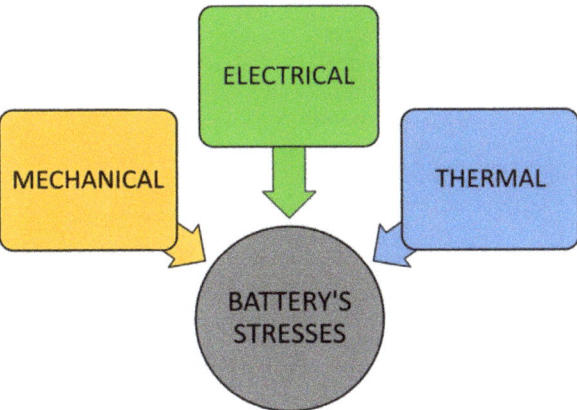

Figure 1. Stresses occurring in lithium-ion batteries during operation and in an impact/road accident.

In various research, the reactions to the safety incidents that occur in the energy sources after overloading have been analysed in detail, and based on this, the following classification, the external stresses on the energy sources have been determined as in [20–22] and as follows:

- Mechanical: impact, deformation, penetration, free fall, mechanical shock, vibrations, and immersion.
- Electrical: overload, overloading, forced download, and high C-rate.
- Thermal: heating, overheating, and thermal runaway.

During the operation/exploitation of an electric vehicle, the battery may undergo deformations due to an external impact (as a result of a traffic accident), which subsequently leads to mechanical damage to the housing (case) and components of the battery systems. To improve the safety and performance of lithium-ion power sources, several international organizations and committees have promulgated standards and test specifications for power sources. These tests evaluate the safety limits of the energy sources, such as overload tests, mechanical deformation, and penetration tests, and the analysis of short-circuit and high-temperature behaviours [21,23]. The problem of the structural optimization of the battery case in electric vehicles has been addressed in various research, both at the micro level (the effect of mechanical impact on the cells) and at the macro level (the construction of the case and the materials used in construction). In most cases, modelling and numerical simulations were carried out, taking into account the immediate advantages offered by these engineering methods.

Recent research carried out by Ahn Y.J. et al. [24] and Muresanu and Dudescu [25] on safety issues in lithium-ion batteries in mechanical impact cases, used a numerical model to evaluate the degree of integrity of each component of the electric cell after being mechanical stressed. The numerical finite element models incorporating the multilayered structure showed good accuracy in the prediction of the cell-level failure mode, being verified through comparison with the experimental results.

Shui et al. proposed and applied a methodology to optimize the design of the battery case in four phases based on various optimization methods (central composite design—CCD, response surface methodology—RSM, artificial neural network—ANN, Latin hypercube sampling—LHS) [26]. Based on these methods, optimization of the structural rigidity was achieved by determining (reducing against an arbitrarily preset initial value) the optimal thickness of the plates (flats) that form the battery box. The material chosen for the construction of the battery case was aluminium and the battery cells considered were 18650-type Li-ion. It should be mentioned that the use of flat plates in the construction of the battery case can lead to an increased weight compared to construction with thinner plates and stiffening ribs.

Optimizing the construction and mechanical strength of the battery case from the point of view of using different materials (e.g., glass fibre composite, carbon fibre composite, and steel) were considered and researched by An Y. et al. [27]. The study was carried out using numerical simulation methods and showed that the use of carbon fibre composite in the construction of the battery case offers advantages in high strength and low weight. A 30% deformation of the casing was achieved at a force of 81 kN using steel compared to a force of 100 kN for carbon fibre composite.

Ruan et al. [28] also used numerical simulation methods to address the constructive optimization problem of the lug connection and the upper cover of a battery case. They considered that the structural stresses were due to acceleration loads in all directions on the battery, and the major conclusion was that the maximum stress occurred in the connection area between the battery pack ears and the box (of 86 mm, for the particular construction of the battery considered in the study). It should be mentioned that in this study, a specific box construction and geometric shape were used, and the plates that formed the box were considered flat and without stiffness ribs.

Qiao et al. added additional EVA (Ethylene-vinyl acetate) foam in the construction of the battery to try and identify the effect on the integrity of the battery in the case of an accident (frontal collision at 50 km/h with a wall) [29]. Based on the imposed conditions and restrictions, they obtained a reduction in deformation of 8.2% for the area of the electrochemical cells and a reduction of 12.65% for the maximum deformation of the casing (by adding EVA foam). However, the study had many limitations: the construction of the battery case and the monomer model were simple, the design, heat dissipation structure, and the external connection of the battery case were not taken into account, and the frontal impact was made with a fixed wall that did not faithfully represent the construction of the body of a vehicle. The plates from which the battery case were made were also flat (without stiffening ribs).

The research carried out by Pan Y. et al. [30] focused on the possibility of reducing the weight of the battery case while maintaining the mechanical resistance properties. The type of material and its optimal thickness were determined using different optimization methods, considering fixed-frequency vibration, and mechanical shock and fatigue life analysis. Further, the obtained optimized design of the battery pack was tested in different tests (crush and crash simulations), which considered the impact of the battery case with a cylindrical pilar (a particular situation that cannot happen because of the design of the car body structure). Regardless, the results obtained showed that the impact protection of the battery was maintained, along with a weight reduction of 10.41% (by changing the construction material of the battery case).

Dong S. et al. [31] addressed the optimization of battery case construction by considering a multitude of goals: to improve the protection level of the battery pack to IP68; to reduce the weight of the battery box frame structure; to simplify the cooling mode of the battery pack; and to improve the battery protection level. In general, the research focused on optimizing the internal resistance structure of the battery (which allowed for the assembly of the modules that made up the battery), and flat plates and covers that made up the battery case (they were not considered in the optimization process). The results obtained showed that it was possible to maintain the mechanical resistance of the battery case by applying a new design and achieved thermal management through a natural air-cooling scheme, which led to a reduction in the total weight of the energy source by 450 kg.

The use of a sandwich-type structure (based on an innovative auxetic structure, located in the core of the structure) to protect the battery system of an electric vehicle in the case of an impact of the load with the road, was designed and studied by Biharta M.A.S. et al. [32]. The protection efficiency of this composite structure was analysed using the nonlinear finite element method and showed that under certain impact conditions, the battery cells were very well protected (a maximum deformation of 1.92 mm was obtained, below the deformation threshold that causes failure of the cells/battery). Another study on the use of sandwich structures in the construction of battery cases was carried out by Pratama L.K. et al. [33]. An optimized form of the lattice structure was determined by considering the structure that offered the highest impact energy absorption value, which was then used further to analyse the protection of a battery in an impact with a body (impactor) at a speed of 42 m/s. Under the conditions imposed on the simulation, the maximum measured deformation of the battery was 2.7 mm, which demonstrated the effectiveness of the construction.

It must be emphasized that with the increasing use of EVs in traffic, there will inevitably be accidents that affect the structural integrity of the batteries, which highlights the need for a specific and effective recycling process. In this context, at the level of the European Union, steps have been taken to establish future directions and policies regarding the need to adopt measures to reintegrate batteries that are no longer used in vehicles into other energy storage systems, or in other industrial processes. In the European Commission document "Green Deal: Sustainable batteries for a circular and climate neutral economy" (2020), it is specified that batteries introduced to the EU market (including those used in the construction of electric vehicles) must be durable, effective, and safe throughout their life cycle [34]. Batteries must be manufactured with the lowest possible environmental impact, using materials obtained with full respect for human rights, as well as social and ecological standards. Batteries must be long-lasting and safe, and at the end of their useful life or purpose, they must be reused, remanufactured, or recycled, putting all or part of them back into the economy. It is expressly stated that " ... *the reuse of batteries from electric vehicles will be facilitated, so that they can have a second life, for example as stationary energy storage systems, or their integration into electrical networks as energy resources*".

Under the conditions and premises presented previously, the topic of this article falls into a topical area of major interest (both on a European and international level) regarding the identification of solutions to increase the degree of inclusion of batteries equipping electric vehicles in the circular economy (reuse of batteries) [35], by optimizing the battery module case to protect against structural mechanical stresses and to protect the electrochemical cells against impact/road accidents. Furthermore, from the analysis of the most recent studies relating to the constructive optimization of the battery case for the protection of the cells upon impact (and already briefly presented), the idea proposed in this article is new. The restriction used for the topographical optimization of the battery casing (box) covers increases the rigidity of the casing through the construction of special ribs. The use of optimized battery case covers provided with stiffening ribs leads to the possibility of using thinner plates (in this case, aluminium), and implicitly, the reduction in the total weight of the battery case (while maintaining the mechanical rigidity properties).

2. Materials and Methods

The research carried out on the possibilities of increasing the integrity of the battery cells (which equip an electric vehicle) to protect against the event of a road accident (through the constructive topographic optimization of the casing) has been carried out in the following directions (see Figure 2):

- Experimental research—The first direction of research was an experimental one and aimed to identify the point where (following mechanical deformation stresses on a Samsung 18650-type cell, with an 80% state of charge) the short-circuit occurs, and the cell becomes unusable.
- Numerical CAD modelling of a 18650-type cell—the creation of a CAD model considering the cell's complex construction/design.
- Validation of a 18650-type cell CAD model—using experimentally obtained data.
- Numerical CAD modelling of a battery case—the creation of a CAD model for a battery (including the internal thermal management system), consisting of 12 modules (a total number of 5760 cells, where each module is made of 480 cells).
- Topographical optimization—the idea of a topographical optimization of the upper and lower surface of the battery case was approached by imposing a restriction relating to the real-world conditions of an electric vehicle. The considered restriction was the range of frequencies (vibrations) due to the tire–road interaction, which are transmitted to the vehicle chassis (and implicitly to the mechanical structure of the battery).
- Numerical simulation of base and optimized cases.
- Comparative analysis of data and conclusions on the efficiency.

2.1. Experimental Research

The experimental tests were performed in a closed room with an average ambient temperature of 20 °C, using an experimental stand and 50 experimental tests (on 50 cells). The experimental stand was configured using a 10-ton hydraulic press, a mini hydraulic group equipped with an electric motor, and the associated hydraulic installation equipment (Figure 3). The hydraulic group was equipped with a simple manual valve that allowed the oil flow to be adjusted between 0.52 and 2.5 L/min, at a working pressure between 50–600 bar. The terminals of the voltmeter were placed at the two terminals of the cell (to identify the moment of short-circuit) and the temperature measuring probe was placed on the cell anode area. Figure 3 shows the configuration of the cylindrical cell bending test system. The linear movement of the piston is driven by the hydraulic pump with an advance of 5 mm/minute and the T-type device bends and deforms the test cell. The obtained results were analysed and validated with statistical tests (removal of erroneous values using Fischer's test). The corresponding values from the 50 tests were averaged and are shown in Figures 4 and 5.

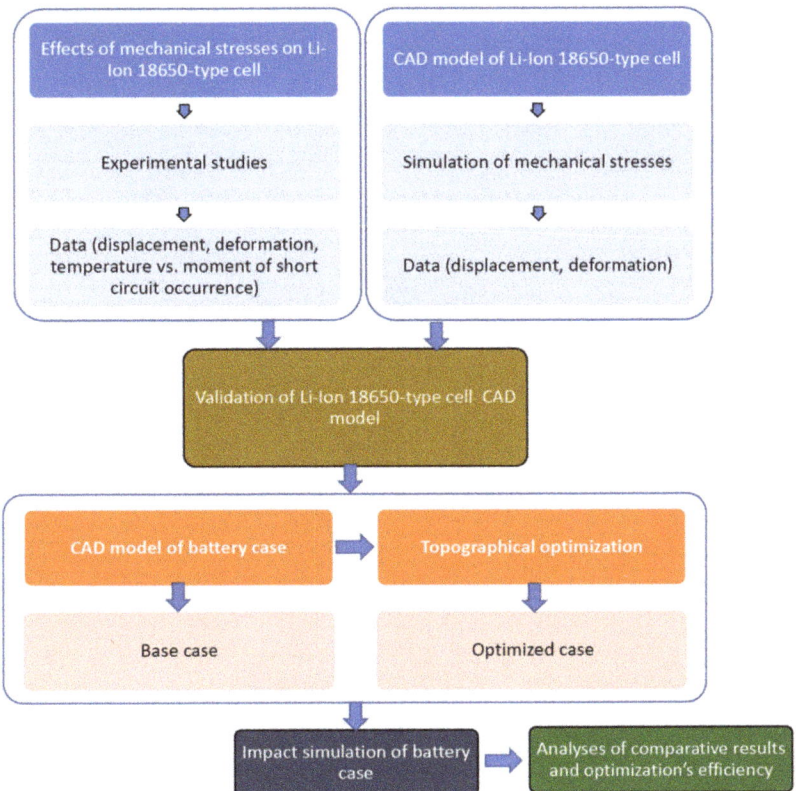

Figure 2. The algorithm of research methodology.

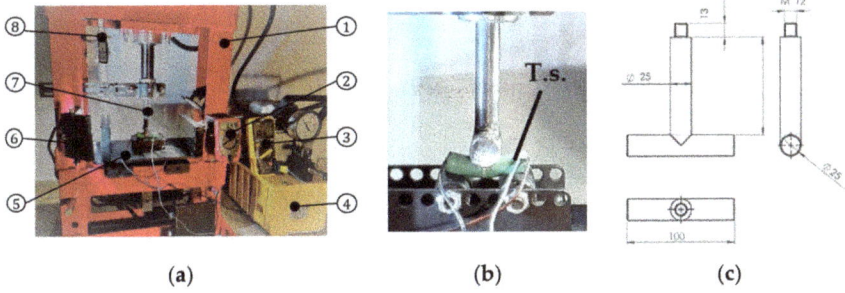

Figure 3. Experimental setup ((**a**)—experimental stand, (**b**)—battery bending test, (**c**)—indentation device geometric dimensions, 1–press; 2–thermometer; 3–voltmeter; 4–hydraulic group; 5–force transducer support; 6–reflector; 7–indentation device; 8–electronic calliper; T.s.—temperature sensor).

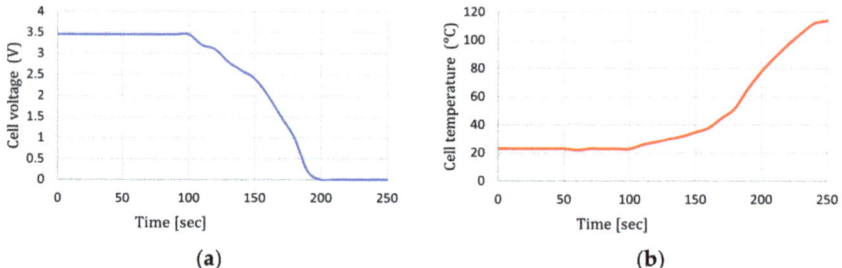

Figure 4. The change in the electric voltage (**a**) and the change in the temperature (**b**) of the 18650-type cell under the action of the deformation force.

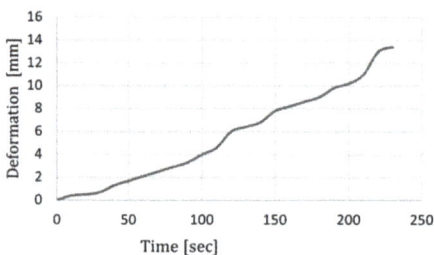

Figure 5. The change in the 18650-type cell deformation over time.

Under the action of the deformation force, the cell's functional parameters started to change at the 100th second. The cell voltage changed from 3.45 V to 3.20 V at a deformation of 4 mm and reached lower than 3 V at the 130th second at a cell deformation of 6.4 mm.

The short-circuit that occurred at the 190th second was also highlighted by the sudden variation in the battery temperature, with an immediate increase from 25 °C to a maximum value of 114 °C (roughly a 5-fold increase). The value of the force at the onset of the short-circuit was 7.18 kN at a cell deformation of 9.8 mm.

2.2. Numerical Modelling and Simulation

2.2.1. Numerical Modelling and Simulation of the 18650-Type Cell

The CAD model of the cylindrical cells was created using SolidWorks computer-aided design software. The geometric dimensions used to create the models corresponded to the real physical dimensions of the cells used in the experimental tests and the cell geometry was imported into HyperMesh and prepared for discretization into finite elements. For each part of the cell, an optimal quality criterion was applied. After the generation of the finite element model, the mechanical structural properties and the materials were assigned, and the last step of this stage was the definition of the load case (forces, contact between components, and impact body velocity). TYPE7 was used as the contact interface, which ensures permanent contact between the model's surfaces. The properties of the materials used are centralized in Table 1 (the material type assigned to the components being M2_PLAS_JOHNS_ZERIL). This is an isotropic elastic–plastic type of material and reproduces internal stresses as a function of strain, strength, and temperature. All components of this study were meshed into mixed shell elements, having assigned type P1_SHELL predefined properties. The preprocessing part was carried out in Altair HyperMesh software in the Radioss solver environment. The results obtained from the simulation process were interpreted using the HyperView and HyperGraph modules of HyperWorks software.

Table 1. Physical–mechanical properties of the 18650-type cell components considered in the modelling and simulation [36].

Battery's Component	Material [-]	Density [kg/m³]	Elasticity Modulus [GPa]	Poisson's Coefficient [-]	Single-Layer Thickness [μm]
Cathode	LiFePO$_4$	4000	100	0.35	90
Anode	Graphite	2300	110	0.23	130
Positive collector	Aluminium	2700	180	0.35	20
Negative collector	Copper	7980	210	0.34	10
Separator	Polyethylene (PE)	1500	20	0.3	10
Case	Steel	7850	210	0.3	560

From the simulation results, when analysing the distribution of von Mises stresses (Figure 6a), the maximum value appeared in the outer areas of the battery case where the impactor has the main point of impact (6.245×10^{-1} GPa).

(a)   (b)

Figure 6. Bending test of the 18650-type cell ((**a**)—von Mises stress variation by simulation; (**b**)—experimental result).

In the graph presented in Figure 7, the results from the experimental tests and the results from the simulation for the bending of the 18650-type cell were superimposed. The analysis of the results shows that the force–deformation variation is almost linear and the experimental results curve follows the simulation results curve through almost the entire deformation range considered. It was also considered that the CAD model of the cell was validated and could be used further in the study since the differences between the simulated and experimental values were approximately 1% (A maximum of 13% difference was recorded for the very final part, which was outside of the range of consideration in this study).

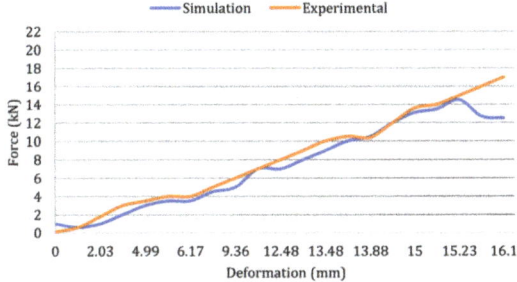

Figure 7. Comparison of experimental results with those obtained by simulation for the bending test of the 18650-type cell.

2.2.2. Topographical Optimization of the Battery Module Case

The batteries that equip electric vehicles must have a certain mechanical rigidity required for the manufacturing process, the handling process, and actual use. Hence, at present, it was found that the necessary rigidity of the battery case is achieved by creating ribs on the surface/surfaces of the case (especially the upper and lower surfaces), ribs whose shapes and placement differ from one manufacturer to another. In general, topographic optimization is applied to flat or already deformed surfaces of a model to reach a geometric shape with reinforcements on various surfaces, so that, through this process, a product can be developed to meet the constructive parameters and ensure a higher level of functionality. Optimization can be performed on the components of an assembly or subassembly, wherever it is considered that there is potential for improvement. In the process of topographical optimization of a basic model, certain configuration levels must be completed in the optimization algorithm: design and development of the model; defining the variables for topographical optimization; defining optimization responses; and postprocessing and analysis of the results. In carrying out this study, we started with a battery module case based on a simple geometric model, with well-defined flat surfaces and with the following geometric dimensions: a length of 500 mm, a width of 240 mm, and a height of 73 mm (Figure 8). The optimization areas chosen were the lower and upper surfaces of the battery module case. The material chosen and defined for the module case model was aluminium, due to its low specific weight, thus reducing the total weight of the battery assembly (thickness, 4 mm; density, 2.7 g/cm^3; Poisson's ratio, 0.33; and Young's modulus, 70 GPa).

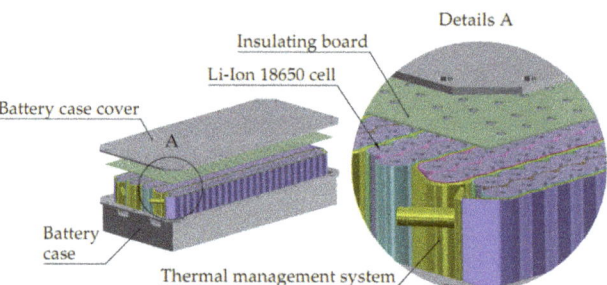

Figure 8. The structure and components of a battery module.

Topographic optimization is an advanced shape optimization, applicable to models discretized into shell elements, and the process is presented in Figure 9. The process of topographic optimization begins by creating the finite element model from the imported basic geometry of the model. The definition of objective functions, design variables, and the creation of the load case is carried out in the second part of the process. The required design parameters are represented by the minimum width of the geometric elements (w_{min} = 28 mm), the height of the elements (h_{max} = 2 mm), and the drawing angle (α = 82°) (Figure 10).

After creating the load case, the model was statically analysed to extract the vibration modes in the frequency range of 22–33 Hz (a frequency range generally considered to be transmitted to the vehicle chassis due to the wheel–road interaction and also to the battery and the battery case, which is fixed to the chassis [37]). In the last stage of the process, it was determined if the results converged towards a feasible solution. If the results converged, then the optimization process was completed by generating the optimized model of the battery case, and in a case where the results did not converge, the process would enter an iterative loop by modifying the geometrics of the beads and reanalysing the pattern until its geometry converged.

After the optimization process was performed on the base case, the topographic optimization algorithm provided an optimized model according to the input data (Figure 11).

Furthermore, the optimized model was completed by the geometric optimization of the areas proposed by the algorithm, for the construction of simple geometric shapes (which are easy to implement in the manufacturing process).

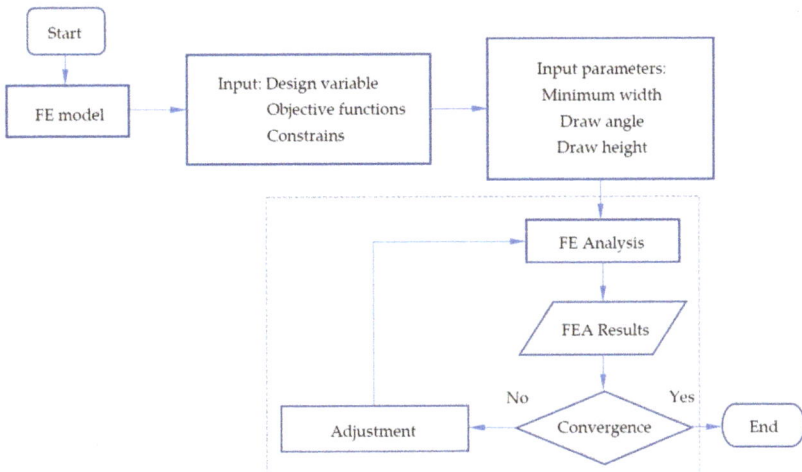

Figure 9. Flowchart of the topography optimization process.

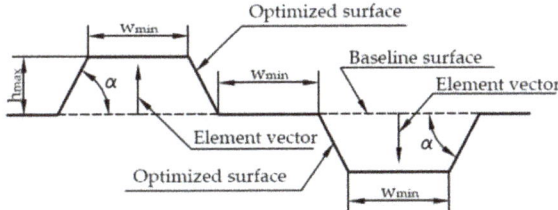

Figure 10. Design parameters of the battery case cover for topographical optimization.

 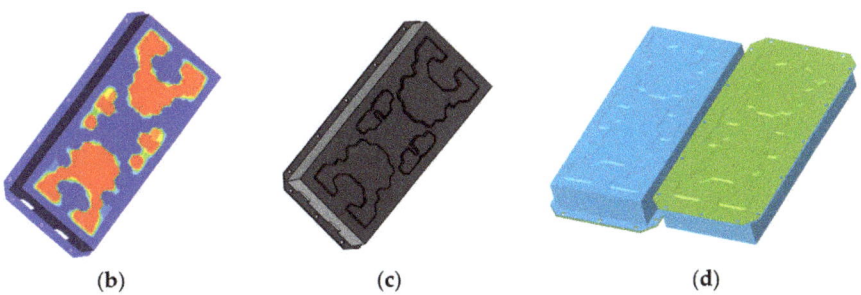

Figure 11. Battery module case model used in simulations (from bottom view): (**a**) base nonoptimized (initial); (**b**) topographically optimized; (**c**) the final shape used in the simulations; (**d**) optimized module case view from top and bottom.

To continue the comparative analysis between the basic case model and the optimized version, an electric vehicle battery pack including 12 modules was modelled, each module having 480 electrochemical cells (18650-type), resulting in a total number of 5760 cells (Figure 8). The discretization was performed in 767,396 shell elements for the basic model (initial shell) and 869,620 shell elements for the topographically optimized shell.

It should also be noted that these simulations did not consider the thermal influences that may occur (temperature variations in the external environment and variations due to the electrochemical processes in operation), and a constant temperature of 25 °C was considered (it was considered that the battery has a functional thermal management system).

The numerical analysis aimed to analyse the behaviour and the mechanical deformation of the cells and the battery module in the event of impact/accident (the effect of impact on the integrity of the battery's cells).

The most unfavourable case was also chosen in terms of the possibility of accidental damage to the battery, namely the side impact to the body of an electric vehicle (the impact area is located between the vehicle's pillars A and B), and the hypothesis considered was that only a maximum of 80% of the total impact force/energy was transmitted to the battery pack (Figure 12). In the case of a frontal impact, the design of the vehicle chassis offers the greater possibility of absorbing and dispersing a large part of the impact energy up to the area where the battery is located; in the case of a side impact, the battery is essentially only protected by the longitudinal beams, the passenger compartment floor, and the structures that form the body (pillars).

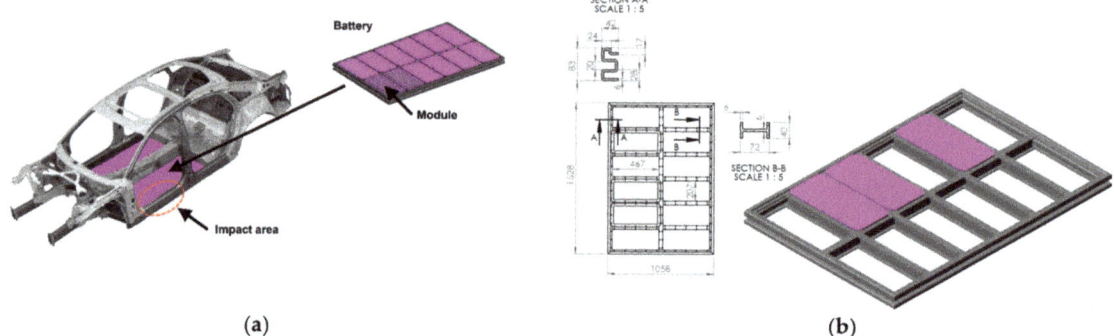

Figure 12. The hypothesis regarding the side impact simulation of the electric vehicle (**a**) and the design details for the battery frame (**b**).

Reporting on the efficiency of the topographical optimization solution for the battery, a scenario was chosen that represented the number of cells that undergo deformation, but which, according to the experimental research carried out, remains in a working condition (a short-circuit does not occur) and can be reused. The chosen value of the maximum deformation that could be suffered by a cell, at which the cell is still considered to be functional, is 2.5 mm (measured on the transversal plane of the cell, i.e., 15% of the size of the diameter of the 18650-type cell). The test impact velocities considered for both the initial/base and topographically optimized cases were 36 km/h (10 m/s), 72 km/h (20 m/s), and 108 km/h (30 m/s).

3. Results and Discussion

The obtained results for the simulation of the deformation of the battery modules (and the effects on the integrity of the cells) under the considered impact conditions are presented below, considering the variation in the mechanical stresses, kinetic energy and internal energy, and von Mises stress values. The energy balance plots represent the method to check the quality of the results. An important indicator for proper analysis is the total energy of the system, which remains constant. It can be observed that the total energy variation remains almost constant in all simulation cases and the higher values of the kinetic energy are absorbed in the frame and the case structure. To obtain an accurate simulation result, more attention was focused on hourglass energy, which must be less than 5% of the overall internal energy.

The results obtained by simulating the battery module case are presented in Figures 13–30, both for the basic version and for the topographically optimized version (considering parameters such as: energy balance, displacement and von Mises stresses). For a better view of the deformation and mechanical stresses that occur in the battery cells area, the module cover is hidden.

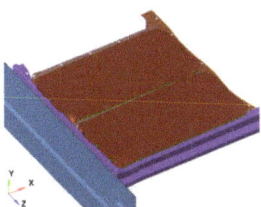 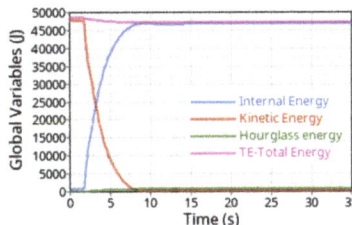

Figure 13. Details of the mechanical stress and energy balance on the modules and base module case at an impact velocity of 10 m/s.

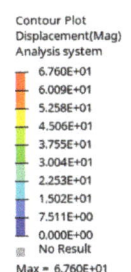 

Figure 14. View of displacement magnitude (mm), base module case, 10 m/s impact velocity.

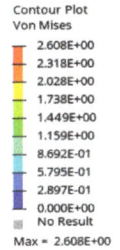 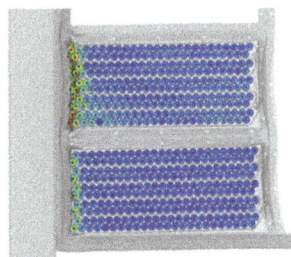

Figure 15. View of von Mises stress (GPa), base module case, 10 m/s impact velocity.

 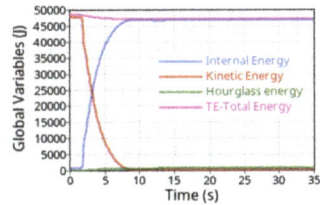

Figure 16. Details of the mechanical stress and energy balance on modules, optimized module case, 10 m/s impact velocity.

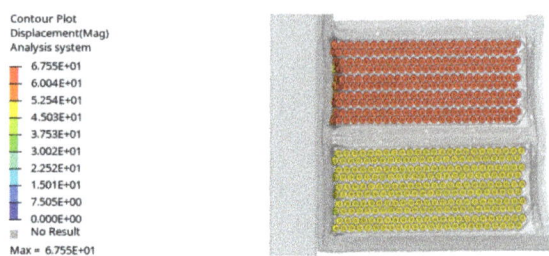

Figure 17. View of displacement magnitude (mm), optimized module case, 10 m/s impact velocity.

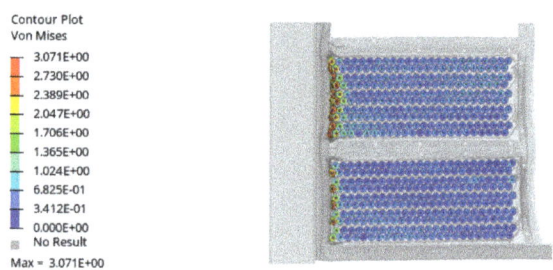

Figure 18. View of von Mises stress (GPa), optimized module case, 10 m/s impact velocity.

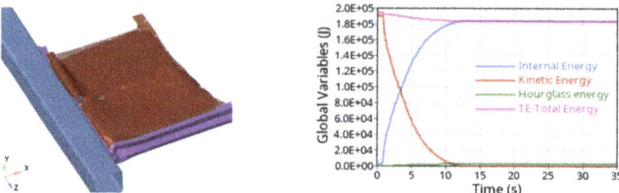

Figure 19. Details of the mechanical stress and energy balance on modules, base module case, 20 m/s impact velocity.

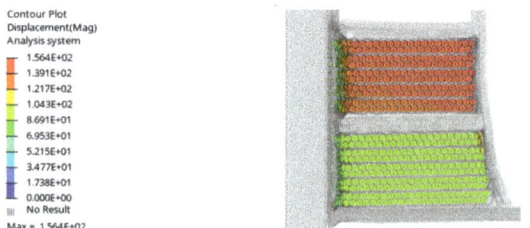

Figure 20. View of displacement magnitude (mm), base module case, 20 m/s impact velocity.

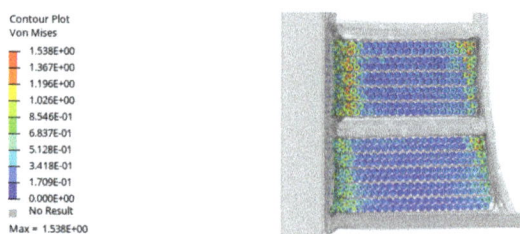

Figure 21. View of von Mises stress (GPa), base module case, 20 m/s impact velocity.

Figure 22. Details of the mechanical stress and energy balance on modules, optimized module case, 20 m/s impact velocity.

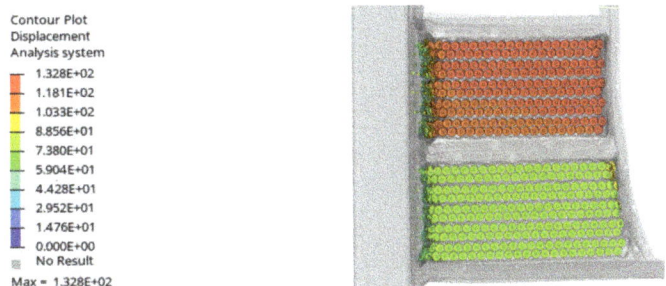

Figure 23. View of displacement magnitude (mm), optimized module case, 20 m/s impact velocity.

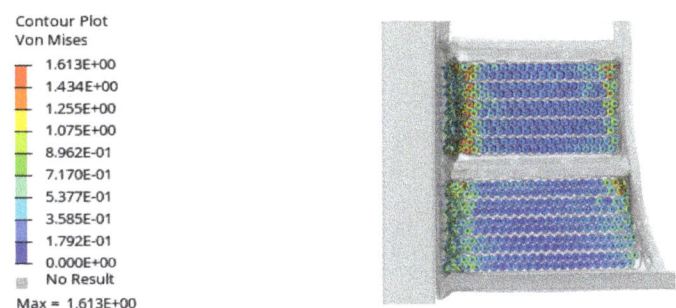

Figure 24. View of von Mises stress (GPa), optimized module case, 20 m/s impact velocity.

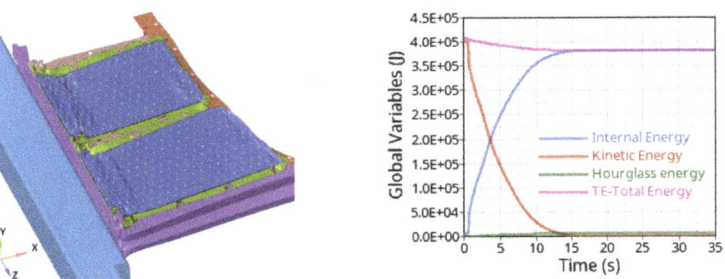

Figure 25. Details of the mechanical stress and energy balance on modules, base module case, 30 m/s impact velocity.

Figure 26. View of displacement magnitude (mm), base module case, 30 m/s impact velocity.

Figure 27. View of von Mises stress (GPa), base module case, 30 m/s impact velocity.

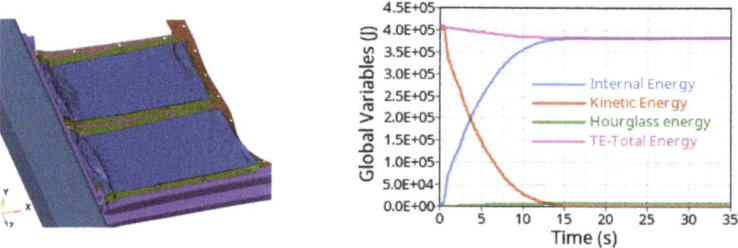

Figure 28. Details of the mechanical stress on modules, optimized module case, 30 m/s impact velocity.

Figure 29. View of displacement magnitude (mm), optimized module case, 30 m/s impact velocity.

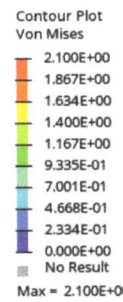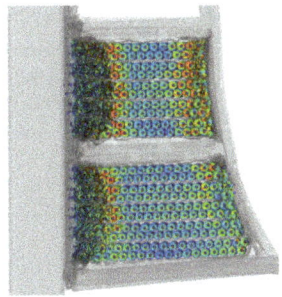

Figure 30. View of von Mises stress (GPa), optimized module case, 30 m/s impact velocity.

The energy balance was plotted after the impact analysis of the battery modules. The kinetic energy variation shown in Figure 13 starts at an initial value of 47,773.4 J and remains constant until the first 1.7 ms when the battery assembly first comes into contact with the longitudinal beam. Thereafter, the kinetic energy drops dramatically, intersecting the internal energy variation at 24,080 J, and continues to decrease to near zero at 8 ms. During the impact, the first row of cells in the battery of the first module is initially damaged by direct contact with the case, while the following rows of cells absorb (and at the same time dissipate) the mechanical shock through contact with the first row of batteries. The deformation of the battery cells due to the impact at a velocity of 10 m/s is shown in Figure 15, where the first row of batteries located in the impact area is partially damaged and the heat exchanger is destroyed during the impact.

The energy balance variation during the rigid side impact of the optimized battery assembly model is presented in Figure 16. The maximum kinetic energy curve starts from the same kinetic energy value as the previous simulation for the base model. The kinetic energy curve intersects the internal energy curve at 23,966.84 J, at 3.185 ms. In Figure 18 it can be observed that the largest deformations are in the impact area of the first rows of battery cells of the battery's modules. The cells are not seriously damaged, compared to the simulation results of the base module simulation. Increasing the impact velocity of the battery packs, the number of damaged batteries is higher, as can be viewed in Figures 20, 23, 26 and 29. In addition to the previous basic simulation, the more affected cells are situated in the impact area and on the opposite side of the battery module frame. The deformation from the opposite of the impact area is due to the mass attached to the battery module's support. In the simulation case presented for optimized module, the number of affected cells is lower than in the previous basic simulation case.

Following the simulations carried out in the three cases with different impact velocities and with different case models (base vs. topographically optimized), a series of values were obtained for the number of batteries deformed by less than 2.5 mm (maximum value at which the cells are considered/alleged to be (still) electrically functional), as shown in Table 2.

Based on the results obtained, it can be stated that the proposed constructive solution offers a reduction in cell deformation with an efficiency of between 16.6–60% (19.7% to 40.7% if the mean value differences are considered for all three impact velocities), which shows that the topographically optimized solution (from a constructive point of view) presents more effective battery case protection against the mechanical stresses from impact phenomena (i.e., road accidents). It should be noted that through topographical optimization, the case tends to be more effective in protecting the cells of the battery, but at the same time, the modification of the optimized surfaces (by increasing them) also leads to modification of the mass of the battery case covers. A larger mass has an immediate effect in the possibility of taking over (and dispersing) more of the impact energy, maintaining the deformation of the structure in the plastic domain for a longer period of time.

Table 2. Comparative results for the considered simulation cases.

	Range of Cells Deformation [mm]	Base Battery Module Case			Optimized Battery Module Case			Difference (Base vs. Optimized) [%]		
		Impact Velocity [m/s]								
		10	20	30	10	20	30	10	20	30
Deformed cells (pcs)	<0.5	5	8	12	2	4	8	−60.0	−50.0	−33.3
	0.5–1.0	6	9	19	3	5	13	−50.0	−44.4	−31.6
	1.0–1.5	8	18	22	4	11	16	−50.0	−38.9	−27.3
	1.5–2.0	9	25	29	5	16	22	−44.4	−36.0	−24.1
	2.0–2.5	13	35	49	8	24	38	−38.5	−31.4	−22.4
	>2.5	18	122	265	13	91	221	−27.8	−25.4	−16.6
Total affected cells		59	217	396	35	151	318	−40.7 *	−30.4 *	−19.7 *

* Mean values.

Considering that, according to the statistics on road accidents in the EU, most accidents take place at low velocity, it can be observed that for the proposed constructive battery case solution, the maximum effect is obtained at low impact velocities (<36 km/h), and at an impact velocity greater than 70 km/h the difference falls below 17%. Thus, the effectiveness of the mechanical protection for the battery cells through the proposed method of topographical optimization could provide a benefit in the application of the circular economy concept by reusing more valid electrochemical cells in other/future constructions of energy sources, reducing (or why not eliminating) costs and economic and environmental pollution due to industrial manufacturing processes.

4. Conclusions

Based on the results obtained from the experimental research activities and numerical analysis presented in this article, it can be stated that research in the field of electric vehicles is necessary, current, and important, considering the continuous growth dynamics of the number of electric vehicles in traffic, with the immediate reality that they will inevitably be involved in road accidents. Considering the major cost that an energy source (battery) is as part of the total construction costs of an electric vehicle, as well as the risks associated with electrochemical cells' behaviour in the event of accidents, this article presents an approach that aims to improve the construction of the energy source protective housing/case for electric vehicles from the design phase to impact, to meet the requirements related to their use and operation in conditions of maximum safety and reliability.

The structural optimization of the battery module case was based on the topographical optimization process (considering the vibrations due to the wheel–road interaction transmitted to the chassis), which allowed for the proposal of a modified but rigid casing structure. In the study, the most unfavourable case of a road accident was considered, namely the side impact (impact area between A and B pillars, for different impact velocities), and the proposed theme was studied through numerical analysis methods (modelling and simulation). Based on the obtained results, it was observed that there was a reduction in the number of damaged cells, which shows that the proposed topographically optimized solution tends to be highly effective at low impact velocities (<36 km/h), when a major number of batteries are protected from impact (with 27.8% more protection). The functional remaining cells can later be used in other constructions/applications of energy sources that do not require strict and highly efficient operating parameters, according to the circular economy concept.

The main directions for future research and development that can offer solutions for the reuse of electrochemical cells in the construction of electric vehicle batteries could be the use of other/complex topographical restrictions to further optimize the design of the casing shape, topographical optimization of all casing surfaces, the development and use of a complex numerical model of an electrochemical cell that also takes into account the mechanics of the internal chemical structure of the battery (friction phenomena for example), and the influences of the electrochemical and thermal phenomena that occur in the operation of the battery.

Author Contributions: Conceptualization, I.S. and F.M.; methodology, I.S. and F.M.; validation, I.S. and L.I.S.; formal analysis, I.S., L.I.S., and H.R.; investigation, I.S.; data analysis and curation, H.R.; writing—original draft preparation, F.M. and H.R.; writing—review and editing, F.M. and H.R.; supervision, F.M. All authors have read and agreed to the published version of the manuscript.

Funding: This research received no external funding.

Data Availability Statement: Not applicable.

Acknowledgments: The authors express their gratitude to the Technical University of Cluj-Napoca, for the financial support in the publication of this article.

Conflicts of Interest: The authors declare no conflict of interest.

References

1. Winkler, S.L.; Anderson, J.E.; Garza, L.; Ruona, W.C.; Vogt, R.; Wallington, T.J. Vehicle criteria pollutant (PM, NOx, CO, HCs) emissions: How low should we go? *Clim. Atmos. Sci.* **2018**, *1*, 26. [CrossRef]
2. Abdel-Rahman, A.A. On the emissions from internal-combustion engines: A review. *Int. J. Energy Res.* **1998**, *22*, 489–513. [CrossRef]
3. Adeyanju, A. Effects of Vehicular Emissions on Human Health. *J. Clean Energy Technol.* **2018**, *6*, 411–420. [CrossRef]
4. Kim, B.-G.; Lee, P.-H.; Lee, S.-H.; Kim, Y.-E.; Shin, M.-Y.; Kang, Y.; Bae, S.-H.; Kim, M.-J.; Rhim, T.; Park, C.-S.; et al. Long-Term Effects of Diesel Exhaust Particles on Airway Inflammation and Remodeling in a Mouse Model. *Allergy Asthma Immunol. Res.* **2016**, *8*, 246–256. [CrossRef]
5. European Environmental Agency (EEA). Greenhouse Gas Emissions from Transport in Europe. 2020. Available online: https://www.eea.europa.eu/data-and-maps/indicators/transport-emissions-of-greenhouse-gases/transport-emissions-of-greenhouse-gases-12 (accessed on 15 September 2022).
6. Chau, K.T.; Chan, C.C. Emerging Energy-Efficient Technologies for Hybrid Electric Vehicles. *Proc. IEEE* **2007**, *95*, 821–835. [CrossRef]
7. Amamra, S.-A.; Tripathy, Y.; Barai, A.; Moore, A.D.; Marco, J. Electric Vehicle Battery Performance Investigation Based on Real World Current Harmonics. *Energies* **2020**, *13*, 489. [CrossRef]
8. Kim, J.-S.; Lee, D.-C.; Lee, J.-J.; Kim, C.-W. Optimization for maximum specific energy density of a lithium-ion battery using progressive quadratic response surface method and design of experiments. *Sci. Rep.* **2020**, *10*, 15586. [CrossRef] [PubMed]
9. Lei, Y.; Zhang, C.; Gao, Y.; Li, T. Charging Optimization of Lithium-ion Batteries Based on Capacity Degradation Speed and Energy Loss. *Energy Procedia* **2018**, *152*, 544–549. [CrossRef]
10. Li, R.; Wei, X.; Sun, H.; Sun, H.; Zhang, X. Fast Charging Optimization for Lithium-Ion Batteries Based on Improved Electro-Thermal Coupling Model. *Energies* **2022**, *15*, 7038. [CrossRef]
11. Liu, K.; Kang Li, K.; Peng, Q.; Zhang, C. A brief review on key technologies in the battery management system of electric vehicles. *Front. Mech. Eng.* **2019**, *14*, 47–64. [CrossRef]
12. Buidin, T.I.C.; Mariasiu, F. Battery Thermal Management Systems: Current Status and Design Approach of Cooling Technologies. *Energies* **2021**, *14*, 4879. [CrossRef]
13. Lai, W.-J.; Ali, M.Y.; Pan, J. Mechanical behavior of representative volume elements of lithium-ion battery modules under various loading conditions. *J. Power Source* **2014**, *248*, 789–808. [CrossRef]
14. Abada, S.; Marlair, G.; Lecocq, A.; Petit, M.; Sauvant-Moynot, V.; Huet, F. Safety focused modeling of lithium-ion batteries: A review. *J. Power Source* **2016**, *306*, 178–192. [CrossRef]
15. Feng, X.; Ouyang, M.; Liu, X.; Lua, L.; Xia, Y.; He, X. Thermal runaway mechanism of lithium ion battery for electric vehicles: A review. *Energy Storage Mater.* **2018**, *10*, 246–267. [CrossRef]
16. Gao, Z.; Zhang, X.; Xiao, Y.; Gao, H.; Wang, H.; Piao, C. Influence of Low-Temperature Charge on the Mechanical Integrity Behavior of 18650 Lithium-Ion Battery Cells Subject to Lateral Compression. *Energies* **2019**, *12*, 797. [CrossRef]
17. Sahraei, E.; Meier, J.; Wierzbicki, T. Characterizing and modeling mechanical properties and onset of short circuit for three types of lithium-ion pouch cells. *J. Power Source* **2014**, *247*, 503–516. [CrossRef]
18. Zhu, J.; Zhang, X.; Sahraei, E.; Wierzbicki, T. Deformation and failure mechanisms of 18650 battery cells under axial compression. *J. Power Source* **2016**, *336*, 332–340. [CrossRef]
19. Xu, J.; Ma, J.; Zhao, X.; Chen, H.; Xu, B.; Wu, X. Detection Technology for Battery Safety in Electric Vehicles: A Review. *Energies* **2020**, *13*, 4636. [CrossRef]
20. Evarts, E.C. Lithium batteries: To the limits of lithium. *Nature* **2015**, *526*, 93–95. [CrossRef]

21. Chombo, P.V.; Laonual, Y. A review of safety strategies of a Li-ion battery. *J. Power Source* **2020**, *478*, 228649. [CrossRef]
22. Klink, J.; Hebenbrock, A.; Grabow, J.; Orazov, N.; Nylén, U.; Benger, R.; Beck, H.-P. Comparison of Model-Based and Sensor-Based Detection of Thermal Runaway in Li-Ion Battery Modules for Automotive Application. *Batteries* **2022**, *8*, 34. [CrossRef]
23. Ruiz, V.; Pfrang, A.; Kriston, A.; Omar, N.; Van den Bossche, P.; Boon-Brett, L. A review of international abuse testing standards and regulations for lithium ion batteries in electric and hybrid electric vehicles. *Renew. Sustain. Energy Rev.* **2018**, *81*, 1427–1452. [CrossRef]
24. Ahn, Y.J.; Lee, Y.-S.; Cho, J.-R. Multi-Layered Numerical Model Development of a Standard Cylindrical Lithium-Ion Battery for the Impact Test. *Energies* **2022**, *15*, 2509. [CrossRef]
25. Muresanu, A.D.; Dudescu, M.C. Numerical and Experimental Evaluation of a Battery Cell under Impact Load. *Batteries* **2022**, *8*, 48. [CrossRef]
26. Shui, L.; Chen, F.; Garg, A.; Peng, X.; Bao, N.; Zhang, J. Design optimization of battery pack enclosure for electric vehicle. *Struct. Multidiscip. Optim.* **2018**, *58*, 331–347. [CrossRef]
27. An, Y.; Wang, X.; Dou, N.; Wu, Z. Strength analysis of the lightweight-designed power battery boxes in electric vehicle. *E3S Web Conf.* **2022**, *341*, 01025. [CrossRef]
28. Ruan, G.; Yu, C.; Hu, X.; Hua, J. Simulation and optimization of a new energy vehicle power battery pack structure. *J. Theor. Appl. Mech.* **2021**, *59*, 565–578. [CrossRef]
29. Qiao, W.; Yu, L.; Zhang, Z.; Pan, T. Study on the Battery Safety in Frontal Collision of Electric Vehicle. *J. Phys.: Conf. Ser.* **2021**, *2137*, 012008. [CrossRef]
30. Pan, Y.; Xiong, Y.; Wu, L.; Diao, K.; Guoa, W. Lightwieght design of an automotive battery-pack enclosure via advanced high-strength steels and size optimization. *Int. J. Automot. Technol.* **2021**, *22*, 1279–1290. [CrossRef]
31. Dong, S.; Lv, J.; Wang, K.; Li, W.; Tian, Y. Design and Optimization for a New Locomotive Power Battery Box. *Sustainability* **2022**, *14*, 12810. [CrossRef]
32. Biharta, M.A.S.; Santosa, S.P.; Widagdo, D.; Gunawan, L. Design and Optimization of Lightweight Lithium-Ion Battery Protector with 3D Auxetic Meta Structures. *World Electr. Veh. J.* **2022**, *13*, 118. [CrossRef]
33. Pratama, L.K.; Santosa, S.P.; Dirgantara, T.; Widagdo, D. Design and Numerical Analysis of Electric Vehicle Li-Ion Battery Protections Using Lattice Structure Undergoing Ground Impact. *World Electr. Veh. J.* **2022**, *13*, 10. [CrossRef]
34. European Commission. Green Deal: Sustainable Batteries for a Circular and Climate Neutral Economy. 2020. Available online: https://digital-strategy.ec.europa.eu/fr/node/475/printable/pdf (accessed on 15 September 2022).
35. Kehl, D.; Jennert, T.; Lienesch, F.; Kurrat, M. Electrical Characterization of Li-Ion Battery Modules for Second-Life Applications. *Batteries* **2021**, *7*, 32. [CrossRef]
36. Szabo, I.; Sirca, A.A.; Scurtu, L.; Kocsis, L.; Hanches, I.N.; Mariasiu, F. Comparative study of Li-ion 18650 cylindrical cell under pinch indentation. *IOP Conf. Ser. Mater. Sci. Eng.* **2022**, *1256*, 012022. [CrossRef]
37. Weber, J. *Automotive Development Process*; Springer: Berlin, Germany, 2009; pp. 142–143.

Disclaimer/Publisher's Note: The statements, opinions and data contained in all publications are solely those of the individual author(s) and contributor(s) and not of MDPI and/or the editor(s). MDPI and/or the editor(s) disclaim responsibility for any injury to people or property resulting from any ideas, methods, instructions or products referred to in the content.

Article

Iterative Nonlinear Fuzzy Modeling of Lithium-Ion Batteries

José M. Andújar *, Antonio J. Barragán, Francisco J. Vivas, Juan M. Enrique and Francisca Segura

Research Centre for Technology, Energy and Sustainability, La Rábida, Palos de la Frontera, 21071 Huelva, Spain
* Correspondence: andujar@diesia.uhu.es

Abstract: Electric vehicles (EVs), in their pure and hybrid variants, have become the main alternative to ensure the decarbonization of the current vehicle fleet. Due to its excellent performance, EV technology is closely linked to lithium-ion battery (LIB) technology. A LIB is a complex dynamic system with extraordinary nonlinear behavior defined by electrical, thermal and electrochemical dynamics. To ensure the proper management of a LIB in such demanding applications as EVs, it is crucial to have an accurate mathematical model that can adequately predict its dynamic behavior. Furthermore, this model must be able to iteratively adapt its parameters to accommodate system disturbances during its operation as well as performance loss in terms of efficiency and nominal capacity during its life cycle. To this end, a methodology that employs the extended Kalman filter to iteratively improve a fuzzy model applied to a real LIB is presented in this paper. This algorithm allows to improve the classical Takagi–Sugeno fuzzy model (TSFM) with each new set of data obtained, adapting the model to the variations of the battery characteristics throughout its operating cycle. Data for modeling and subsequent validation were collected during experimental tests on a real LIB under EVs driving cycle conditions according to the "worldwide harmonised light vehicle test procedure" (WLTP) standard. The TSFM results allow the creation of an accurate nonlinear dynamic model of the LIB, even under fluctuating operating conditions, demonstrating its suitability for modeling and design of model-based control systems for LIBs used in EVs applications.

Keywords: adaptation; batteries; fuzzy; intelligent system; iterative; Kalman; lithium-ion; modeling; WLTP

Citation: Andújar, J.M.; Barragán, A.J.; Vivas, F.J.; Enrique, J.M.; Segura, F. Iterative Nonlinear Fuzzy Modeling of Lithium-Ion Batteries. *Batteries* **2023**, *9*, 100. https://doi.org/10.3390/batteries9020100

Academic Editor: Carlos Ziebert

Received: 30 December 2022
Revised: 20 January 2023
Accepted: 25 January 2023
Published: 1 February 2023

Copyright: © 2023 by the authors. Licensee MDPI, Basel, Switzerland. This article is an open access article distributed under the terms and conditions of the Creative Commons Attribution (CC BY) license (https:// creativecommons.org/licenses/by/ 4.0/).

1. Introduction

The growing popularity in recent decades of devices that require reliable and efficient storage of electrical energy, such as portable devices (phones, laptops, etc.) and electric vehicles (EVs) [1], logically entails a development of battery technology and the devices that manage them.

Currently, the decarbonization of the transport industry involves the use of lithium-ion batteries (LIBs) [2]. The need to know precisely and in real time the autonomy of EVs requires not only reliable models but also control systems to increase this autonomy [3,4]. To design an efficient control system, it is desirable to have an accurate model at all times, regardless of conditions in which the battery is working. The model must provide reliable predictions about the state of the battery, available energy and remaining operating time for given conditions and power requirements. However, the behavior of batteries changes with time, charge and discharge cycles, temperature, degradation and even depending on the instantaneous power required from them [5,6]. Therefore, it is necessary to develop modeling solutions that allow the dynamic model to be adapted or modified to provide a computationally cost-effective, efficient and robust solution to the modeling problem for any instantaneous battery operating condition.

Thus, many models have been suggested in the specialized scientific literature, with their corresponding modifications and subsequent improvements, which try to describe the behavior of the different battery technologies [7]. However, given that what underlies the operation of these energy accumulators are chemical reactions, more or less complex, that depend on various factors and variables, and that show highly nonlinear behavior, obtaining accurate models is not an easy task. Therefore, it is currently a scientific field in full activity.

In summary, most of the models developed to date can be grouped as follows [8,9], (summarized in a simple way in Figure 1):

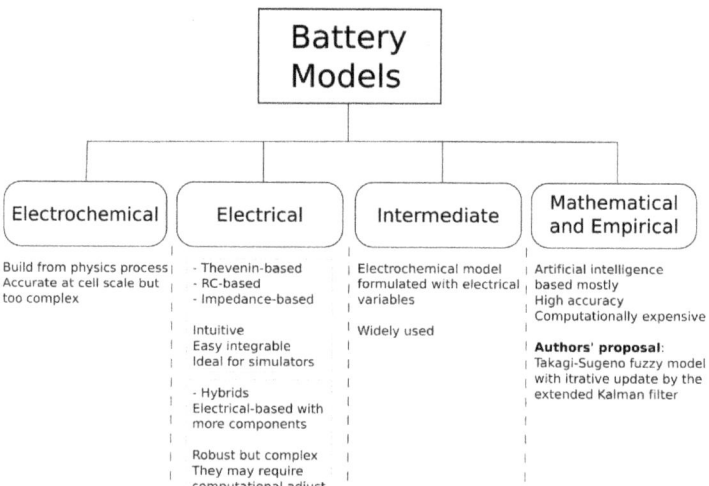

Figure 1. Brief classification of battery models.

(a) Electrochemical models:

These models are built from the physics of electrochemical processes that occur at the battery cell level. They are quite accurate at this level [10–16], although they present significant errors at larger scales. In view of the above, its usefulness is justified to optimize the physical design of the cells [17]. These models are highly complex to define and implement, requiring a large computational burden, since they are usually described by nonlinear differential equations [10,18,19] that model microscopic transport phenomena or the chemical kinetics of the reactions involved [20,21]. However, recent research seeks to obtain low-order models for use in real-time applications [22]. The most common are usually single particle models [23], porous electrodes [24] and pseudo-two-dimensional models [9].

(b) Electrical models:

They are implemented by means of electrical circuits (resistors, capacitors and sources), to describe the behavior of a battery. They are usually intuitive models, and by their nature, they are directly applicable to the use of simulators where it is easy to integrate them with the rest of the participating systems. That is why their uses, in their different versions, are widespread. Most of these models can be grouped into a few categories:

- Thevenin-based models [25–28]. These models basically consist of a real voltage source followed by a concatenation of n RC cells in series. The voltage source represents the open circuit voltage, V_{OC}, related to the state of charge (SOC) of the battery [29]. This model has a relatively uncomplicated design and demonstrates good accuracy in simulations [30]. In addition, it does not consider nonlinear behavior and temperature effects in batteries, although recent works are making improvements in this aspect [31]. A Thevenin-based model of n-RC order is shown in Figure 2, and its transfer function is shown in (1).

$$\frac{V_{OC}(s) - V(s)}{I(s)} = R_0 + \sum_{i=1}^{n}\left(\frac{R_i}{R_i C_i s + 1}\right) \tag{1}$$

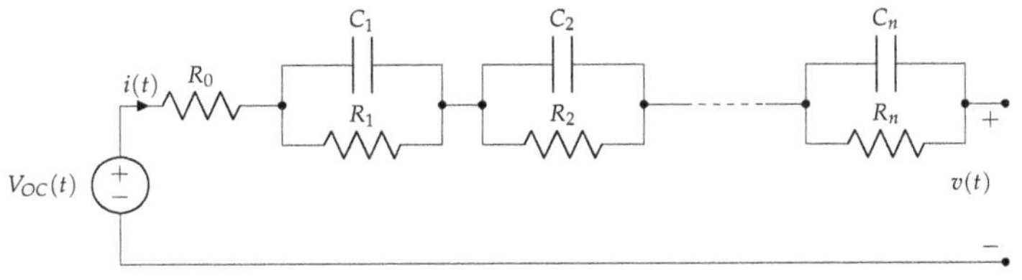

Figure 2. Thevenin-based model of nth order RC cells.

- RC-based models [32–34]. These are simpler electronics models that basically consist of a network formed by two capacitors and several resistors. Generally, a large capacitor models the energy storage capacity of the battery, and another, smaller than the previous one, models its transient effects.
- Impedance-based models [35–37]. These types of models change the RC cells of the Thevenin model by impedances determined by electrochemical impedance spectroscopy methods. They are usually suitable models for slow discharges but not for high currents.

Obviously, in the scientific literature, there are models that, trying to obtain better performance, hybridize the previous ones and introduce controlled sources and nonlinear elements [38,39] and even operators or mathematical blocks such as integrators and filters [40], obtaining general and robust models, although due to their complexity, they may require a lot of computational burden to adjust their parameters.

We should also mention the so-called intermediate models, which are located between purely electrochemical and equivalent circuit models, although they are basically formulated with electrical variables [32]. The simplest is the Peukert model from 1897 [41], and among the most famous is the Shepherd model from 1965 [42], which is still widely used today, as well as their respective improvements in the Unnewehr, Nertst and Plett models [32,43–45]. Plett's model can be considered an improved compendium of the previous ones; however, its improvement comes at the cost of increasing its complexity and requiring several parameters to be experimentally determined.

(c) Mathematical and empirical models: There are multiple techniques in the literature, most of them employing artificial intelligence techniques, that allow obtaining nonlinear models of dynamic systems with high accuracy, such as neural networks [46–48], neuro-fuzzy models [49], particle swarm optimization [50], or the most recent hybrid models [51,52], among others [53]. Many of these models (algorithms) have been used to estimate the LIB performance from operating data [50,54–57]. The goodness of these algorithms is more than proved, especially when they are based on a large set of data, but they do not allow these models to adapt to subsequent LIB changes.

The previous methodologies allow obtaining more or less accurate models, but they are unable to adapt to the changes in the dynamics that a LIB undergoes during its life cycle. The main objective of this work is to present an iterative modeling methodology, capable of improving an initial model with each new set of data obtained and, therefore, capable of adapting to such changes. In order to respond to these problems, the main novelty of the paper is the use of an iterative modeling technique, which allows a balanced solution in terms of modeling error and computational cost, as well as adaptability to changes in battery performance in demanding applications such as the one studied. In this paper, we propose the use of an extended Kalman filter (EKF) algorithm [58] as an iterative modeling algorithm to adapt a Takagi–Sugeno fuzzy model (TSFM) [59]. This combination provides the accuracy and interpretability of a fuzzy model, together with the adaptability and computational efficiency of the EKF [60], so that it is possible to maintain model accuracy in the face of dynamic variations experienced by the batteries during their operating cycle. In addition, it should also be noted that the EKF is particularly suitable for systems where noise or inaccuracies are present, making it ideal for real-world applications [61].

Following this introduction, in Section 2, the materials and methods used during the experiments are presented. The iterative battery modeling process is described in Section 3, and the results obtained are discussed in Section 4. Finally, some conclusions are presented.

2. Materials and Methods

To evaluate and validate the adaptive capacity of the developed algorithm to model the LIB, two experimental tests (charging and discharging tests) were carried out on a 59.2 VDC and 120 Ah LIB (The battery used has the following characteristics: Pack lithium ion Samsung 50 59.2 V 120 Ah + BMS 80 A + aluminum box). Specifications of lithium Ion battery 59.2 V 120 Ah—Samsung 50E. Technology: NMC. Battery pack nominal voltage: 59.2 V. Battery pack charging voltage: 67.2 V. Capacity: 120 Ah (7.10 kWh). Continuous maximum discharge current: 240 A. In this case, the setting is 16 cells of 3.7 V in series, and 24 cells of 5 Ah in parallel to obtain the 120 Ah., specially designed and manufactured from Samsung, model INR21700-50S cells by Batesur©. For this purpose, simulations results of the proposed algorithm were compared with the experimental data obtained on the real battery. The model simulations were performed in the MATLAB© environment. The sampling time during all tests was set at 1 s.

To evaluate the battery performance, in particular its voltage, with respect to its operating temperature, a temperature-controlled high-capacity thermal chamber was used. Specifically, during the two tests, different heating cycles were programmed up to a maximum internal temperature of 60 °C, and a subsequent forced cooling of the battery, always taking into account the safety temperature limits established by the manufacturer. The cooling stage was carried out using a high-flow axial fan. The objective of the temperature control was never to reach a target temperature but to provoke substantial temperature variations in a short period of time, essential to appreciate the effect of the temperature on the voltage both in the training and validation phases of the model.

For the experimental tests, a 32 kW programmable source and sink of the Regatron© TC.GSS series were used. This device allows operation in the first (source) and fourth (load) quadrant depending on the voltage set point and the current or power limits established. The electrical connection between the battery and the programmable source was made by direct connection using fuses. Due to the internal protections of the device, it was not necessary to add additional elements such as antireverse diodes. The charge and discharge profiles (current and power profiles, respectively) were previously defined in the software that includes the programmable source and executed directly on the final assembly. The measurement of the variables of interest, voltage, current and operating temperature of the battery was performed by means of specific sensors and a signal conditioning circuit designed ad hoc. For voltage and current measurements, Hall Effect sensors model LV25-P and LA25-P from LEM©, respectively, were used. For temperature measurement, an NTC type thermoresistance and a conditioning circuit based on a resistive divider and zero correction circuit were used. To ensure an optimal and homogeneous measurement, the temperature sensor was placed in the center of the side face of the battery, at the opposite end of the thermal chamber heat source. The measurement obtained was compared at all times with a reference contact thermometer. The acquisition of the variables was carried out by means of a 12-bit resolution data acquisition card (DAQ) NI-USB-6008 from National Instruments©. Their storage, processing and representation was carried out by means of a data acquisition and control system (SCADA) designed ad hoc and programmed in LabVIEW©. The experimental setup carried out can be seen in Figure 3.

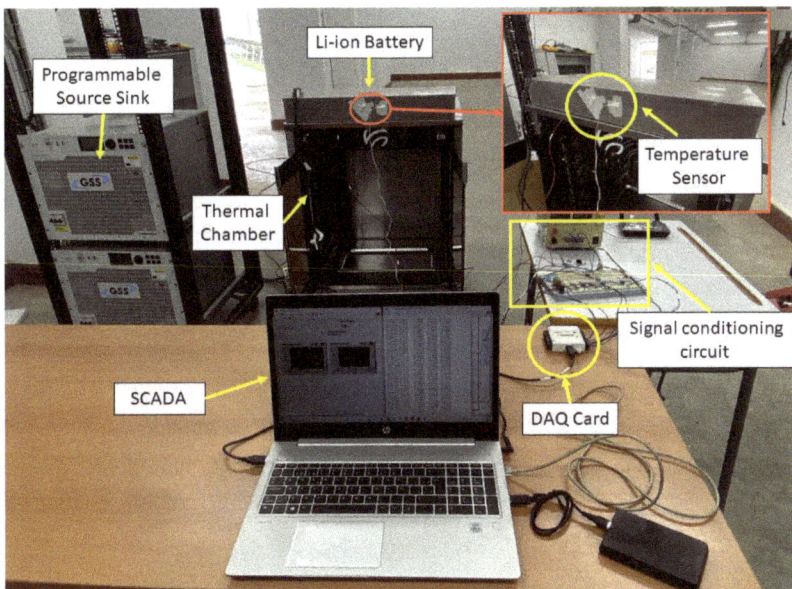

Figure 3. Laboratory experimental setup.

2.1. Takagi–Sugeno Fuzzy Model

Fuzzy logic, unlike bi-evaluated logic, is able to work with infinite possible values, not only with 'TRUE' and 'FALSE'. It is therefore very close to human reasoning, able to infer by vague predicates, i.e., able to logically integrate the infinite shades of gray that exist between absolute black and white. Fuzzy models are based on IF-THEN rules. The condition that must be fulfilled (the 'IF'), is called the antecedent, and the one that is activated if the antecedent is fulfilled (the 'THEN'), is the consequent. Between the different fuzzy models that exist, TSFM are those whose consequent is a linear polynomial [59,62], see (2).

$$R^{(l,i)}: \text{If } x_1(k) \text{ is } A^l_{1i} \text{ and} \ldots \text{and } x_n(k) \text{ is } A^l_{ni}$$
$$\text{Then } y^l_i(k) = a^l_{0i} + \sum_{j=1}^{n} a^l_{ji} x_j(k), \tag{2}$$

where n and m are respectively the number of input (x_j) and output (y_i) variables of a system to be modeled; $l = 1, \ldots, M_i$ represent the index of the rule and M_i the total number of rules for the i-th output ($y_i(k), i = 1, \ldots, m$). On the other hand, $a^l_{ij}, j = 0, \ldots, n$ is the parameters of the consequents, and k is the sampling time. A^l_{ji} are the fuzzy sets that define the antecedents of the fuzzy rules [63].

Obtaining an accurate model is a fundamental step to control any system. TSFM are universal approximators [64], allowing approximate reasoning [65], and have proven to be very accurate in modeling nonlinear systems [66–69]. Usually, the number of rules increases if the modeling error is to be reduced [70]; therefore, this process is crucial to obtain a good, accurate and manageable model for both analysis [71–73] and control system design [74,75].

To work with a more compact expression, the input vector $\mathbf{x} = (x_1, x_2, \ldots, x_n)^T$ could be extend [67,74] by the coordinate $\tilde{x}_0 = 1$:

$$\tilde{\mathbf{x}} = (\tilde{x}_0, \tilde{x}_1, \ldots, \tilde{x}_n)^T = (1, x_1, \ldots, x_n)^T \tag{3}$$

Thus, the output y_i can be calculated by [76]:

$$y_i(k) = h_i(\mathbf{x}(k)) = \sum_{j=0}^{n} a_{ji}(\mathbf{x})\tilde{x}_j(k), \tag{4}$$

where $a_{ji}(\mathbf{x})$ are variables coefficients [77]

$$a_{ji}(\mathbf{x}) = \frac{\sum_{l=1}^{M_i} w_i^l(\mathbf{x}) a_{ji}^l}{\sum_{l=1}^{M_i} w_i^l(\mathbf{x})}, \tag{5}$$

and $w_i^l(\mathbf{x})$ is the degree of activation of the fuzzy rules:

$$w_i^l(\mathbf{x}) = \prod_{j=1}^{n} \mu_{ji}^l(x_j(k), \sigma_{ji}^l). \tag{6}$$

$\mu_{ji}^l(x_j(k), \sigma_{ji}^l)$ is the jth membership function in the l rule, for the ith output (A_{ji}^l fuzzy set). The σ_{ji}^l are the parameters of this membership functions. So, the modeling process consists of determining σ_{ji}^l and a_{ji}^l parameters to obtain a TSFM.

2.2. Extended Kalman Filter and Its Application to Takagi–Sugeno Fuzzy Modeling

The EKF allows to construct a pseudo-optimum observer for nonlinear systems [58,78,79] (pseudo because it is based on a linear approximation). The filter is based on the assumption that the noises (uncertainties in the model and inaccuracies of the sensors) are zero mean white Gaussian noise, although there are also variants that allow working with other types of noise. It has shown that the EKF is an excellent algorithm for iteratively modeling complex systems based on data [60,61,80–84]. Therefore, in this work, it is used as an iterative adjustment method for the TSFM.

Be a nonlinear dynamical system in discrete time:

$$\begin{aligned} \mathbf{x}(k+1) &= \mathbf{f}(\mathbf{x}(k), \mathbf{u}(k)) + \mathbf{v}(k) \\ \mathbf{y}(k) &= \mathbf{g}(\mathbf{x}(k)) + \mathbf{e}(k), \end{aligned} \tag{7}$$

where $x(k)$ is the state vector, $u(k)$ is the input vector, and $\mathbf{v}(k)$ and $\mathbf{e}(k)$ are white noise vectors. The system's Jacobian matrices are:

$$\mathbf{\Phi}(k) = \left.\frac{\partial \mathbf{f}}{\partial \mathbf{x}}\right|_{\mathbf{x}=\mathbf{x}(k), \mathbf{u}=\mathbf{u}(k)}, \tag{8}$$

$$\mathbf{\Gamma}(k) = \left.\frac{\partial \mathbf{f}}{\partial \mathbf{u}}\right|_{\mathbf{x}=\mathbf{x}(k), \mathbf{u}=\mathbf{u}(k)}, \tag{9}$$

and

$$\mathbf{C}(k) = \left.\frac{\partial \mathbf{g}}{\partial \mathbf{x}}\right|_{\mathbf{x}=\mathbf{x}(k)}. \tag{10}$$

The EKF is considered as the execution of a prediction phase followed by an update phase. The prediction phase uses the state model and past data, while the update phase adds sensor information to the prediction—all this taking into account the quality of this information known through its covariance matrices. To solve the EKF, the following set of equations must be used iteratively:

Prediction (*a priori* estimation):

$$\hat{\mathbf{x}}(k|k-1) = \mathbf{\Phi}(k)\hat{\mathbf{x}}(k-1|k-1) + \mathbf{\Gamma}(k)\mathbf{u}(k) \tag{11}$$

$$\mathbf{P}(k|k-1) = \mathbf{\Phi}(k)\mathbf{P}(k-1|k-1)\mathbf{\Phi}^T(k) + \mathbf{R}_v \tag{12}$$

Update (*a posteriori* estimation):

$$\mathbf{K}(k) = \left(\mathbf{\Phi}(k)\mathbf{P}(k|k-1)\mathbf{C}^T(k) + \mathbf{R}_{ve}\right)\left(\mathbf{C}(k)\mathbf{P}(k|k-1)\mathbf{C}^T(k) + \mathbf{R}_e\right)^{-1} \tag{13}$$

$$\hat{\mathbf{x}}(k|k) = \hat{\mathbf{x}}(k|k-1) + \mathbf{K}(k)(\mathbf{y}(k) - \hat{\mathbf{y}}(k)) \tag{14}$$

$$\mathbf{P}(k|k) = \mathbf{\Phi}(k)\mathbf{P}(k|k-1)\mathbf{\Phi}^T(k) + \mathbf{R}_v - \mathbf{K}(k)\left(\mathbf{C}(k)\mathbf{P}(k|k-1)\mathbf{\Phi}^T(k) + \mathbf{R}_{ve}^T\right), \tag{15}$$

From (11) to (15), $\hat{\mathbf{x}}(k|k-1)$ represents the *priori* estimations of the true state $\mathbf{x}$, and $\hat{\mathbf{x}}(k|k)$ represents the *posteriori* one.

$\mathbf{P}(k|k-1)$ and $\mathbf{P}(k|k)$ are, respectively, the *a priori* and the *a posteriori* estimate covariance matrices. $\hat{\mathbf{y}}(k)$ are the estimated output vector, $\mathbf{R}_v$, $\mathbf{R}_{ve}$ and $\mathbf{R}_e$, are the noise covariance matrices, and $\mathbf{K}(k)$ is the Kalman gain.

Starting from an initial prediction $\hat{\mathbf{x}}(0)$, and its corresponding covariance matrix, $\mathbf{P}(0)$, the algorithm evolves iteratively, minimizing the estimation error and its covariance matrix for the linearization obtained at each instant.

As shown in [60,61], the EKF algorithm allows to estimate the parameters of a TSFM. To do it, it is necessary to define a new model whose states are the parameters to be estimated [85], $\mathbf{p}(k)$ in (16), which will be adjusted iteratively to reduce the estimation error on the output to be modeled, $\mathbf{y}(k)$. $\mathbf{e}(k)$ represents the uncertainty of the output measurements of the system, which are assumed to be a zero mean Gaussian white noise with $\mathbf{R}_e$ covariance. So, Equations (11) to (15) can be iteratively applied to estimate the model's parameters, $\mathbf{p}(k)$.

$$\begin{aligned} \mathbf{p}(k+1) &= \mathbf{p}(k) \\ \mathbf{y}(k) &= \mathbf{h}(\mathbf{x}(k), \mathbf{p}(k)) + \mathbf{e}(k). \end{aligned} \tag{16}$$

Applying (8), (9) and (10) on (16):

$$\mathbf{\Phi}(\mathbf{p}(k)) = \mathbf{I}, \mathbf{\Gamma}(\mathbf{p}(k)) = \mathbf{0}, \text{ and } \mathbf{C}(\mathbf{p}(k)) = \left.\frac{\partial \mathbf{h}}{\partial \mathbf{p}}\right|_{\mathbf{p}=\hat{\mathbf{p}}(k)}. \tag{17}$$

In the above expressions, $\hat{\mathbf{p}}(k)$ is the current estimation of $\mathbf{p}(k)$, which includes all the TSFM parameters, both antecedent (σ_{ji}^l) and consequent (a_{ji}^l). $\mathbf{C}(\mathbf{p}(k))$ is the derivative of the TSFM with respect to its parameters, which was obtained in [60], and its calculation was implemented in the Fuzzy Logic Tools library (FLT) [86].

3. Iterative Modeling of a Lithium-Ion Batteries

A LIB is a system whose parameters change over time due to degradation of its components or internal chemical processes, so a good prediction of its behavior requires the constant adjustment of the model. The conjunction of the Kalman algorithm with TSFM has demonstrated the ability to obtain good models while allowing to improve and adapt these models with each new data obtained from the operation of the system [60,61]. The use of such an algorithm for modeling the consequents of a TSFM based on operating data from a real LIB is presented below. It has been decided not to also adapt the antecedents of the TSFM in order to obtain a more efficient algorithm that can be executed iteratively at run time without requiring particularly fast hardware. Of course, if a more accurate model is required, it is possible to adapt the model antecedents as well [87].

The modeling process consisted of 4 phases:

3.1. Phase 1. Obtaining Discharge and Charge Data of a Real Battery

The first step is to obtain a sufficiently large and representative dataset of the LIB, taking that the data must be obtained in an operating environment as close as possible to that of a real EV into account. For this purpose, using the test bench presented in section 2, the battery was discharged from its maximum capacity until it was completely discharged, and then it was subjected to a charging process. To do that, the "Worldwide Harmonised Light Vehicle Test Procedure" [88,89] (WLTP) was used for the discharge test. This procedure is used as a global standard to determine the CO_2 emissions, fuel consumption and pollutant levels of traditional and hybrid cars, as well as the range of fully EVs. Specifically, based on the performance of the LIB, the WLTP standard was used assuming a low-power class 2 electric vehicle (with a power-to-mass ratio between 22 and 34 W/kg and a upper speed of less than 90 km/h). This driving profile is characterized by a speed and acceleration profile composed of three phases associated with low, medium, and high-speed driving, each with a defined duration, Figure 4. The associated discharge power profile was calculated from the acceleration profile, assuming a direct relationship between power and acceleration, matching the peak acceleration with the maximum allowable discharge power of the LIB. Due to the type of electric vehicle under study, in this test, only the discharge of the LIB has been considered. In Figure 4b, the negative accelerations present in the acceleration profile will not have any effect, as the battery recharging by means of regenerative systems is not considered. To evaluate the performance of the model, this power profile was evaluated over the entire operating range of the battery (0% ≤ SOC ≤ 100%). The described power profile is shown in Figure 5.

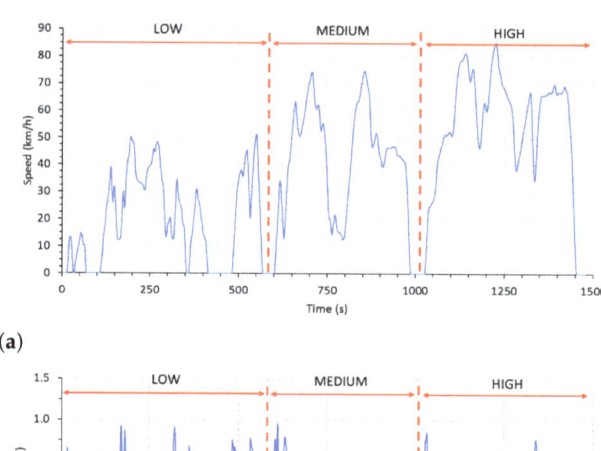

(a)

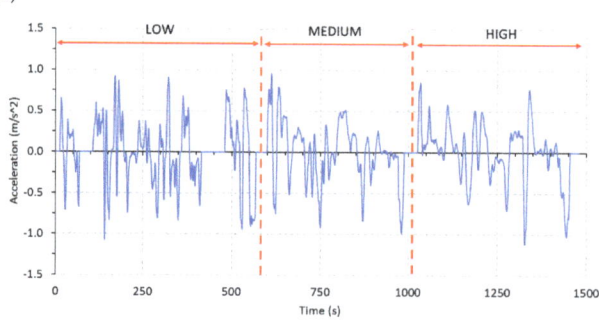

(b)

Figure 4. WLTP class 2 driving profile: (**a**) speed; (**b**) acceleration.

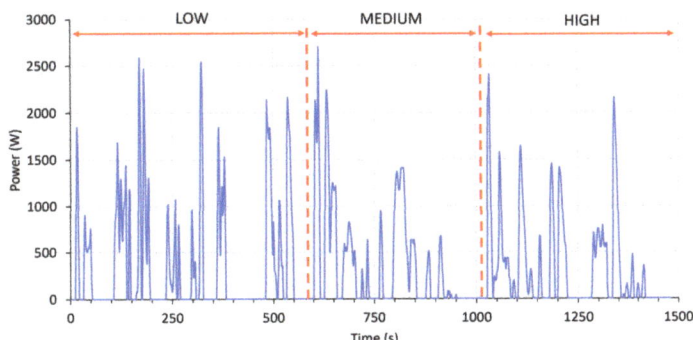

Figure 5. Power profile for WLTP class 2 driving profile.

In accordance with the application and with the aim of evaluating the response of the modeling algorithm throughout the battery's operating range, the LIB charging process was carried out by simulating the use of a commercial type 2 semifast charger. For this purpose, a charging test was performed for a maximum current of 35 A, according to the well-known two-phase charging protocol, constant current (Bulk Phase) and constant voltage (Absorption Phase).

In this first phase, a set of 86508 data sampled every second was obtained from the battery voltage, current, temperature and SOC (The battery SOC was estimated by integration of the current according to the Coulomb counting method [90].). These data are shown in Figure 6.

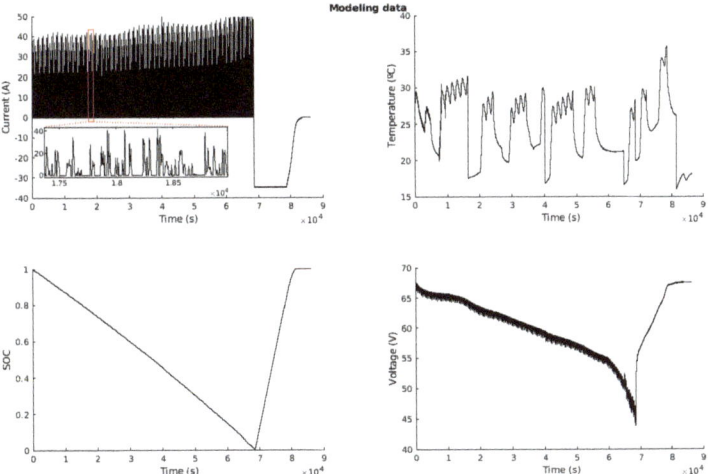

Figure 6. Modeling data.

3.2. Phase 2. Initial TSFM

To obtain an initial structure of the dynamic model for the LIB similar to (7), a series of tests were performed on a reduced dataset consisting of the first 1000 data. Through these tests, it was determined to perform a third-order discrete dynamic model, i.e., including as inputs three memories of the previous outputs, together with the inputs (current, temperature, and SOC), and a memory of the current. That is, the discrete dynamic model of the system will have the form:

$$v(k+1) = f(i(k), i(k-1), T(k), SOC(k), v(k), v(k-1), v(k-2)) \tag{18}$$

where k is the discrete time, $v(k+1)$ is the estimate future voltage, $v(k \ldots k-2)$ are the previous voltages, $i(k)$ and $i(k-1)$ are the current and its previous value, $T(k)$ is the temperature, and $SOC(k)$ is the estimated SOC. f is the TSFM to be fitted.

Once the model structure was determined, the dataset was used to obtain an initial TSFM using the well-known subtractive clustering algorithm of Chiu [91]. Using a Cluster center's range of influence (RADII) of 0.6, six rules with Gaussian antecedents were obtained.

3.3. Phase 3. Iterative Modeling

This is the most relevant phase of the experiment, since the objective of the present work is to demonstrate that it is possible to perform an iterative modeling that allows readjusting an existing model based on the new data obtained.

With the initial model obtained in the previous phase, the EKF was applied to improve the estimation of the parameters with each of the data, i.e., iteratively as it would be improved in a working system. The initial covariance matrix of the EKF, $\mathbf{P}(0)$, was initialized as $2\mathbf{I}$, this is $\alpha = 2$, being $\mathbf{I}$ an identity matrix. The fitting process is shown in Algorithm 1, where α is a value indicating the reliability on the initial parameters of the model.

Algorithm 1 EKF algorithm for the adaptation of consequents.

1: $\tilde{\mathbf{p}}(0|-1) = \mathbf{0}$
2: $\mathbf{P}(0|-1) = \mathbf{I}\alpha$
3: **for** $k = 0..k_{end}$ **do**
4:     Calculate $\mathbf{P}(k|k-1)$ by (12)
5:     Estimate $\tilde{\mathbf{y}}(k)$ using the fuzzy model
6:     Calculate $\mathbf{C}(k)$ by (17)
7:     Get $\mathbf{K}(k)$ by (13)
8:     Update $\tilde{\mathbf{p}}(k|k)$ by (14)
9:     Update $\mathbf{P}(k|k)$ by (15)
10: **end for** k

A Mean Absolute Error (MAE) of 31.40 mV, or the equivalent Root Mean Square Error (RMSE) of 53.00 mV, was obtained during the modeling phase.

The prediction of the model at each iteration can be seen in Figure 7, and the error obtained at each instant can be seen in Figure 8.

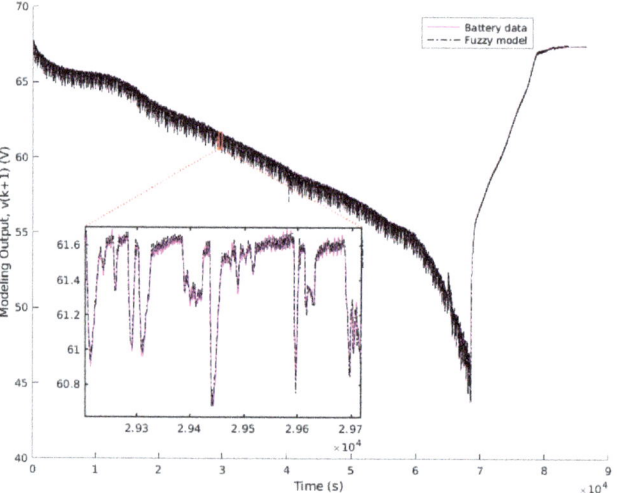

Figure 7. Modeling output.

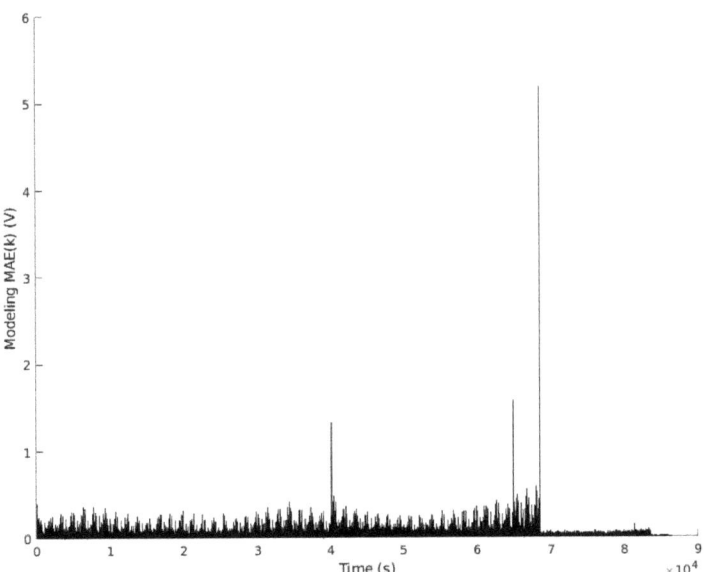

Figure 8. Modeling error.

The model improves as new data appears. When a new behavior not yet modeled occurs, an error is raised, as can be seen in Figure 8. However, the algorithm will learn from this error and will improve the model.

The fuzzy model obtained after running through all the experimental data is reflected by the following rules, where $GAUSS(c, \sigma)$ is the Gauss function (this antecedents can be seen in Figure 9):

IF $i(k)$ is GAUSS(2.317; 25.41) and $i(k-1)$ is GAUSS(3.925; 25.41)
    and $T(k)$ is GAUSS(24.8; 5.94) and $SOC(k)$ is GAUSS(0.526; 0.3)
    and $v(k)$ is GAUSS(60.39; 7.187) and $v(k-1)$ is GAUSS(60.33; 7.187) and $v(k-2)$ is GAUSS(60.3; 7.187)
THEN $v(k+1) = 9.007 - 0.03385\ i(k) + 0.02963\ i(k-1) + 0.0007178\ T(k) + 2.638\ SOC(k) + 0.6987\ v(k) - 0.3983\ v(k-1) + 0.527\ v(k-2)$

IF $i(k)$ is GAUSS(1.501; 25.41) and $i(k-1)$ is GAUSS(3.032; 25.41)
    and $T(k)$ is GAUSS(27.22; 5.94) and $SOC(k)$ is GAUSS(0.878; 0.3)
    and $v(k)$ is GAUSS(65.51; 7.187) and $v(k-1)$ is GAUSS(65.47; 7.187) and $v(k-2)$ is GAUSS(65.38; 7.187)
THEN $v(k+1) = 0.08457 - 0.03393\ i(k) + 0.03365\ i(k-1) + 0.00247\ T(k) + 0.08013\ SOC(k) + 0.8829\ v(k) - 0.4889\ v(k-1) + 0.6025\ v(k-2)$

IF $i(k)$ is GAUSS(3.032; 25.41) and $i(k-1)$ is GAUSS(4.027; 25.41)
    and $T(k)$ is GAUSS(22.84; 5.94) and $SOC(k)$ is GAUSS(0.197; 0.3)
    and $v(k)$ is GAUSS(55.41; 7.187) and $v(k-1)$ is GAUSS(55.42; 7.187) and $v(k-2)$ is GAUSS(55.33; 7.187)
THEN $v(k+1) = 0.1629 - 0.03028\ i(k) + 0.03077\ i(k-1) + 0.00001303\ T(k) + 0.1531\ SOC(k) + 0.8291\ v(k) - 0.03837\ v(k-1) + 0.2058\ v(k-2)$

IF $i(k)$ is GAUSS(−0.158; 25.41) and $i(k-1)$ is GAUSS(−0.158; 25.41)
    and $T(k)$ is GAUSS(17.56; 5.94) and $SOC(k)$ is GAUSS(1; 0.3)
    and $v(k)$ is GAUSS(67.44; 7.187) and $v(k-1)$ is GAUSS(67.44; 7.187) and $v(k-2)$ is GAUSS(67.44; 7.187)
THEN $v(k+1) = 1.453 - 0.04581\ i(k) + 0.04429\ i(k-1) - 0.0007327\ T(k) + 0.3583\ SOC(k) + 0.6178\ v(k) - 0.3521\ v(k-1) + 0.7076\ v(k-2)$

IF $i(k)$ is GAUSS(-34.87; 25.41) and $i(k-1)$ is GAUSS(-34.87; 25.41)
and $T(k)$ is GAUSS(25.5; 5.94) and $SOC(k)$ is GAUSS(0.318; 0.3)
and $v(k)$ is GAUSS(59.38; 7.187) and $v(k-1)$ is GAUSS(59.43; 7.187) and $v(k-2)$ is GAUSS(59.43; 7.187)
THEN $v(k+1) = 1.598 + 0.1012\ i(k) - 0.08297\ i(k-1) + 0.0002574\ T(k) + 0.2328\ SOC(k) + 1.248\ v(k) - 1.106\ v(k-1) + 0.8397\ v(k-2)$

IF $i(k)$ is GAUSS(-34.77; 25.41) and $i(k-1)$ is GAUSS(-34.9; 25.41)
and $T(k)$ is GAUSS(32.82; 5.94) and $SOC(k)$ is GAUSS(0.774; 0.3)
and $v(k)$ is GAUSS(65.61; 7.187) and $v(k-1)$ is GAUSS(65.61; 7.187) and $v(k-2)$ is GAUSS(65.64; 7.187)
THEN $v(k+1) = 0.2383 + 0.08311\ i(k) - 0.08389\ i(k-1) - 0.0001479\ T(k) + 0.06106\ SOC(k) + 1.446\ v(k) - 1.274\ v(k-1) + 0.8237\ v(k-2)$

This model allows us to interpret the relationships between the variables linguistically within each rule. In this case, the initial model was not considered to be easily interpretable as a whole, so the antecedents appear overlapped, but there are algorithms that allow this to be done [92–94].

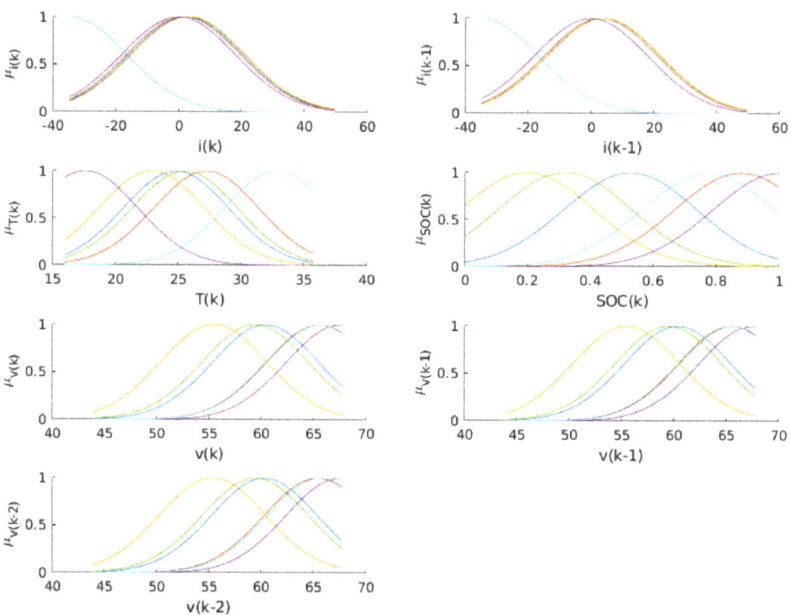

Figure 9. TSFM antecedents.

3.4. Phase 4. Validation Data

Using the same philosophy as previous modeling data, a new charge/discharge cycle was performed to obtain another dataset to validate the resulting TSFM. On this occasion, 14746 data were obtained with the same sample time of 1 s, which are shown in Figure 10. The final TSFM obtained in the previous phase was validated to check the performance of the iterative algorithm.

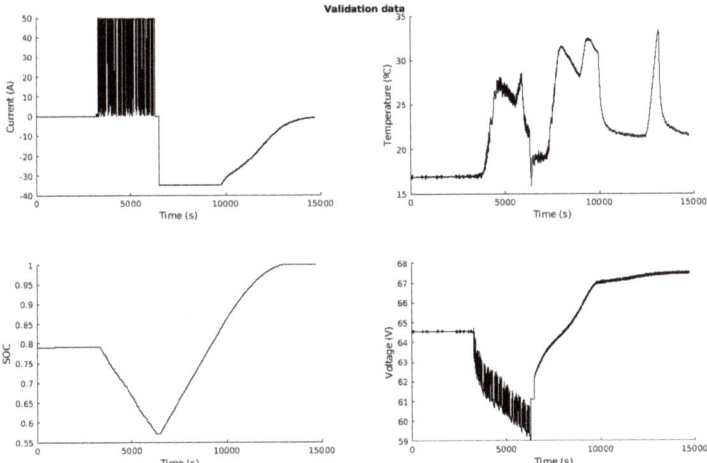

Figure 10. Validation data.

4. Analysis of the Results

Finally, once all the data were run through, it was checked that the model obtained adequately represented the dynamic behavior of the battery under study. Note that the proposed algorithm does not perform the modeling with all the data, but iteratively adjusts the model with each new data obtained. Therefore, the main objective is not focused on obtaining a lower error rate than those obtained by other techniques presented in the scientific literature, but to demonstrate that the algorithm is able to perform the iterative adjustment, with good performance considering the model error, against diverse and highly fluctuating power profiles, regardless of the battery temperature, SOC or operating condition (charge or discharge). Obviously, the performance of the modeling process will be highly dependent on the initial model. In this case, for the test performed, the initial model was obtained from the clustering, without any additional tuning. Despite this, the model validation obtained a MAE of 69.90 mV, and a RMSE of 21.20 mV. The model output during the validation process is shown in Figure 11, and the error in Figure 12.

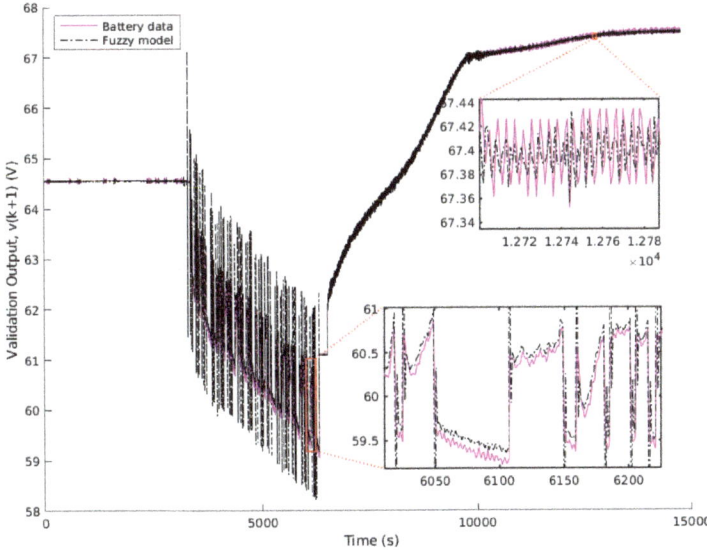

Figure 11. Validation output.

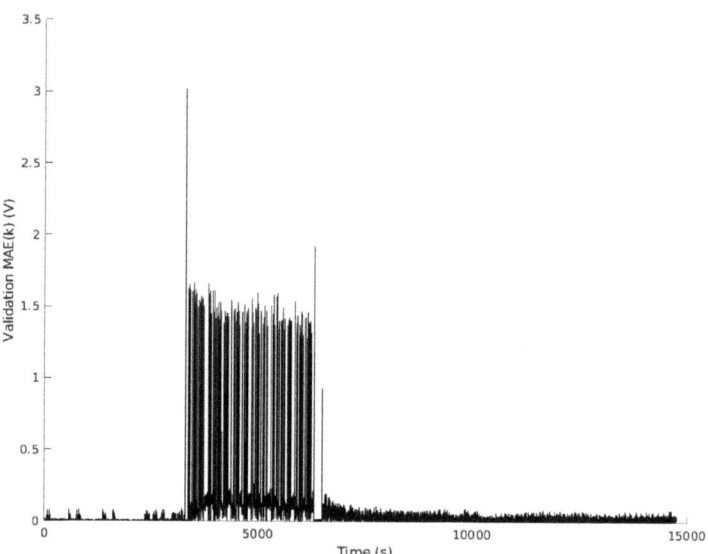

Figure 12. Validation error.

From the results obtained, it is verified that the EKF works properly, and that it allows modeling iteratively a real LIB with a reduced error. This will allow continuous improvement and adaptation of the TSFM to the variations that the LIB may undergo.

5. Conclusions and Future Works

The modeling process of a LIB is a complex task that requires a deep knowledge of the laws of electrochemistry but is essential for its efficient use and proper management. For EVs application, the high fluctuation of the power profile means that the dynamic behavior of the LIB is strongly influenced by intrinsic nonlinearities. The main proposals found in the scientific literature for these applications are based on equivalent electrical models that present reduced performance in terms of dynamic behavior and in many cases, for the sake of simplicity, ignore the dependence on certain parameters such as temperature.

To address to the shortcomings found in terms of modeling, this research has applied an iterative fuzzy modeling methodology based on the EKF to a real LIB for use in EVs. The proposed method is not a closed algorithm and can be easily applied to other types of batteries or applications. The model parameters were fitted iteratively from an initial unfitted model, using a 59.2 VDC and 120 Ah commercial LIB on a profile under WLTP standard conditions for a class 2 driving cycle. Experimental results for both charging and discharging, for the entire useful operating range of the LIB and different operating SOC and temperatures, validate the effectiveness of the model in a WLTP test resulting a MAE below 70 mV (about 21 mV RMSE) and taking into account that the starting model was unadjusted.

Based on the results, it is found that the designed algorithm and the obtained dynamic model are sufficiently accurate to iteratively model nonlinear LIBs even under such demanding operating profiles. Similarly, thanks to the iterative run-time fuzzy modeling methodology proposed in this work, the model can adapt to changes in the LIB's performance or rated capacity due to usage time. This will contribute to the design and development of model-based energy management strategies and controllers for the proper operation of LIBs in EVs applications throughout their lifetime.

Further experiments and verifications will be carried out in the future to test the effectiveness of the proposed method for different driving cycle profiles and operating conditions, as well as its application for the design of iterative model-based SOC estimators. Finally, as a natural outcome of the research process, these models will be applied in the design of new model-based energy management and control strategies for use in hybrid

EVs powered by LIBs hybridized with traditional combustion engines and/or hydrogen fuel cells.

Author Contributions: Conceptualization, F.J.V., J.M.E. and F.S.; Methodology, A.J.B., F.J.V., J.M.E. and F.S.; Software, A.J.B.; Validation, A.J.B., F.J.V., J.M.E.; Formal analysis, A.J.B., F.J.V., J.M.E. and F.S.; Investigation, A.J.B., F.J.V., J.M.E. and F.S.; Data curation, A.J.B., F.J.V., J.M.E.; Writing—original draft, A.J.B., F.J.V., J.M.E.; Writing—review & editing, A.J.B., F.J.V., J.M.E. and F.S.; Visualization, F.J.V. and J.M.E.; Supervision, J.M.A.; Project administration, J.M.A.; Funding acquisition, J.M.A. All authors have read and agreed to the published version of the manuscript.

Funding: This research was funded by "H2Integration&Control. Integration and Control of a hydrogen-based pilot plant in residential applications for energy supply" from the Spanish Government (PID2020-116616RB-C31), "SALTES: Smartgrid with reconfigurable Architecture for testing control Techniques and Energy Storage priority" by Andalusian Regional Program of R+D+i (P20-00730), and by the project "The green hydrogen vector. Residential and mobility application", approved in the call for research projects of the Cepsa Foundation Chair of the University of Huelva.

Data Availability Statement: The data is available upon request to the authors.

Conflicts of Interest: The authors declare no conflict of interest.

Abbreviations

The following abbreviations are used in this manuscript:

Acronym or Parameter	Description
A_{ji}^{l}	Fuzzy antecedent for rule l, derivative i and input j.
$\mathbf{a}_{ji}^{l}$	Consequents parameters of the fuzzy model
DAQ	data acquisition
$\mathbf{C}(\mathbf{p}(k))$	Derivative of the fuzzy model (output matrix of the EKF model)
EKF	Extended Kalman filter
EVs	Electric vehicles
LIB	Lithium-ion battery
MAE	Mean Absolute Error
n	System order (length of the state vector)
P	Covariance matrix of the Kalman filter
$\mathbf{p}(k)$	Vector of parameters to be estimated
$\hat{\mathbf{p}}(k)$	Vector of estimated parameters
RADII	Cluster center's range of influence parameter
RMSE	Root Mean Square Error
SCADA	Supervisory control and data acquisition system
SOC	Battery State of Charge
TSFM	Takagi–Sugeno fuzzy model
WLTP	Worldwide Harmonised Light Vehicle Test Procedure standard
x	State vector
x̃	Extended state vector
$\mathbf{y}(k)$	Output of the actual system
$\hat{\mathbf{y}}(k)$	Estimated output

References

1. Philippot, M.; Alvarez, G.; Ayerbe, E.; Mierlo, J.V.; Messagie, M. Eco-efficiency of a lithium-ion battery for electric vehicles: Influence of manufacturing country and commodity prices on ghg emissions and costs. *Batteries* **2019**, *5*, 23. [CrossRef]
2. Lowe, I. How Planning Can Address the Challenge of Transitioning to Low-Carbon Urban Economies. In *Proceedings of the The Routledge Handbook of Australian Urban and Regional Planning*; Routledge: New York, NY, USA, 2017.
3. Sajadi-Alamdari, S.A.; Voos, H.; Darouach, M. Nonlinear model predictive extended eco-cruise control for battery electric vehicles. In Proceedings of the 2016 24th Mediterranean Conference on Control and Automation (MED), Athens, Greece, 21–24 June 2016; pp. 467–472. [CrossRef]
4. Tehrani, K. A smart cyber physical multi-source energy system for an electric vehicle prototype. *J. Syst. Archit.* **2020**, *111*, 101804. [CrossRef]

5. Mao, N.; Wang, Z.R.; Chung, Y.H.; Shu, C.M. Overcharge cycling effect on the thermal behavior, structure, and material of lithium-ion batteries. *Appl. Therm. Eng.* **2019**, *163*, 114147. [CrossRef]
6. Ren, D.; Hsu, H.; Li, R.; Feng, X.; Guo, D.; Han, X.; Lu, L.; He, X.; Gao, S.; Hou, J.; et al. A comparative investigation of aging effects on thermal runaway behavior of lithium-ion batteries. *eTransportation* **2019**, *2*, 100034. [CrossRef]
7. Tran, M.K.; Dacosta, A.; Mevawalla, A.; Panchal, S.; Fowler, M. Comparative study of equivalent circuit models performance in four common lithium-ion batteries: LFP, NMC, LMO, NCA. *Batteries* **2021**, *7*, 51. [CrossRef]
8. Cittanti, D.; Ferraris, A.; Airale, A.; Fiorot, S.; Scavuzzo, S.; Carello, M. Modeling Li-ion batteries for automotive application: A trade-off between accuracy and complexity. In Proceedings of the 2017 International Conference of Electrical and Electronic Technologies for Automotive, Antalya, Turkey, 21–23 August 2017; pp. 1–8. [CrossRef]
9. Jokar, A.; Rajabloo, B.; Désilets, M.; Lacroix, M. Review of simplified Pseudo-two-Dimensional models of lithium-ion batteries. *J. Power Sources* **2016**, *327*, 44–55. [CrossRef]
10. Hu, Y.; Yurkovich, S.; Guezennec, Y.; Yurkovich, B. A technique for dynamic battery model identification in automotive applications using linear parameter varying structures. *Control Eng. Pract.* **2009**, *17*, 1190–1201. [CrossRef]
11. Liu, K.; Gao, Y.; Zhu, C.; Li, K.; Fei, M.; Peng, C.; Zhang, X.; Han, Q.L. Electrochemical modeling and parameterization towards control-oriented management of lithium-ion batteries. *Control Eng. Pract.* **2022**, *124*, 105176. [CrossRef]
12. Madani, S.S.; Schaltz, E.; Kær, S.K. Review of parameter determination for thermal modeling of lithium ion batteries. *Batteries* **2018**, *4*, 20. [CrossRef]
13. Mathew, M.; Mastali, M.; Catton, J.; Samadani, E.; Janhunen, S.; Fowler, M. Development of an electro-thermal model for electric vehicles using a design of experiments approach. *Batteries* **2018**, *4*, 29. [CrossRef]
14. Smith, K.A.; Rahn, C.D.; Wang, C.Y. Model-Based Electrochemical Estimation and Constraint Management for Pulse Operation of Lithium Ion Batteries. *IEEE Trans. Control. Syst. Technol.* **2010**, *18*, 654–663. [CrossRef]
15. Sikha, G.; White, R.E.; Popov, B.N. A Mathematical Model for a Lithium-Ion Battery/Electrochemical Capacitor Hybrid System. *J. Electrochem. Soc.* **2005**, *152*, A1682. [CrossRef]
16. Torchio, M.; Magni, L.; Gopaluni, R.B.; Braatz, R.D.; Raimondo, D.M. LIONSIMBA: A Matlab Framework Based on a Finite Volume Model Suitable for Li-Ion Battery Design, Simulation, and Control. *J. Electrochem. Soc.* **2016**, *163*, A1192–A1205. [CrossRef]
17. Dees, D.W.; Battaglia, V.S.; Bélanger, A. Electrochemical modeling of lithium polymer batteries. *J. Power Sources* **2002**, *110*, 310–320. [CrossRef]
18. Shafiei, A.; Momeni, A.; Williamson, S.S. Battery modeling approaches and management techniques for Plug-in Hybrid Electric Vehicles. In Proceedings of the 2011 IEEE Vehicle Power and Propulsion Conference, Chicago, IL, USA, 6–9 September 2011; pp. 1–5. [CrossRef]
19. Kim, T.; Qiao, W. A Hybrid Battery Model Capable of Capturing Dynamic Circuit Characteristics and Nonlinear Capacity Effects. *IEEE Trans. Energy Convers.* **2011**, *26*, 1172–1180. [CrossRef]
20. Chu, Z.; Jobman, R.; Rodríguez, A.; Plett, G.L.; Trimboli, M.S.; Feng, X.; Ouyang, M. A control-oriented electrochemical model for lithium-ion battery. Part II: Parameter identification based on reference electrode. *J. Energy Storage* **2020**, *27*, 101101. [CrossRef]
21. Song, L.; Evans, J.W. Electrochemical-Thermal Model of Lithium Polymer Batteries. *J. Electrochem. Soc.* **2000**, *147*, 2086. [CrossRef]
22. Hu, Y.; Yin, Y.; Bi, Y.; Choe, S.Y. A control oriented reduced order electrochemical model considering variable diffusivity of lithium ions in solid. *J. Power Sources* **2020**, *468*, 228322. [CrossRef]
23. Tran, N.T.; Vilathgamuwa, M.; Farrell, T.; Choi, S.S. Matlab simulation of lithium ion cell using electrochemical single particle model. In Proceedings of the 2016 IEEE 2nd Annual Southern Power Electronics Conference (SPEC), Auckland, New Zealand, 5–8 December 2016. [CrossRef]
24. Lai, W.; Ciucci, F. Mathematical modeling of porous battery electrodes—Revisit of Newman's model. *Electrochim. Acta* **2011**, *56*, 4369–4377. [CrossRef]
25. Sockeel, N.; Shahverdi, M.; Mazzola, M.; Meadows, W. High-fidelity battery model for model predictive control implemented into a plug-in hybrid electric vehicle. *Batteries* **2017**, *3*, 13. [CrossRef]
26. Salameh, Z.; Casacca, M.; Lynch, W. A mathematical model for lead-acid batteries. *IEEE Trans. Energy Convers.* **1992**, *7*, 93–98. [CrossRef]
27. Barletta, G.; DiPrima, P.; Papurello, D. Thevenin's Battery Model Parameter Estimation Based on Simulink. *Energies* **2022**, *15*, 6207. [CrossRef]
28. Wang, C.; Xu, M.; Zhang, Q.; Feng, J.; Jiang, R.; Wei, Y.; Liu, Y. Parameters identification of Thevenin model for lithium-ion batteries using self-adaptive Particle Swarm Optimization Differential Evolution algorithm to estimate state of charge. *J. Energy Storage* **2021**, *44*, 103244. [CrossRef]
29. Baczyńska, A.; Niewiadomski, W.; Gonçalves, A.; Almeida, P.; Luís, R. LI-NMC batteries model evaluation with experimental data for electric vehicle application. *Batteries* **2018**, *4*, 11. [CrossRef]
30. Zhou, W.; Zheng, Y.; Pan, Z.; Lu, Q. Review on the Battery Model and SOC Estimation Method. *Processes* **2021**, *9*, 1685. [CrossRef]
31. Ding, X.; Zhang, D.; Cheng, J.; Wang, B.; Luk, P.C.K. An improved Thevenin model of lithium-ion battery with high accuracy for electric vehicles. *Appl. Energy* **2019**, *254*, 113615. [CrossRef]
32. Jiang, J.; Zhang, C. *Fundamentals and Applications of Lithium-Ion Batteries in Electric Drive Vehicles*; Wiley: Hoboken, NJ, USA, 2015. [CrossRef]

33. He, H.; Xiong, R.; Fan, J. Evaluation of Lithium-Ion Battery Equivalent Circuit Models for State of Charge Estimation by an Experimental Approach. *Energies* **2011**, *4*, 582–598. [CrossRef]
34. Bhangu, B.; Bentley, P.; Stone, D.; Bingham, C. Nonlinear Observers for Predicting State-of-Charge and State-of-Health of Lead-Acid Batteries for Hybrid-Electric Vehicles. *IEEE Trans. Veh. Technol.* **2005**, *54*, 783–794. [CrossRef]
35. Dubarry, M.; Liaw, B.Y. Development of a universal modeling tool for rechargeable lithium batteries. *J. Power Sources* **2007**, *174*, 856–860. [CrossRef]
36. Danzer, M.; Liebau, V.; Maglia, F. Aging of lithium-ion batteries for electric vehicles. In *Advances in Battery Technologies for Electric Vehicles*; Elsevier: Amsterdam, The Netherlands, 2015; pp. 359–387. [CrossRef]
37. Liaw, B.Y.; Jungst, R.G.; Nagasubramanian, G.; Case, H.L.; Doughty, D.H. Modeling capacity fade in lithium-ion cells. *J. Power Sources* **2005**, *140*, 157–161. [CrossRef]
38. Chen, M.; Rincon-Mora, G. Accurate electrical battery model capable of predicting runtime and I-V performance. *IEEE Trans. Energy Convers.* **2006**, *21*, 504–511. [CrossRef]
39. Li, C.; Cui, N.; Cui, Z.; Wang, C.; Zhang, C. Novel equivalent circuit model for high-energy lithium-ion batteries considering the effect of nonlinear solid-phase diffusion. *J. Power Sources* **2022**, *523*, 230993. [CrossRef]
40. Tremblay, O.; Dessaint, L.A. Experimental Validation of a Battery Dynamic Model for EV Applications. *World Electr. Veh. J.* **2009**, *3*, 289–298. [CrossRef]
41. Peukert, W. Über die Abhängigkeit der Kapazität von der Entladestromstärke bei Bleiakkumulatoren. *Elektrotechnisch Z.* **1897**, *20*, 287–288. (In German)
42. Shepherd, C.M. Design of Primary and Secondary Cells. *J. Electrochem. Soc.* **1965**, *112*, 657. [CrossRef]
43. Degla, A.; Chikh, M.; Mahrane, A.; Arab, A.H. Improved lithium-ion battery model for photovoltaic applications based on comparative analysis and experimental tests. *Int. J. Energy Res.* **2022**, *46*, 10965–10988. [CrossRef]
44. Hussein, A.A.H.; Batarseh, I. An overview of generic battery models. In Proceedings of the 2011 IEEE Power and Energy Society General Meeting, Detroit, MI, USA, 24–29 July 2011. [CrossRef]
45. He, H.; Xiong, R.; Guo, H.; Li, S. Comparison study on the battery models used for the energy management of batteries in electric vehicles. *Energy Convers. Manag.* **2012**, *64*, 113–121. [CrossRef]
46. Denaï, M.A.; Palis, F.; Zeghbib, A.H. Modeling and control of non-linear systems using soft computing techniques. *Appl. Soft Comput.* **2007**, *7*, 728–738. [CrossRef]
47. Jang, J.S.R. ANFIS: Adaptive–network-based fuzzy inference system. *IEEE Trans. Syst. Man Cybern.* **1993**, *23*, 665–685. [CrossRef]
48. Zhang, C.W.; Chen, S.R.; Gao, H.B.; Xu, K.J.; Yang, M.Y. State of charge estimation of power battery using improved back propagation neural network. *Batteries* **2018**, *4*, 69. [CrossRef]
49. Jang, J.S.R.; Sun, C.T. Neuro-fuzzy modeling and control. *Proc. IEEE* **1995**, *83*, 378–406. [CrossRef]
50. Xiong, R.; He, H.; Guo, H.; Ding, Y. Modeling for Lithium-Ion Battery used in Electric Vehicles. *Procedia Eng.* **2011**, *15*, 2869–2874. [CrossRef]
51. Casteleiro-Roca, J.L.; Barragán, A.J.; Segura, F.; Calvo-Rolle, J.L.; Andújar, J.M. Intelligent hybrid system for the prediction of the voltage-current characteristic curve of a hydrogen-based fuel cell. *Rev. Iberoam. Autom. ÁTica Inform. ÁTica Ind.* **2019**, *16*, 492–501. [CrossRef]
52. Casteleiro-Roca, J.; Barragán, A.J.; Segura, F.; Calvo-Rolle, J.L.; Andújar, J.M. Fuel Cell Output Current Prediction with a Hybrid Intelligent System. *Complexity* **2019**, *2019*, 10. [CrossRef]
53. Miao, J.; Tong, Z.; Tong, S.; Zhang, J.; Mao, J. State of Charge Estimation of Lithium-Ion Battery for Electric Vehicles under Extreme Operating Temperatures Based on an Adaptive Temporal Convolutional Network. *Batteries* **2022**, *8*, 145. [CrossRef]
54. He, W.; Williard, N.; Chen, C.; Pecht, M. State of charge estimation for Li-ion batteries using neural network modeling and unscented Kalman filter-based error cancellation. *Int. J. Electr. Power Energy Syst.* **2014**, *62*, 783–791. [CrossRef]
55. How, D.N.T.; Hannan, M.A.; Hossain Lipu, M.S.; Ker, P.J. State of Charge Estimation for Lithium-Ion Batteries Using Model-Based and Data-Driven Methods: A Review. *IEEE Access* **2019**, *7*, 136116–136136. [CrossRef]
56. Singh, P.; Vinjamuri, R.; Wang, X.; Reisner, D. Design and implementation of a fuzzy logic-based state-of-charge meter for Li-ion batteries used in portable defibrillators. *J. Power Sources* **2006**, *162*, 829–836. [CrossRef]
57. Tong, S.; Lacap, J.H.; Park, J.W. Battery state of charge estimation using a load-classifying neural network. *J. Energy Storage* **2016**, *7*, 236–243. [CrossRef]
58. Kalman, R.E. A new approach to linear filtering and prediction problems. *Trans. -Asme-J. Basic Eng.* **1960**, *82*, 35–45. [CrossRef]
59. Takagi, T.; Sugeno, M. Fuzzy identification of systems and its applications to modeling and control. *IEEE Trans. Syst. Man Cybern.* **1985**, *15*, 116–132. [CrossRef]
60. Barragán, A.J.; Al-Hadithi, B.M.; Jiménez, A.; Andújar, J.M. A general methodology for online TS fuzzy modeling by the extended Kalman filter. *Appl. Soft Comput.* **2014**, *18*, 277–289. [CrossRef]
61. Barragán, A.; Enrique, J.; Segura, F.; Andújar, J.M. Iterative Fuzzy Modeling Of Hydrogen Fuel Cells By The Extended Kalman Filter. *IEEE Access* **2020**, *8*, 180280–180294. [CrossRef]
62. Babuška, R. Fuzzy modeling-A control engineering perspective. In Proceedings of the 1995 IEEE International Conference on Fuzzy Systems, Yokohama, Japan, 20–24 March 1995; Volume 4, pp. 1897–1902. [CrossRef]

63. Barragán, A.J.; Andújar, J.M.; Aznar, M.; Jiménez, A. Methodology for adapting the parameters of a fuzzy system using the extended Kalman filter. In *European Society for Fuzzy Logic and Technology (EUSFLAT-2011) and LFA-2011*; Galichet, S., Montero, J., Mauris, G., Eds.; Number 1 in Advances in Intelligent Systems Research: Dordrecht, The Netherlands, 2011; pp. 686–690. [CrossRef]
64. Zeng, K.; Zhang, N.Y.; Xu, W.L. A comparative study on sufficient conditions for Takagi–Sugeno fuzzy systems as universal approximators. *IEEE Trans. Fuzzy Syst.* **2000**, *8*, 773–780. [CrossRef]
65. Zadeh, L.A. Fuzzy sets. *Inf. Control* **1965**, *8*, 338–353. [CrossRef]
66. Sanz, R.; Matia, F.; de Antonio, A.; Segarra, M.J. Fuzzy agents for ICa. In Proceedings of the 1998 IEEE International Conference on Fuzzy Systems, IEEE World Congress on Computational Intelligence, Kota Kinabalu, Malaysia, 3–5 December 2014; Anon, Ed.; IEEE: Piscataway, NJ, USA; Anchorage, AK, USA, 1998; Volume 1, pp. 545–550. [CrossRef]
67. Andújar, J.M.; Barragán, A.J. A methodology to design stable nonlinear fuzzy control systems. *Fuzzy Sets Syst.* **2005**, *154*, 157–181. [CrossRef]
68. Al-Hadithi, B.M.; Jiménez, A.; Matía, F. A new approach to fuzzy estimation of Takagi–Sugeno model and its applications to optimal control for nonlinear systems. *Appl. Soft Comput.* **2012**, *12*, 280–290. [CrossRef]
69. Doubabi, H.; Ezzara, A.; Salhi, I. Simulation and Experimental Validation for Takagi-Sugeno Fuzzy-Based Li-ion Battery Model. *Int. J. Renew. Energy Res.* **2022**, *12*, 339–348. [CrossRef]
70. Kóczy, L.T.; Hirota, K. Size reduction by interpolation in fuzzy rule bases. *IEEE Trans. Syst. Man -Cybern. -Part B Cybern.* **1997**, *27*, 14–25. [CrossRef] [PubMed]
71. Andújar, J.M.; Barragán, A.J. Hybridization of fuzzy systems for modeling and control. *Rev. Iberoam. AutomáTica InformáTica Ind. {RIAI}* **2014**, *11*, 127–141. [CrossRef]
72. Barragán, A.J.; Al-Hadithi, B.M.; Andújar, J.M.; Jiménez, A. Formal methodology for analyzing the dynamic behavior of nonlinear systems using fuzzy logic. *Rev. Iberoam. AutomÁTica InformÁTica Ind. (Riai)* **2015**, *12*, 434–445. [CrossRef]
73. Barragán, A.J.; Enrique, J.M.; Calderón, A.J.; Andújar, J.M. Discovering the dynamic behavior of unknown systems using fuzzy logic. *Fuzzy Optim. Decis. Mak.* **2018**, 1–25. [CrossRef]
74. Márquez, J.M.A.; Piña, A.J.B.; Arias, M.E.G. A general and formal methodology for designing stable nonlinear fuzzy control systems. *IEEE Trans. Fuzzy Syst.* **2009**, *17*, 1081–1091. [CrossRef]
75. Al-Hadithi, B.M.; Barragán, A.J.; Andújar, J.M.; Jiménez, A. Fuzzy Optimal Control for Double Inverted Pendulum. In Proceedings of the 7th IEEE Conference on Industrial Electronics and Applications (ICIEA 2012), Singapore, 18–20 July 2012.
76. Wang, L.X. *A Course in Fuzzy Systems and Control*; Prentice Hall: Hoboken, NJ, USA, 1997.
77. Wong, L.; Leung, F.; Tam, P. Stability design of TS model based fuzzy systems. In Proceedings of the IEEE International Conference on Fuzzy Systems, Barcelona, Spain, 5 July 1997; Volume 1, pp. 83–86. [CrossRef]
78. Kalman, R.E. New Methods in Wiener Filtering Theory. In *Proceedings of the 1st Symposium On Engineering Applications of Random Function Theory and Probability*; Bogdanoff, J.L., Kozin, F., Eds.; John Wiley and Sons: New York, NY, USA, 1963.
79. Maybeck, P.S. *Stochastic Models, Estimation, and Control*; Mathematics in Science and Engineering; Academyc Press: New York, NY, USA, 1979; Volume 141.
80. Al-Hadithi, B.M.; Jiménez, A.; Matía, F.; Andújar, J.M.; Barragán, A.J. New Concepts for the Estimation of Takagi–Sugeno Model Based on Extended Kalman Filter. In *Fuzzy Modeling and Control: Theory and Applications*; Matía, F., Marichal, G.N., Jiménez, E., Eds.; Atlantis Computational Intelligence Systems; Atlantis Press: Dordrecht, The Netherlands, 2014; Chapter 1; Volume 9, pp. 3–24. [CrossRef]
81. Al-Hadithi, B.M.; Jiménez Avello, A.; Matía, F. New Methods for the Estimation of Takagi–Sugeno Model Based Extended Kalman Filter and its Applications to Optimal Control for Nonlinear Systems. *Optim. Control. Appl. Methods* **2012**, *33*, 552–575. [CrossRef]
82. Chafaa, K.; Ghanaï, M.; Benmahammed, K. Fuzzy modelling using Kalman filter. *IET Control Theory Appl.* **2007**, *1*, 58–64. [CrossRef]
83. Simon, D. Kalman filtering with state constraints: A survey of linear and nonlinear algorithms. *IET Control Theory Appl.* **2010**, *4*, 1303–1318. [CrossRef]
84. Ketabipour, S.; Samet, H.; Vafamand, N. TS Fuzzy Prediction-based SVC Compensation of Wind Farms Flicker: A Dual-UKF Approach. *CSEE J. Power Energy Syst.* **2022**, *8*, 1594–1602. [CrossRef]
85. Simon, D. Training fuzzy systems with the extended Kalman filter. *Fuzzy Sets Syst.* **2002**, *132*, 189–199. [CrossRef]
86. Barragán, A.J.; Andújar, J.M. *Fuzzy Logic Tools Reference Manual v1.0*; University of Huelva: Huelva, Spain, 2012.
87. Matía, F.; Marichal, G.N.; Jiménez, E. (Eds.) *Fuzzy Modeling and Control: Theory and Applications*; Atlantis Computational Intelligence Systems; Atlantis Press: Dordrecht, The Netherlands, 2014; Volume 9. [CrossRef]
88. Economic and Social Council, United Nations Economic Commission for Europe. *Proposal for a New Global Technical Regulation on the Worldwide Harmonized Light Vehicles Test Procedure (WLTP)*; World Forum for Harmonization of Vehicle Regulations: Geneva, Switzerland, 2013.
89. (UNECE), U.N.E.C.f.E. Parameter List for RLD-Validation (WLTP-DTP-10-08, WLTP-DTP). Available online: http://www.unece.org/fileadmin/DAM/trans/doc/2012/wp29grpe/WLTP-DHC-12-07e.xls (accessed on 27 December 2022).
90. Vivas, F.; Segura, F.; Andújar, J.; Caparrós, J. A suitable state-space model for renewable source-based microgrids with hydrogen as backup for the design of energy management systems. *Energy Convers. Manag.* **2020**, *219*, 113053. [CrossRef]

91. Chiu, S. Fuzzy model identification based on cluster estimation. *J. Intell. Fuzzy Syst.* **1994**, *2*, 267–278. [CrossRef]
92. Johansen, T.A.; Shorten, R.N.; Murray-Smith, R. On the interpretation and identification of dynamic Takagi–Sugeno fuzzy models. *IEEE Trans. Fuzzy Syst.* **2000**, *8*, 297–313. [CrossRef] [PubMed]
93. Vélez, M.A.; Sánchez, O.; Romero, S.; Manuel, A.J. A new methodology to improve interpretability in neuro-fuzzy TSK models. *Appl. Soft Comput.* **2010**, *10*, 578–591. [CrossRef]
94. Díez, J.L.; Navarro, J.L.; Sala, A. A fuzzy clustering algorithm enhancing local model interpretability. *Soft Comput.* **2007**, *11*, 973–983. [CrossRef]

Disclaimer/Publisher's Note: The statements, opinions and data contained in all publications are solely those of the individual author(s) and contributor(s) and not of MDPI and/or the editor(s). MDPI and/or the editor(s) disclaim responsibility for any injury to people or property resulting from any ideas, methods, instructions or products referred to in the content.

Review

Critical Analysis of Simulation of Misalignment in Wireless Charging of Electric Vehicles Batteries

Saeid Ghazizadeh *, Kafeel Ahmed, Mehdi Seyedmahmoudian, Saad Mekhilef, Jaideep Chandran and Alex Stojcevski *

School of Science, Computing and Engineering Technologies, Swinburne University of Technology, Melbourne, VIC 3122, Australia
* Correspondence: sa.gh74@yahoo.com (S.G.); astojcevski@swin.edu.au (A.S.)

Abstract: The transition from conventional to electric transportation has become inevitable in recent years owing to the significant impact of electric vehicles (EVs) on energy sustainability, reduction of global warming and carbon emission reduction. Despite the rapidly growing global adoption of EVs in today's electrical and transportation networks, energy storage in EVs, particularly in regards to bulky size and charging process, still remains a major bottleneck. As a result, wireless charging of EVs via inductively coupled power transfer (ICPT) through coupled coils is becoming a promising solution. However, the efficiency of charging EV batteries via wireless charging is hugely affected by misalignment between the primary and secondary coils. This paper presents an in-depth analysis of various key factors affecting the efficiency of EV battery charging. Finite element analysis (FEA) using Ansys Maxwell® is performed on commonly used coil designs such as circular and rectangular coils under various misalignment conditions. In addition, various reactive power compensation topologies applied in ICPT are investigated and the behavior of each topology is observed in simulation. It is revealed that circular structures with S–S compensation topology show more robustness in misalignment conditions and maintain the desired efficiency for a wider range of displacement. A critical analysis of coil designs, compensation techniques and the combination of both factors is accomplished and conclusions are presented.

Keywords: electric vehicles; inductively coupled power transfer; EV battery charging; misalignment; coil design; compensation topology

Citation: Ghazizadeh, S.; Ahmed, K.; Seyedmahmoudian, M.; Mekhilef, S.; Chandran, J.; Stojcevski, A. Critical Analysis of Simulation of Misalignment in Wireless Charging of Electric Vehicles Batteries. *Batteries* **2023**, *9*, 106. https://doi.org/10.3390/batteries9020106

Academic Editors: Federico Baronti and Carlos Ziebert

Received: 30 November 2022
Revised: 20 January 2023
Accepted: 27 January 2023
Published: 3 February 2023

Copyright: © 2023 by the authors. Licensee MDPI, Basel, Switzerland. This article is an open access article distributed under the terms and conditions of the Creative Commons Attribution (CC BY) license (https://creativecommons.org/licenses/by/4.0/).

1. Introduction

Transportation accounts for a large percentage of global energy consumption and greenhouse gas emissions [1,2]. EVs are becoming more commercialized due to the global transition from internal combustion engine vehicles (ICEVs) to EVs. Studies suggest that the number of EVs is growing significantly due to their environment-friendly merits and high efficiency [3]. In this regard, batteries are considered as the heart of electric transportation [4,5]. Given that EVs essentially work on the basis of electricity, one vital issue is to address the challenges of charging the electric reservoir of the vehicles, namely, the batteries. Conventional conductive charging methods are still widely used; however, their drawbacks such as wiring, trip hazards, contact wear and risk of tearing [6,7] are motivating researchers to pursue the idea of wireless power transfer (WPT). Compared to conductive charging, wireless charging offers safer, more convenient and more robust power transfer [8,9]. In addition, wireless charging benefits from the advantages of automated charging processes [10], i.e., the charging process becomes driver-independent. In the 1990s, the idea of plug-in inductive charging was proposed [11–17], but the issue of plugging in the charger cable remained unsolved until early 2000s, when the idea of contact-less charging was significantly redeveloped [16]. Therefore, the design and production of efficient wireless charging methods became more vital.

Power is essentially transferred between two coils through the air medium [18] via the ICPT system. The coupling between the two coils is evaluated by measuring the coupling coefficient (k). The system is termed as tightly coupled if k > 0.5, and loosely coupled if k < 0.5 [19]. The efficiency of the charging process is affected by the coupling coefficient, which itself depends on the linkage flux and alignment between the coils [18,20–22]. Once displacement is applied to the relative position of coils, the linkage flux reduces, resulting in lower coupling and lower efficiency, and causing disturbance to the output power stability [23]. Numerous solutions have been proposed by scholars to improve the misalignment tolerance, e.g., enlarging the transmitter coil size or using multicoil structures rather than single coil systems [23–26], albeit at an extra cost. The tolerance of WPT system against displacement possess more importance in Dynamic wireless charging (DWC) systems than static wireless charging (SWC) due to higher vulnerability of efficiency against misalignment [27]. SWC refers to the wireless charging of an EV while the vehicle is stationary, such as at home or in office parking, whereas DWC refers to charging an EV while it is moving on the road, e.g., on highways. The emergence of DWC may revolutionize the EV industry by alleviating the energy storage issue, which is a critical bottleneck of today's costly EVs [28]. The driving range of EVs can be extended by increasing the size and capacity of the batteries, but this increases the price of the product for the customer [29]. Thus, the idea of DWC could tackle the issue. Having said this, despite SWC being commercially feasible, DWC has not been commercialized yet due to its significant challenges [10]. As implied in [16], the investment cost of DWC may be as high as 10% of a highway lane. To solve the misalignment issue, Ahmad et al. investigated the reduction of efficiency due to misalignment by simulating different displacements applied to coils, and proposed the concept of an "automatic alignment receiver system" to address the problem by employing Hall effect sensors and electromechanical motors to adjust the receiver coil to the best possible aligned position [6]. However, this approach could make the charging process slow and mechanically complex [30]. Therefore, in order to achieve an efficient fast-charging system, it is necessary to have a highly misalignment-tolerant coupling circuit. One factor that affects the coupling between the two coils is the use of shielding in the couplers. Kushwaha et al. investigated variations of mutual inductance and efficiency of a WPT system with and without shielding in various types of misalignment [18]. The effect of shielding on the performance of the system depends on the structure of couplers, which will be elaborated on later in this paper.

Section 2 demonstrates a general classification of different misalignment types. Since the theoretical calculations of mutual inductance of coils under misaligned conditions are relatively complex [31], the behavior of a WPT system in misalignment is analyzed using finite element analysis (FEA) simulation with Ansys Maxwell®, while the effect of coil shape on magnetic coupling is investigated in Section 3. Section 4 includes the analysis of power transfer efficiency under different compensation topologies during misalignment conditions using ANSYS Simplorer® software. Based on the investigations on different types of coil designs for different misalignment conditions, a conclusion is presented in Section 5.

2. Classification of Different Types of Misalignments

Fundamentally the function of a WPT system relies on two separate coils forming mutual inductance. Any alterations in the shape and position of these two coils is known as misalignment and leads to a deforming airgap, which alters the mutual inductance and thus affects the transferred power [32]. This can put the stability of the system at risk, as well as producing oscillations in output voltage [33]. Moreover, misalignment creates challenging safety issues due to increased chances of electromagnetic exposure [34]. In [33], a classification of different misalignments is presented for different positions of two square coils. The two coils are perfectly aligned when the planes are close enough and both axes are parallel to each other, as shown in Figure 1. There are different classifications of misalignment in the literature. In this paper, displacements are categorized by the type of movement, which can be either translational or angular.

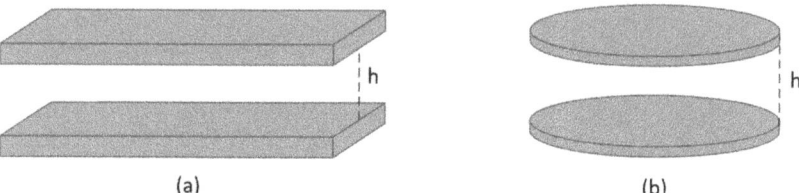

Figure 1. Perfectly aligned coils: (**a**) rectangular coils, (**b**) circular coils.

Various structures of coils have been introduced by scholars. The design of the coil impacts the magnetic flux distribution and the linked flux between the two coils [35,36]. Moreover, mutual inductance, self-inductance and the coupling coefficient are affected by the design of coils [35]. Mostak et al. [37] has shown the impact of the core on the coupling coefficient in horizontal misalignment. As reported, the core significantly increases the coupling coefficient. Misalignment in cases of circular- and rectangular-shaped coils have are further elaborated in this paper.

The misalignment phenomenon occurs when either translational or rotational displacement occurs in one or both of the coils; thus, there are two main types of misalignment:

(1) Translational misalignment;
(2) Angular and rotational misalignment.

Translational misalignment refers to either lateral or longitudinal displacement of planes with respect to each other, as shown in Figure 2a,b. Therefore, the linkage flux is reduced, resulting in reduction of MI, and the system enters loose coupling mode, with diminished efficiency [38,39]. In [16], longitudinal and lateral misalignments are referred to as X-axis and Y-axis misalignments, respectively.

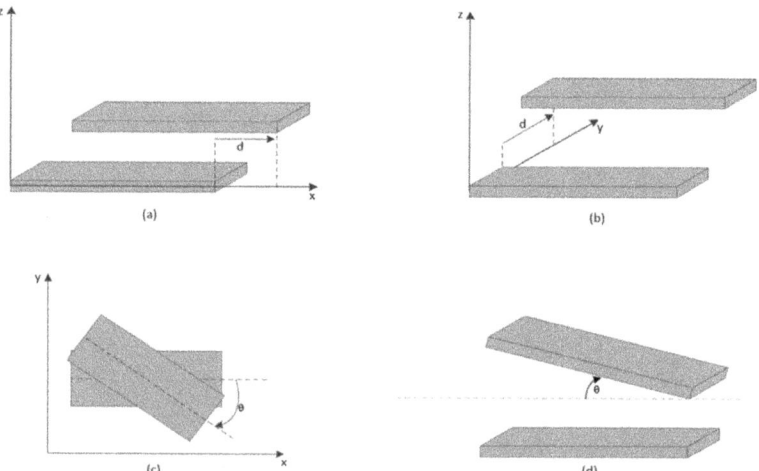

Figure 2. Misalignment forms of rectangular coils: (**a**) longitudinal, (**b**) lateral, (**c**) rotational, (**d**) angular.

In Figure 2c,d rotational misalignment is depicted where one plane rotates around its axis either in-plane or out of the parallel plane, classified as rotational and angular misalignment, respectively. In practice, there is a high possibility of a combination of different types of misalignments occurring. Although perfect alignment is desirable, placing the vehicle at the exact point of perfect alignment does not necessarily guarantee the best performance, since the load of the vehicle alters the height of the airgap; thus, changing the airgap between the two coils results in a different MI and different efficiency. This situation is not considered to be misalignment and is categorized as vertical variation [18].

In addition to rectangular structures, circular coils are also employed in practice. Although both systems are based on mutual inductance and magnetic interaction between the two coils, their tolerance to misalignment differs hugely. As shown in Figure 1b, in the circular design, as long as the two planes are parallel, any flat rotation in the position of the coils does not change the linked flux density. In addition, due to the intrinsic symmetrical geometry, there is no difference between lateral or longitudinal displacement in circular pads. Therefore, as depicted in Figure 3, the possible displacement scenarios for circular structures diminishes to two classes, i.e., translational and angular misalignment.

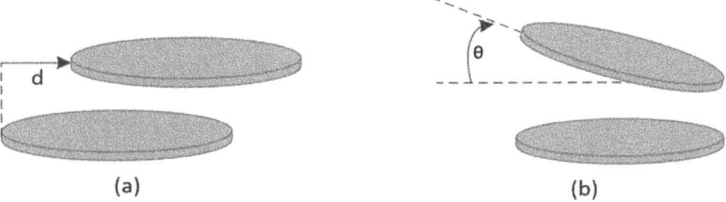

Figure 3. Misalignment in circular coils: (**a**) translational, (**b**) angular.

3. The Effect of Coupler Structure on the Efficiency of a WPT System

As described earlier, the performance of a WPT system is affected by the coupling coefficient [40]. The coupling coefficient is considerably affected by coupler design; thus, the structure of the magnetic couplers is one of the most substantial factors in a WPT system [41]. More precisely, the coupling coefficient determines what portion of the flux created by the primary coil is linked with the secondary coil [10]. The rest of the flux that is not linked with the secondary is termed as flux leakage. As a result, the voltage induced in the secondary coil depends on the coupling coefficient. The relation between the coupling coefficient and mutual inductance is shown in (1):

$$M = k\sqrt{L_p L_s} \qquad (1)$$

Equation (1) indicates the linear relationship between mutual inductance (M) and coupling coefficient (k). The results of the simulation can compute M, which then can be used to find the efficiency of the system [18]. Thus, in order to investigate the impact on the efficiency of a WPT system, the mutual inductance and the coupling coefficient can be analyzed; then, the efficiency can be derived from the results.

Couplers are the most substantial part of a WPT system and should have high coupling coefficient and good misalignment tolerance [10]. Research on square coil design [42–44] and circular design [45,46] has been presented in the literature. Utilizing different coil geometries can change the magnetic coupling considerably [47]. Various structures have been proposed by scholars to achieve the desired characteristics. Conventionally, couplers used in EV charging have either a flat spiral circular shape or a cylindrical spiral solenoid [16]. Compared to solenoidal types, the flat spiral coils offer various advantages including compactness, robustness and light weight [31]. Coil geometries vary from basic structures such as circular and square-shaped pads to more complex designs such as double D (DD) and DD-quadrature (DDQ) [48]. Covic and Boys [16] presented a review of different coupler structures and analyzed the performance of interoperation of different coupler designs, showing the tolerance of each design to displacements in the x- and y-axes. In a comparison between circular pads and polarized solenoidal pads, circular pads were found to have poorer coupling ability. However, polarized pads are restricted in terms of the direction in which the vehicle approaches the pad, while circular pads can be approached in any direction [16], proving that circular pads are strongly robust against rotational misalignment. Li et al. argued that for similar size, DD coils provide a better coupling, with a charging area almost two times larger than that of a circular pad, making them a possible solution for DWC [41].

Although using DD coils can improve the lateral misalignment tolerance, in the EV industry, circular-shaped and square-shaped coils are mostly used for static wireless charging [47]. Both circular and square-shaped (including rectangular) geometries offer a single-sided field that goes through the coil from its front [10,49]. The higher coupling coefficient of circular geometries compared to square-shaped or rectangular coils has been pointed out by scholars [50]. However, there still remains a lack of analysis of their performance in misalignment. The proposed assessment is based on the comparison of coupling coefficients computed by the simulator. Apart from the coil structure and displacement, there are other factors such as shielding that could affect the result. Patil et al. compared the coupling coefficient and misalignment tolerance of various structures such as circular, combined circular and square, circular with ferrite bar, bipolar and three-circular coils [10]. It was concluded that the bipolar structure demonstrated the best performance without shielding, whereas with aluminum shielding, the coupling coefficient of circular coils was less affected and thus achieved better performance than bipolar pads. As a general conclusion, the effect of shielding on couplers with a single-sided flux path is low, while it has a considerable impact on couplers with a double-sided flux path. This is because the double-sided coils have half of their flux on the front side and half on the rear side, while single-sided pads offer the advantage of having most of their flux on one side, with only a small portion making its way through the back [41]. Nevertheless, given that the computations without considering the effects of shielding are still valid [51], it is not included in this paper. Moreover, different coil shapes can be used interchangeably, e.g., a circular coil can be placed as a secondary coupled with a square-shaped primary. However, the following simulations are executed on identical coupled coils.

3.1. Circular Coils

Circular pads are the most reported structure and the most widely used couplers for SWC charging in EVs [10,50,52,53]. Due to the nature of circles, there is no difference between lateral and longitudinal displacement in this design, i.e., it is non-directional [52]. In addition, due to the symmetrical shape, rotational displacement is not applicable for this type of coils. In this section, the behavior of a coil set during lateral and angular misalignment is examined. Moreover, the effect of height alteration between the two coils is also investigated. To evaluate the behavior of the couplers in misalignment, various parameters such as magnetic field density, coupling coefficient and mutual inductance were computed by simulation using ANSYS Maxwell ®. The simulation was executed in a boundary area with relative permeability of 1 and conductivity of 0; this boundary emulates ambient conditions identical to air.

In order to calculate the coupling coefficient, this simulation was performed in magnetostatic mode and a parametric sweep was executed to analyze the alterations experienced by the coupling system during misalignment conditions. The values of the parameters used for the simulation are presented in Table 1. The airgap between the coils was set to one quarter of the coils' average diameter [54].

Table 1. Simulation parameters for circular coupler set.

Parameters	Values
Current in Rx	10 A
Coil inner radius (Tx and Rx)	50 mm
Coil outer radius (Tx and Rx)	70 mm
Vertical distance between coils	30 mm
Tx turns	20
Rx turns	20

To better represent the issue, relevant figures are provided where needed. Given that mutual inductance is not a function of current, there is no load applied to the secondary coil, i.e., the secondary coil is an open circuit. The effects of induced voltage and current (in order to assess the efficiency rate) in the secondary coil are analyzed in detail in further sections.

Figure 4 depicts the distribution of magnetic flux density around the transmitter. To analyze the effect of misalignment on the coupling coefficient and mutual inductance, the receiver coil was first moved both vertically and horizontally and the output parameters were computed at certain distances; the same approach was implemented for the angular (tilted) displacement. In the final stage, the values were plotted against different variables to assess tolerance against each type of misalignment. Figure 5a shows that the coupling coefficient decreases hyperbolically when the height of the secondary coil increases from 5 to 200 mm; in this case, the coupling coefficient drops from around 0.6 to close to zero. In Figure 5b, lateral displacement in applied to the secondary coil. As can be seen, the curve plunges to zero at a point regarded as the "null coupling position" or "flux cancellation" area, and rises again as the displacement increases [40,52]. The reason for this local zero coupling area is that at this particular distance, the amount of flux intersected by the secondary pad is equal to the intersected flux in the reverse direction, resulting in a net flux equal to zero. Ke and Chen et al. developed a method to correct this phenomenon [40]. In this case, the flux cancellation occurs at around 80 mm lateral offset, confirming the results of the empirical formula derived by [40] and implying that the flux cancellation zone of circular pads occurs at a displacement of approximately 50% of the coil outer diameter. However, this can vary based on the airgap between the two coils. Figure 5c represents the alterations of the coupling factor with angular misalignment from 0 to 20 degrees.

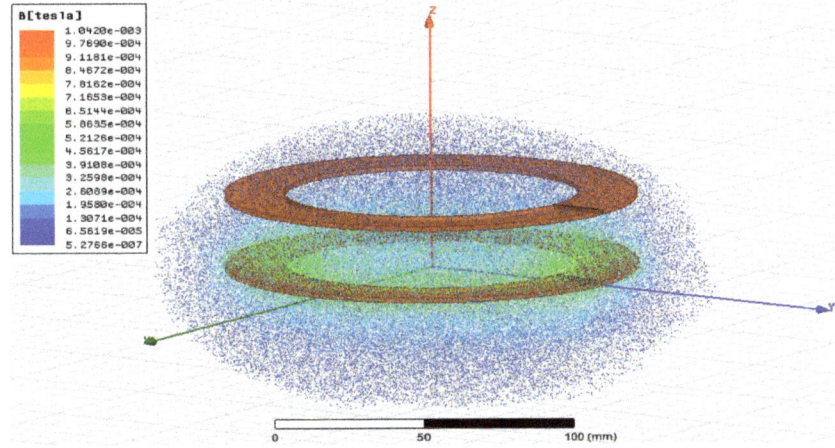

Figure 4. Magnetic flux density of perfectly aligned circular coils with 30 mm vertical distance.

3.2. Rectangular Coils

Another basic shape for couplers is rectangular. To obtain fair comparison results, the surface area of the rectangular coil in the x-y plane (7520 mm^2) is kept close to that of circular coil (7540 mm^2). This means that the quantity of copper used for manufacturing both structures is the same. The simulation parameters are shown in the Table 2. The magnetic flux distribution around the rectangular coil is shown in Figure 6.

Rectangular and square-shaped geometries have been widely investigated by scholars [18,31,33,55,56]. The rectangular design creates a polarized magnetic flux spread. Thus, the flux linkage is better than circular design coils [57] and is suitable for mid-range ICPT systems [56]. Wang et al. investigated a rectangular coupler set at a fixed distance of 5 cm to analyze lateral and angular misalignment effects [56]. However, rotational misalignment was not covered. Kushwaha et al. examined the misalignment tolerance of rectangular coils and confirmed that the results of analytical and experimental methods are in close agreement with simulation results [18]. In [33], a detailed analysis was conducted on square-shaped coils with all types of misalignment. However, the null coupling zone was not analyzed. The layout of the magnetic flux density of rectangular coil geometry is depicted in Figure 6. As can be seen, under vertical and angular misalignment, the rectangular coil experiences a similar pattern to that of a circular structure. Figure 7b shows sensitive this structure is to lateral displacement, while longitudinal offset has less of an impact on the linked flux. The null coupling zone occurs at approximately 60 and 105 mm for lateral and longitudinal displacements, respectively. In addition, such coils are vulnerable to rotation of the secondary coil, i.e., when the vehicle approaches the ground pad at an angle. Figure 7d shows that rotational misalignment leads to a significant reduction in coupling coefficient from around 0.45 to below 0.15. However, this alteration varies based on the airgap between the windings. For an airgap of 30 mm, the alteration range is significantly reduced.

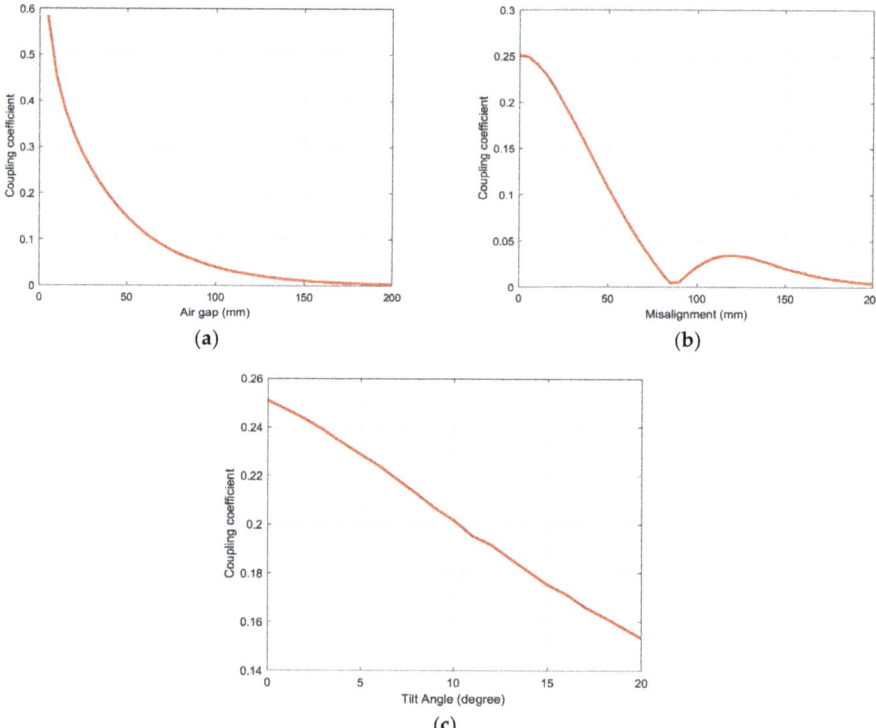

Figure 5. Circular coil under misalignment conditions: (**a**) against height of the secondary coil, (**b**) lateral misalignment, (**c**) angular misalignment.

Table 2. Simulation parameters for rectangular coupler set.

Parameter	Value
Current in Rx	10 A
Coil length (Tx and Rx)	140 mm
Coil width (Tx and Rx)	88 mm
Vertical distance between coils	30 mm
Tx turns	20
Rx turns	20

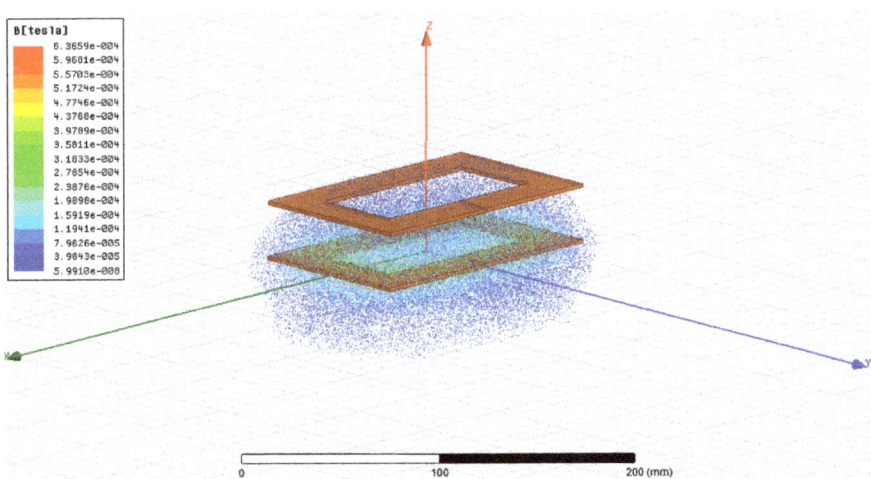

Figure 6. Magnetic flux density of perfectly aligned rectangular coils with 30 mm vertical distance.

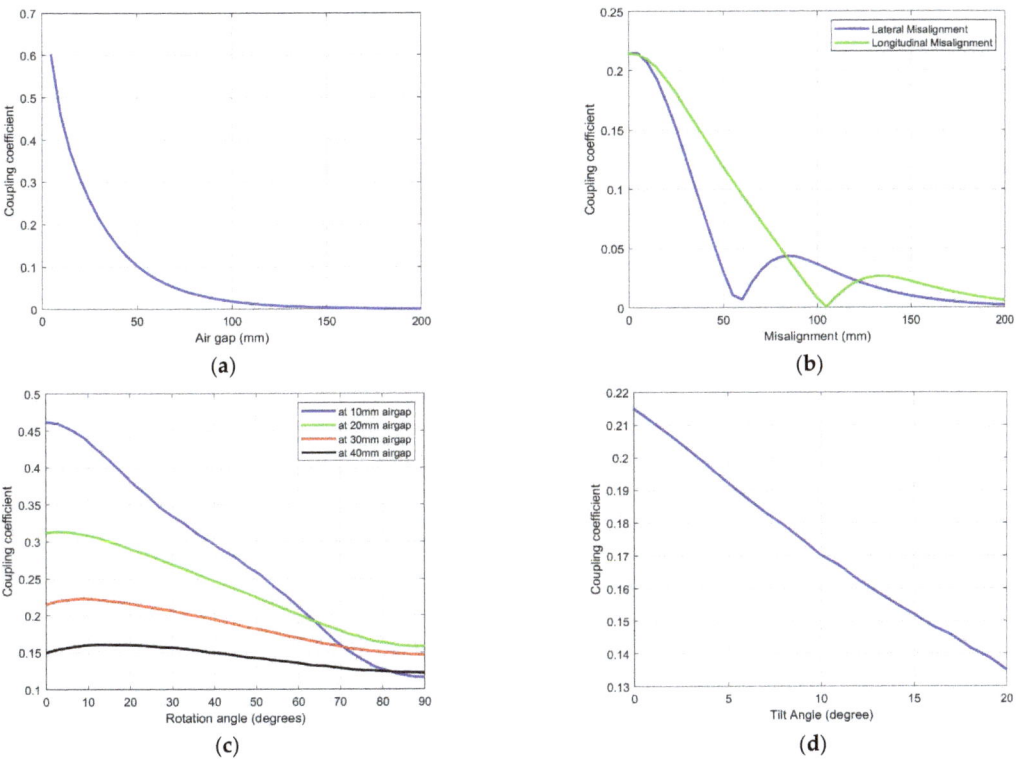

Figure 7. Rectangular coil under misalignment conditions: (**a**) against height of the secondary coil, (**b**) lateral and longitudinal misalignment, (**c**) angular misalignment, (**d**) rotational misalignment.

3.3. Comparative Analysis

In order to visualize the differences between the two coil shapes, in this section, the results of both structures are superimposed. As depicted in Figure 8a, a circular pad offers better tolerance to vertical displacement, as the rectangular coil experiences a steeper slope. At 50 mm height, the circular pad considerably outperforms the rectangular coil with a

50% higher coupling coefficient. Figure 8b indicates that the coupling coefficient of the rectangular coil plunges more drastically when lateral displacement is applied. However, due to the lengthened geometry of the rectangular coil, it achieves better performance than the circular coil in terms of longitudinal misalignment. This could make such coils a suitable choice for DWC, where high tolerance to longitudinal misalignment is more desired. The behavior of the system against angular misalignment is shown in Figure 8c, where the circular coil outperforms the rectangular pad slightly. Given that both structures require the same amount of copper for production, circular coils achieve greater tolerance to against misalignment overall. Moreover, such couplers offer better performance under rotational displacement due to their intrinsic neutrality to flat rotation.

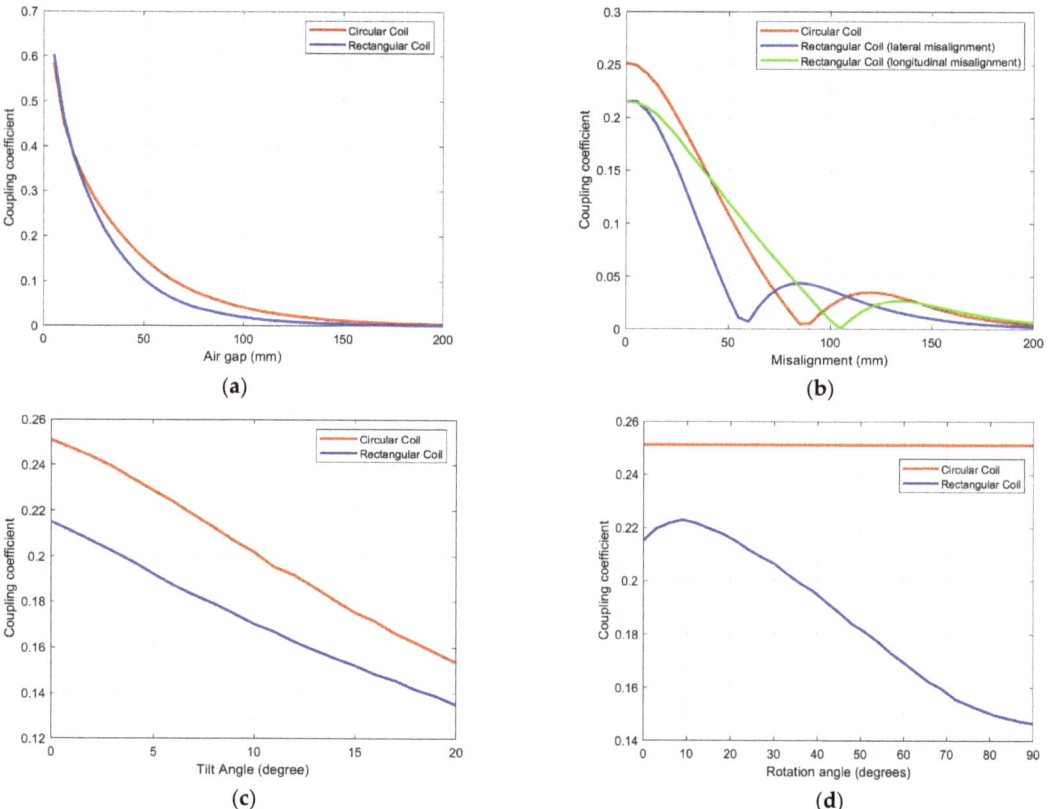

Figure 8. Circular and rectangular pads against different types of misalignments: (**a**) vertical distance, (**b**) lateral and longitudinal displacement, (**c**) angular misalignment, (**d**) rotational misalignment.

4. Impact of Compensation Topology on System Performance

Inductive power transfer can be efficient when the two coils are at close relative distance of a few centimeters, but its efficiency plunges drastically as a result of increased leakage inductance when the distance is increased [58,59]. The inductive nature of the circuit and high leakage flux leads to a high reactive current and, thus, a high VA rating of the power source. In addition, the excessive current results in higher losses and subsequently lower efficiency. Therefore, the presence of a compensatory capacitor is vital [10]. The simultaneous use of capacitive and inductive elements results in a resonance state, which realizes the concept of resonant coupling. The value of the resonant capacitors depends on the desired frequency of resonance. There is a trade-off between choosing the desired frequency and reactive components X_L, X_C, in that higher frequencies lead to a higher quality factor, but may cause higher losses in the switching stage [60]. By operating at the resonance frequency, the capacitor cancels out the effect of the inductive component, i.e., cancelling out the leakage inductance effect; thus, the volt–amp capacity of the inverter on the primary side is optimized. When reducing the phase angle between the primary voltage and current to zero, known as the zero phase angle (ZPA), the power factor (PF) hovers around 1, which is desirable. ZPA maintains the necessary conditions for soft switching of electronic components [41]. Depending on the application, the tuned frequency at which the circuit is designed to operate is determined. Therefore, a precise control system is needed to keep the circuit operating at the resonance frequency even when the coupling is changed due to the misalignment. This frequency control is significantly dependent on the compensation topology [30]. In compliance with the SAE J2954/1 standard, the nominal frequency should be 85 kHz [10]. Nevertheless, the frequency can range between 81.38–90 kHz, which is the agreed frequency spectrum internationally, as it has the least interference with other facilities [10]. One widely used method of compensation comprises only two capacitors; this is also the most economical approach [36]. Therefore, there are four principal compensation networks classified based on the position of compensatory capacitors in the circuit. The capacitors are placed either in series or parallel on both the primary and secondary sides; hence, there are four basic designs of compensation network—series–series (S–S), series–parallel (S–P), parallel–series (P–S) and parallel–parallel (P–P). Each of these topologies has its own merits and drawbacks, which results in a trade-off based on the desired application. Sohn et al. investigated all four schemes in terms of maximum efficiency, maximum transferred power, load-independency, k-independency and no magnetic coupling performance, concluding that current-sourced S–S and S–P topologies have better performance [61]. However, the author mentions that there remains uncertainty regarding which compensation is the most suitable choice depending on the application. In another study, Shevchenko et al. conducted a comprehensive review of basic compensation topologies, as well as a comparison of more complex hybrid compensation approaches [47]. It was concluded that using series primary compensation leads to higher efficiency amongst the four abovementioned topologies [47]. Apart from these four basic compensation strategies, many scholars have proposed more complex compensation topologies. For instance, Villa et al. investigated the performance of four basic topologies under misalignment conditions to find a high-misalignment tolerant design, and presented the use of SPS topology in an ICPT system [30]. In [47], it is concluded that amongst complex topologies, LCL and LCC are more advantageous, with an efficiency of more than 95%. In this paper, however, the four basic topologies are investigated and other more complex designs are excluded. Although in some designs the circuit is driven by a current source, in this paper, a voltage-sourced circuit is evaluated.

The impact of different compensation topologies on the efficiency of a WPT system under misalignment conditions is essentially based on how the circuit reacts to changes applied to the circuit. Different parameters such as mutual inductance (equivalent to the coupling coefficient) [62], operating frequency and load variations eventually result in changes in system efficiency. Here, the aim of the analysis is to determine the impact of mutual inductance alterations on the efficiency of power transfer. It has been determined that each compensation topology has a certain behavior to mutual inductance alterations, which will be analyzed in this section.

In order to investigate the circuit behavior, an equivalent model of the system is required (see Figure 9). The load impedance can be modeled as reflected impedance to the primary side, as shown in Figure 9c.

$$Z_r = -\frac{j\omega M I_2}{I_1} \quad (2)$$

$$I_2 = \frac{j\omega M I_1}{Z_{eq}^2} \quad (3)$$

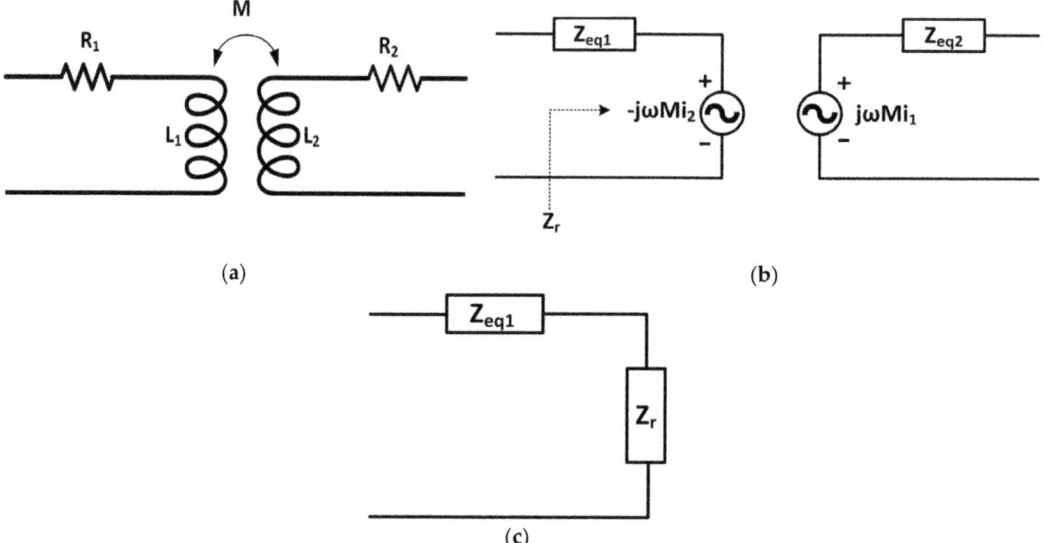

Figure 9. Coupled circuit: (**a**) basic model, (**b**) equivalent model of induced voltages, (**c**) reflected impedance to the primary side.

Substituting (2) in (3) returns the reflected impedance relation:

$$Z_r = \frac{\omega^2 M^2}{Z_{eq}^2} \qquad (4)$$

The reflected secondary impedance to the primary side is derived in (4), where ω is the resonance frequency, Z_{eq}^2 is the equivalent impedance of the secondary side and Z_r is the reflected Z_{eq}^2 to the primary side. The formation of capacitor and load determines the value of Z_{eq}^2 and, consequently, Z_r. Thus, it is vital to calculate the equivalent impedance based on the compensation topologies used in primary and secondary sides (Figure 10). The equivalent impedance of the secondary side (neglecting the parasitic winding resistance) can be found using (5) and (6):

$$\text{Series secondary compensation}: Z_{eq}^2 = j\omega L_S + \frac{1}{j\omega C_S} + Z_L; \qquad (5)$$

$$\text{Parallel secondary compensation}: Z_{eq}^2 = j\omega L_S + \frac{1}{j\omega C_S + \frac{1}{Z_L}}. \qquad (6)$$

Using (4) to transfer Z_{eq}^2 to the primary side, the input impedance seen by the source is determined (Table 3). The circuit can be fed by either a voltage source or current source. In Figure 11, the circuit is driven by voltage source Vs, where the load is represented by Z_L, which is considered as a purely resistive load. The parasitic resistances of the primary and secondary windings are modeled by R1 and R2, while L1 and L2 represent the self-inductances of the respective windings. The secondary capacitor is selected such that the imaginary part of the secondary impedance becomes zero to achieve the maximum transfer capability [63]. To achieve the maximum efficiency, the secondary capacitor is determined by (7) [28].

$$\text{Secondary side capacitor}: C_2 = \frac{1}{\omega^2 L_2} \qquad (7)$$

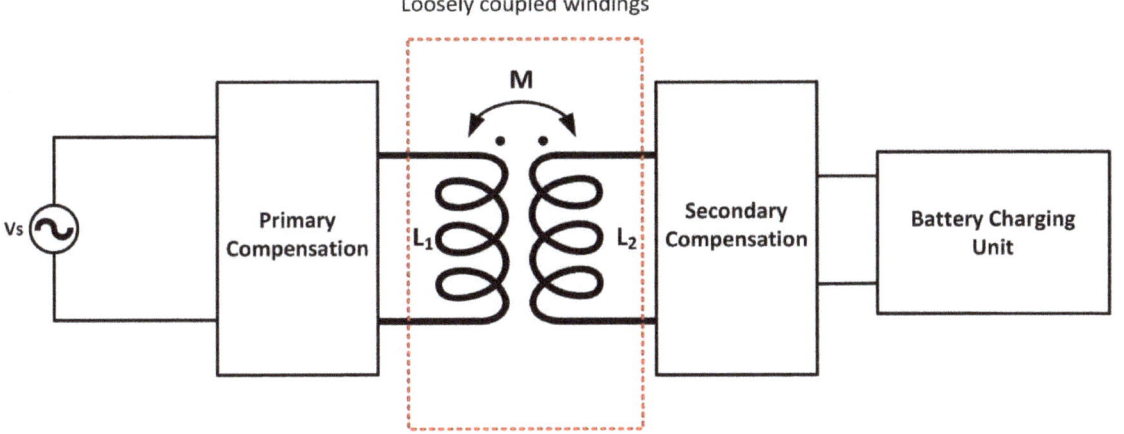

Figure 10. Inductive power transfer system.

Table 3. Total impedance seen by the source [10].

Compensation Topology	Reflected Impedance
S–S	$\left(R_1 + j\left(\omega L_1 - \frac{1}{\omega C_1}\right)\right) + \frac{\omega^2 M^2}{R_2 + R_L + j\left(L_2\omega - \frac{1}{C_2\omega}\right)}$
S–P	$\left(R_1 + j\left(\omega L_1 - \frac{1}{\omega C_1}\right)\right) + \frac{\omega^2 M^2}{R_2 + \frac{R_L}{1+jR_L C_2\omega} + jL_2\omega}$
P–P	$\dfrac{1}{j\omega C_1 + \dfrac{1}{R_1 + jL_1\omega + \dfrac{\omega^2 M^2 \,(1+jR_L C_2\omega)}{R_L + (R_2+jL_2\omega)(1+jR_L C_2\omega)}}}$
P–S	$\dfrac{1}{R_1 + jL_1\omega + \dfrac{\omega^2 M^2}{R_2 + R_L + j\left(L_2\omega - \frac{1}{C_2\omega}\right)}} + jC_1\omega$

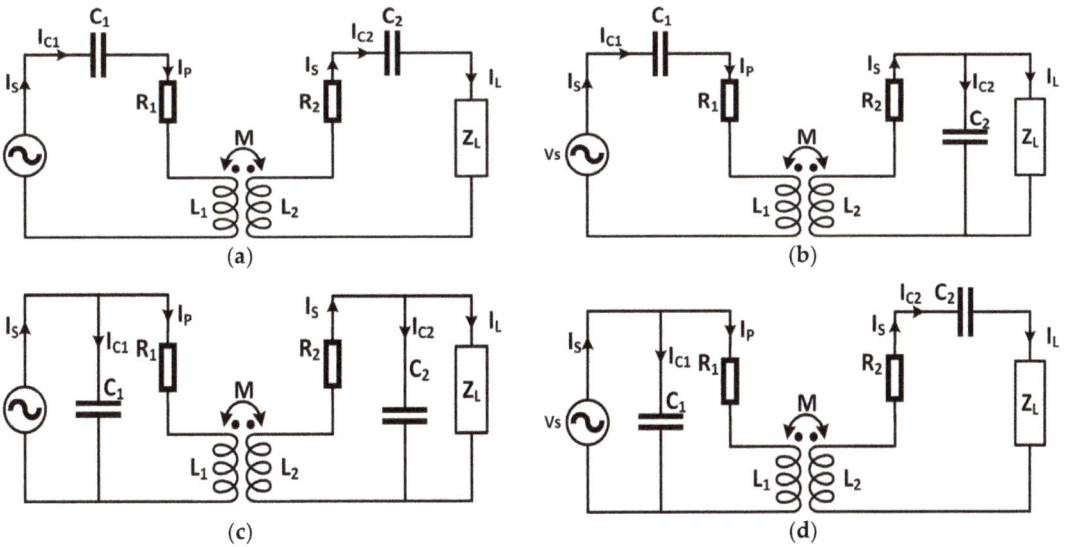

Figure 11. Basic compensation topologies. (**a**) series–series, (**b**) series–parallel, (**c**) parallel–parallel, (**d**) parallel–series.

The resultant impedance is reflected to the primary side, which is seen by the source. The primary capacitor compensates for the reactive power and, thus, the imaginary part of the equivalent impedance seen by the source (including the reflected impedance) is eliminated to ensure that the circuit is operating in ZPA mode; this achieves the minimum VA rating of the power source [63]. Depending on what topology is used, the capacitance of the primary capacitor is determined [29,61,63,64]. The values of the primary and secondary compensating capacitors are determined such that the system resonates at the desired frequency of 85 kHz. The value of the primary capacitor, depending on the corresponding compensation topology, is given by Table 4.

Table 4. Primary capacitance for different topologies [64].

Compensation Topology	Primary Capacitance
S–S	$\frac{1}{\omega_0^2 L_1}$
S–P	$\frac{1}{\omega_0^2 \left(L_1 - \frac{M^2}{L_2}\right)}$
P–P	$\frac{L_1 - \frac{M^2}{L_2}}{\left(\frac{M^2 R}{L_2^2}\right)^2 + \omega_0^2 \left(L_1 - \frac{M^2}{L_2}\right)^2}$
P–S	$\frac{L_1}{\left(\omega_0^2 \frac{M^2}{R}\right)^2 + \omega_0^2 L_1^2}$

The performance of the compensation network is assessed by computing the efficiency of the power transmitted by the transmitter and acquired by the load. Although in practice, the load is often a rectifier, filter and switched controller, it is possible to model the load with an equivalent resistor [31]. To evaluate the efficiency, simulations were executed in ANSYS Simplorer®. The results previously achieved using ANSYS Maxwell® (in Section 3) were used to determine the mutual inductance for simulation in this section. The Maxwell results indicate that the primary and secondary inductances remain constant at 59 uH, whereas the mutual inductance changes as a result of changing the relative coil positions. A constant resistance is placed at the secondary terminals as a load. The simulation was executed in AC-analysis mode in Simplorer to investigate the system efficiency across a wide range of frequencies, including the resonant frequency. The efficiency was then found by computing the ratio of the real power received on the secondary side to the transmitted real power on primary side [65]. The efficiency was plotted in the frequency domain to compare the consistency of the resonance frequency.

$$\eta = \frac{P_{Received}}{P_{transmitted}} \qquad (8)$$

To investigate the overall behavior of various combinations of different compensation methods with different coil designs, simulations were executed for all four basic compensation topologies with both circular and rectangular coils. A lateral displacement of 0–200 mm was then applied to the secondary coil and the efficiency of the system was measured during misalignment. It is desirable that the system maintain its maximum efficiency for a wide range of misalignment.

4.1. S–S (Series–Series)

S–S compensation is widely used in WPT charging for EVs [23]. In terms of copper mass, S–S is the best choice among all four options [47,51]. Another merit of this topology is that the required capacitor for compensation is independent from the coupling coefficient and load variations [10,28,58,59,63]. Therefore, the frequency of resonance remains constant when misalignment occurs (see Figure 12a) [28]. This stability of resonant frequency during misalignment makes WPT suitable for applications with high misalignment requirements. Furthermore, the current drawn on the secondary side depends upon the AC source regardless of the voltage induced at the secondary coil, which is beneficial for battery charging purposes [66]. Another merit of this design is that since the imaginary part of the reflected impedance remains zero (see Table 3), the unity power factor is maintained during misalignment [28,67]. As a result, S–S-compensated systems are the most appropriate for EV charging applications [28,68,69]. Moreover, in terms of charging requirements, this topology can provide constant voltage and current for battery charging applications [28,70,71]. Nevertheless, one major drawback of S–S topology is its performance at light-load conditions and when no receiver is on the secondary side [28]. This is because the equivalent impedance seen by the primary side is reduced when the coupling is weakened, which results in a tremendous current on the primary side and an extreme voltage at the secondary terminal when the circuit is supplied by a voltage source [10,41]. Therefore, working in zero-coupling mode must be avoided [47].

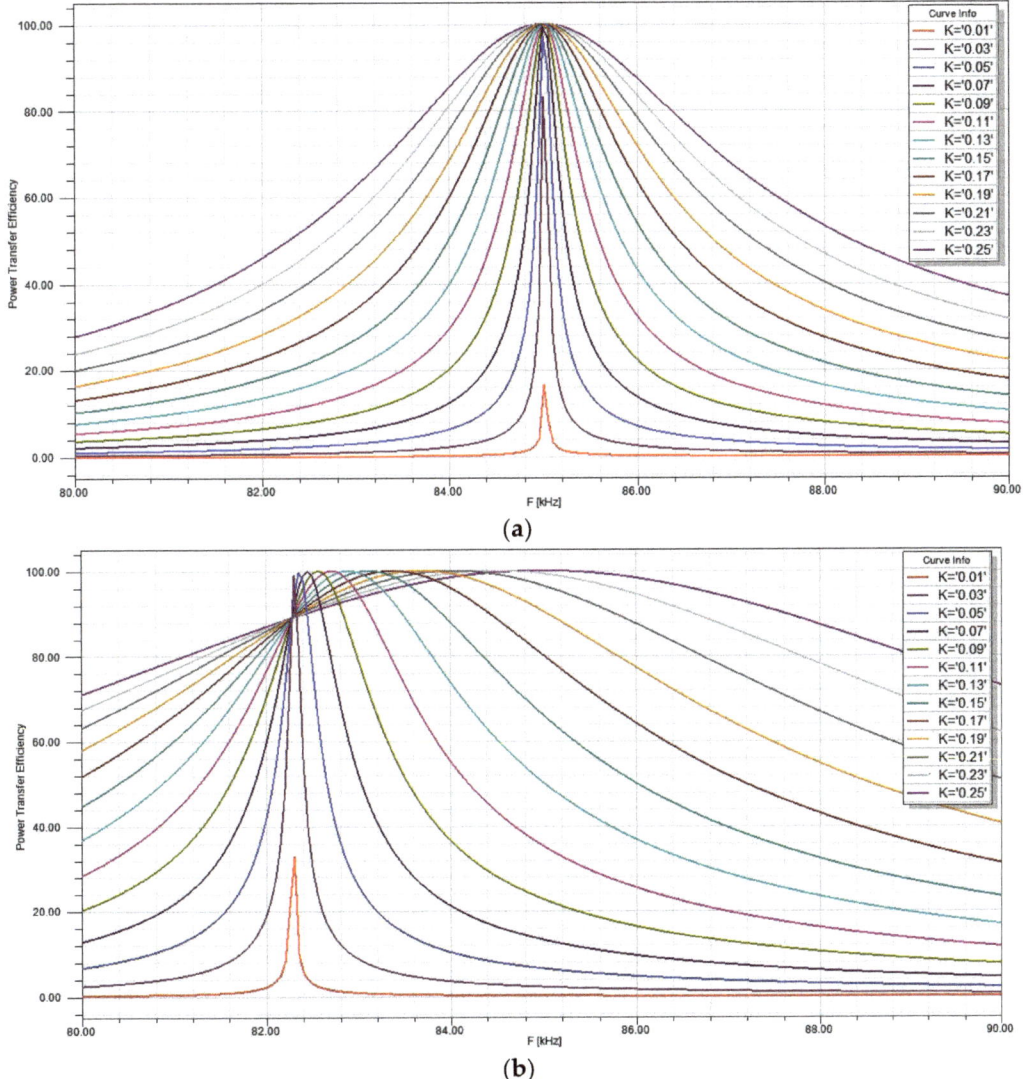

Figure 12. *Cont.*

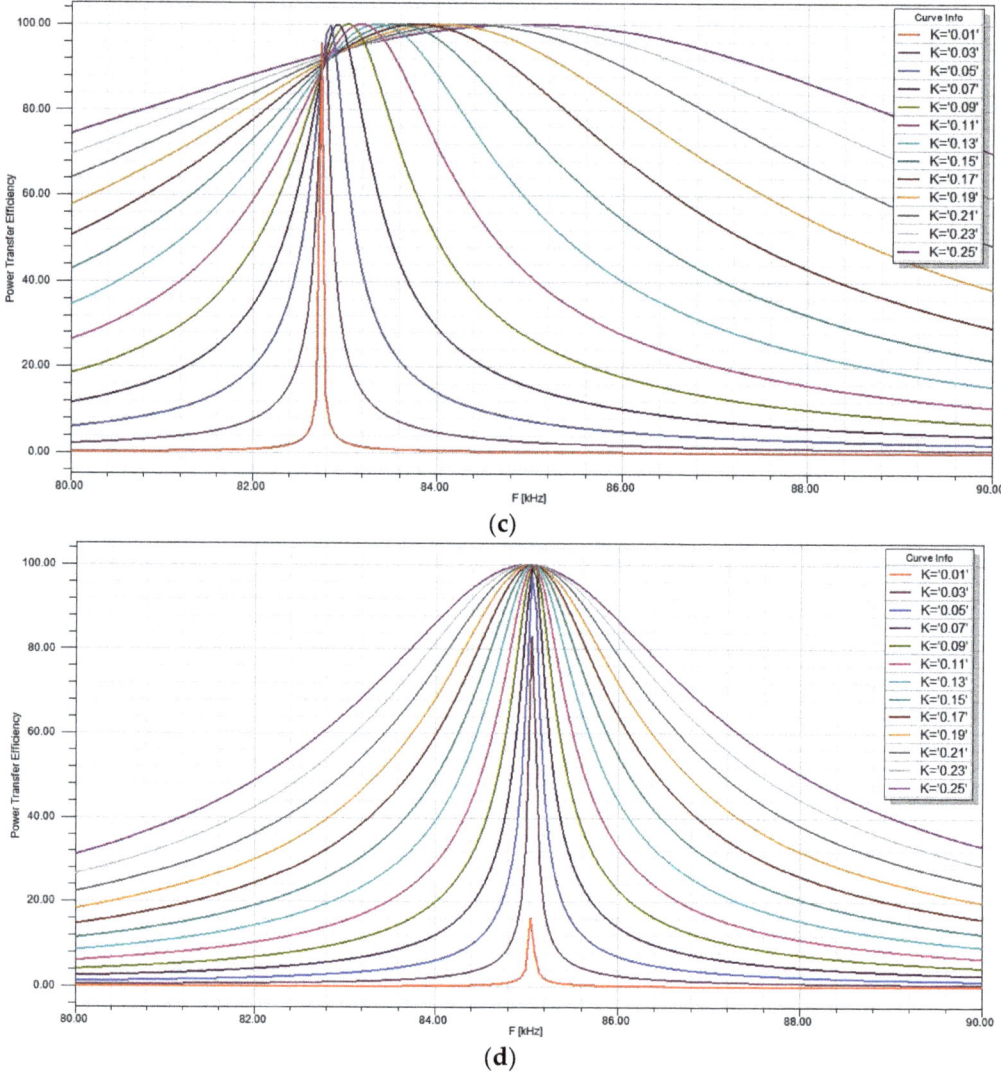

Figure 12. Resonant frequency mutations of different compensation topologies: (**a**) S–S, (**b**) S–P, (**c**) P–P, (**d**) P–S.

4.2. S–P (Series–Parallel)

Due to the series primary connection, both S–S and S–P designs have the advantage of transferring higher power than rated [59]. S–P topology, however, lacks the benefit of independence from mutual inductance, in contrast to S–S topology. In addition, it requires a larger capacitor on the primary side. S–P compensation is usually employed to maintain a constant voltage output [63]. Nevertheless, it has the hazard of excessive current when the secondary side is not in proximity due to the negligible impedance seen by the source [10]; therefore, a current limit mechanism is needed. As depicted in Table 4, the primary capacitance changes with mutual inductance alterations, and given that the capacitor is fixed in practice, a change in the mutual inductance leads to significant resonance frequency mutation (see Figure 12b).

4.3. P–P (Parallel–Parallel)

The relationships of primary capacitance and impedance in this topology are relatively complex. Using a parallel capacitor in the primary side is safe in case of the absence of a secondary coil [59]. However, this also leads to large input impedance, which requires high voltage to transfer power sufficiently [47]. A merit of this technique is that if an appropriate alignment is guaranteed by the coupling unit, it offers a high efficiency with suitable robustness against frequency mutations (see Figure 12c). However, since the value of the primary capacitor depends on both mutual inductance and load impedance (see Table 4), this method has a low misalignment tolerance, as shown in Figure 12c [28]. Another disadvantage of this technique is its low power factor [10]. This topology is not widely investigated due to its numerous disadvantages [10,65,72].

4.4. P–S (Parallel–Series)

Due to the parallel connection on the primary side, the relationship of the primary capacitor in this topology is also relatively complex. The primary capacitor is strongly affected by mutual inductance and load. Compared to P–P, the primary capacitor for P–S has a steeper relation with mutual inductance, as it decreases with the fourth power of M. Another drawback of this design is that it requires more copper than the other three topologies (30% more than S–S) [47]. Unlike other topologies, the reflected impedance in P–S is inversely correlated with the mutual inductance, meaning that the impedance increases when misalignment occurs; thus, the current is limited during misalignment (Table 3). A major advantage of using P–S topology is that it provides soft-switching for semiconductors [47]. However, it is not capable of transferring sufficient power when misalignment happens [28,70]. It also has a low tolerance to frequency mutations [28], especially in misalignment conditions (see Figure 12d).

4.5. Comparative Analysis

Figure 12a–d reveals that there is a trade-off between misalignment tolerance and robustness to frequency mutations. Using series compensation on the secondary side results in higher efficiency during misalignment, but lower tolerance to frequency alterations. This can increase costs, since it requires a high frequency control system to maintain the exact optimal frequency with high precision. On the contrary, using a parallel compensation on the secondary side leads to higher frequency tolerance, but misalignment tolerance is compromised. Each compensation technique has been simulated under the same parametric conditions (see Figure 12a–d). In [10], a comparison indicates that S–S and P–S topologies demonstrate higher efficiency at low mutual inductances than S–P and P–P, meaning that having series compensation on the secondary side offers better tolerance under misalignment [73]. A reconfigurable resonant and hybrid circuit for the ICPT system has been proposed to sustain the changes in load and misalignment variation [74]. However, in terms of economic analysis, using a series capacitor on the primary side, namely, S–S or S–P, is more suitable for high-power applications [51]. Sohn et al. [61] concluded that in terms of maximum efficiency, S–S and P–S schemes offer complete k-independency, meaning that S–S and P–S offer better performance in misalignment conditions.

To compare the performance of all four schemes, the efficiency of each topology was measured at the frequency of 85 kHz while lateral misalignment was applied to the secondary position at certain steps.

Due to the changes in coupling, the resonant frequency was shifted when the misalignment was applied to the system. Figure 12a,d indicates that using a series capacitor on the secondary side (S–S and P–S) mitigates the frequency shift caused by misalignment; hence, better performance is achieved during misalignment. It should be noted that although the efficiency of P–S drops more sharply than S–S, the frequency shift is not considerable and the efficiency can be maintained around the desired area by using a frequency control circuit. However, this can increase the production costs. Figure 13 implies that S–S has the most robust curve under misalignment, whereas P–P and S–P experience more deterioration in efficiency as the mutual inductance decreases. S–P demonstrates the weakest misalignment tolerance of all the topologies.

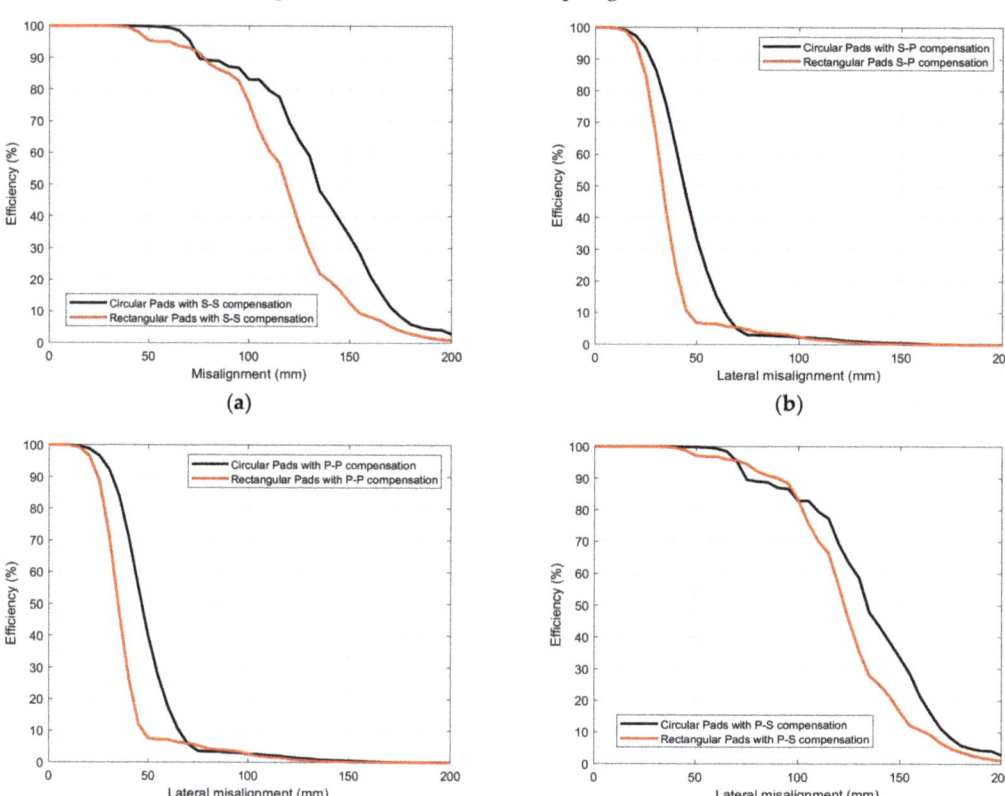

Figure 13. Misalignment tolerance of different combinations of coils and compensation topologies: (**a**) S–S, (**b**) S–P, (**c**) P–P, (**d**) P–S.

5. Limitations and Future Trends

Various limitations still persist in EV wireless charging systems. Conventional control methods of WPT charging systems use communication devices; however, there can be delays data transmission causing unnecessary energy loss. In addition, despite extensive research in achieving high-misalignment-tolerant designs; a high risk of electromagnetic field exposure remains during misalignment. Moreover, the interoperability of different charging coil designs and circuit configurations lacks careful consideration of misalignment.

The future prospects of this technology demand further investigation of the following topics:

- A reconfigurable arrangement of coil design that can switch different arrangements of coils to offer high coupling coefficients between EVs and chargers for each misalignment condition;
- Instead of using communication between the EV and the charging circuit, a new control using system parameters whose changes reflect misalignment between the charging coils, e.g., mutual inductance deviations, should be a candidate for future research;
- Current studies on the thermal failure of the ferrite core employed in coil construction are inadequate, making it difficult to draw any firm conclusions;
- Although dynamic wireless charging is a promising solution to the challenges of EVs, there are limited studies considering misalignment with this type of EV charging.

6. Conclusions

A comparison of non-polarized coupler structures and basic compensation topologies used in EV battery charging is presented in this paper. It was found that with regards to misalignment tolerance, circular pads outperform rectangular structures due to their maintaining a higher coupling coefficient, as well as offering freedom of approach direction. In terms of compensation topology, it was concluded that using series compensation on the secondary side leads to higher misalignment tolerance, while system robustness against frequency alterations is compromised. The S–S topology demonstrates the best performance considering its independency from load and coupling mutations and its ability to provide suitable charging conditions for EV battery charging applications. Given that there are various trade-offs between different designs, selection of the most optimal mixture of coil structure and compensation topology must be determined based on the EV specifications, e.g., chassis height, size, dimensions and battery capacity.

Author Contributions: Conceptualization and development, S.G., M.S., S.M. and A.S.; methodology, S.G. and K.A.; software, S.G.; validation, K.A. and S.M., J.C.; formal analysis, S.M., M.S., J.C. and A.S.; investigation, S.G. and K.A.; data curation, S.G. and K.A.; writing—original draft preparation, S.G. and K.A.; writing—review, analysis and editing, M.S. and A.S.; supervision, A.S.; project administration, M.S. and S.M. All authors have read and agreed to the published version of the manuscript.

Funding: This research received no external funding.

Data Availability Statement: There was no data gathered for the purpose of this study.

Conflicts of Interest: The authors declare no conflict of interest.

References

1. Fachrizal, R.; Shepero, M.; van der Meer, D.; Munkhammar, J.; Widén, J. Smart charging of electric vehicles considering photovoltaic power production and electricity consumption: A review. *eTransportation* **2020**, *4*, 100056.
2. Madani, S.S.; Soghrati, R.; Ziebert, C. A Regression-Based Technique for Capacity Estimation of Lithium-Ion Batteries. *Batteries* **2022**, *8*, 31. [CrossRef]
3. Hovet, S.; Farley, B.; Perry, J.; Kirsche, K.; Jerue, M.; Tse, Z.T.H. Introduction of electric vehicle charging stations to university campuses: A case study for the university of Georgia from 2014 to 2017. *Batteries* **2018**, *4*, 27. [CrossRef]
4. Lipu, M.S.H.; Mamun, A.A.; Ansari, S.; Miah, M.S.; Hasan, K.; Meraj, S.T.; Abdolrasol, M.G.; Rahman, T.; Maruf, M.H.; Sarker, M.R. Battery Management, Key Technologies, Methods, Issues, and Future Trends of Electric Vehicles: A Pathway toward Achieving Sustainable Development Goals. *Batteries* **2022**, *8*, 119. [CrossRef]
5. Gandoman, F.H.; El-Shahat, A.; Alaas, Z.M.; Ali, Z.M.; Berecibar, M.; Abdel Aleem, S.H. Understanding Voltage Behavior of Lithium-Ion Batteries in Electric Vehicles Applications. *Batteries* **2022**, *8*, 130. [CrossRef]
6. Ahmad, A.; Alam, M.S.; Rafat, Y.; Shariff, S. Designing and demonstration of misalignment reduction for wireless charging of autonomous electric vehicle. *eTransportation* **2020**, *4*, 100052. [CrossRef]
7. Hasanzadeh, S.; Vaez-Zadeh, S.; Isfahani, A.H. Optimization of a contactless power transfer system for electric vehicles. *IEEE Trans. Veh. Technol.* **2012**, *61*, 3566–3573. [CrossRef]
8. Li, W.; Zhao, H.; Li, S.; Deng, J.; Kan, T.; Mi, C.C. Integrated LCC compensation topology for wireless charger in electric and plug-in electric vehicles. *IEEE Trans. Ind. Electron.* **2014**, *62*, 4215–4225. [CrossRef]
9. Zhao, L.; Thrimawithana, D.J.; Madawala, U.K.; Hu, A.P.; Mi, C.C. A misalignment-tolerant series-hybrid wireless EV charging system with integrated magnetics. *IEEE Trans. Power Electron.* **2018**, *34*, 1276–1285. [CrossRef]
10. Patil, D.; Mcdonough, M.K.; Miller, J.M.; Fahimi, B.; Balsara, P.T. Wireless power transfer for vehicular applications: Overview and challenges. *IEEE Trans. Transp. Electrif.* **2017**, *4*, 3–37. [CrossRef]
11. Sakamoto, H.; Harada, K.; Washimiya, S.; Takehara, K.; Matsuo, Y.; Nakao, F. Large air-gap coupler for inductive charger [for electric vehicles]. *IEEE Trans. Magn.* **1999**, *35*, 3526–3528. [CrossRef]
12. Hayes, J.G.; Egan, M.G.; Murphy, J.M.; Schulz, S.E.; Hall, J.T. Wide-load-range resonant converter supplying the SAE J-1773 electric vehicle inductive charging interface. *IEEE Trans. Ind. Appl.* **1999**, *35*, 884–895. [CrossRef]
13. Severns, R.; Yeow, E.; Woody, G.; Hall, J.; Hayes, J. An Ultra-Compact Transformer for a 100 W to 120 kW Inductive Coupler for Electric Vehicle Battery Charging. In Proceedings of the Applied Power Electronics Conference, APEC'96, San Jose, CA, USA, 3–7 March 1996; IEEE: New York, NY, USA, 1996; pp. 32–38.
14. Klontz, K.W.; Divan, D.M.; Novotny, D.W. An actively cooled 120 kW coaxial winding transformer for fast charging electric vehicles. *IEEE Trans. Ind. Appl.* **1995**, *31*, 1257–1263. [CrossRef]

15. Klontz, K.; Esser, A.; Bacon, R.; Divan, D.; Novotny, D.; Lorenz, R. An electric vehicle charging system with 'universal' inductive interface. In Proceedings of the Conference Record of the Power Conversion Conference-Yokohama 1993, Yokohama, Japan, 19–21 April 1993; IEEE: New York, NY, USA, 1993; pp. 227–232.
16. Covic, G.A.; Boys, J.T. Modern trends in inductive power transfer for transportation applications. *IEEE J. Emerg. Sel. Top. Power Electron.* **2013**, *1*, 28–41. [CrossRef]
17. Proceedings of the IECON '97: 23rd International Conference on Industrial Electronics, Control, and Instrumentation, New Orleans, LA, USA, 14 November 1997; IEEE: Piscataway, NJ, USA.
18. Kushwaha, B.K.; Rituraj, G.; Kumar, P. 3-D analytical model for computation of mutual inductance for different misalignments with shielding in wireless power transfer system. *IEEE Trans. Transp. Electrif.* **2017**, *3*, 332–342. [CrossRef]
19. Kalwar, K.A.; Aamir, M.; Mekhilef, S. A design method for developing a high misalignment tolerant wireless charging system for electric vehicles. *Measurement* **2018**, *118*, 237–245. [CrossRef]
20. Choi, S.Y.; Huh, J.; Lee, W.Y.; Rim, C.T. Asymmetric coil sets for wireless stationary EV chargers with large lateral tolerance by dominant field analysis. *IEEE Trans. Power Electron.* **2014**, *29*, 6406–6420. [CrossRef]
21. Nadakuduti, J.; Douglas, M.; Lu, L.; Christ, A.; Guckian, P.; Kuster, N. Compliance testing methodology for wireless power transfer systems. *IEEE Trans. Power Electron.* **2015**, *30*, 6264–6273. [CrossRef]
22. Zheng, C.; Ma, H.; Lai, J.-S.; Zhang, L. Design considerations to reduce gap variation and misalignment effects for the inductive power transfer system. *IEEE Trans. Power Electron.* **2015**, *30*, 6108–6119. [CrossRef]
23. Yuan, Z.; Saeedifard, M.; Cai, C.; Yang, Q.; Zhang, P.; Lin, H. A Misalignment Tolerant Design for a Dual-Coupled LCC-S-Compensated WPT System with Load-Independent CC Output. *IEEE Trans. Power Electron.* **2022**, *37*, 7480–7492. [CrossRef]
24. Zhao, L.; Thrimawithana, D.J.; Madawala, U.K. Hybrid bidirectional wireless EV charging system tolerant to pad misalignment. *IEEE Trans. Ind. Electron.* **2017**, *64*, 7079–7086. [CrossRef]
25. Chu, S.Y.; Cui, X.; Avestruz, A.-T. Accurate transfer-power measurement for wireless charging of electric vehicles under misalignment. In Proceedings of the 2018 IEEE PELS Workshop on Emerging Technologies: Wireless Power Transfer (Wow), Montreal, QC, Canada, 3–7 June 2018; IEEE: New York, NY, USA, 2018; pp. 1–6.
26. Ramezani, A.; Narimani, M. Optimized electric vehicle wireless chargers with reduced output voltage sensitivity to misalignment. *IEEE J. Emerg. Sel. Top. Power Electron.* **2019**, *8*, 3569–3581. [CrossRef]
27. Tavakoli, R.; Shabanian, T.; Dede, E.M.; Chou, C.; Pantic, Z. EV Misalignment Estimation in DWPT Systems Utilizing the Roadside Charging Pads. *IEEE Trans. Transp. Electrif.* **2021**, *8*, 752–766. [CrossRef]
28. Panchal, C.; Stegen, S.; Lu, J. Review of static and dynamic wireless electric vehicle charging system. *Eng. Sci. Technol. Int. J.* **2018**, *21*, 922–937. [CrossRef]
29. Zhang, Z.; Pang, H.; Georgiadis, A.; Cecati, C. Wireless power transfer—An overview. *IEEE Trans. Ind. Electron.* **2018**, *66*, 1044–1058. [CrossRef]
30. Villa, J.L.; Sallan, J.; Osorio, J.F.S.; Llombart, A. High-misalignment tolerant compensation topology for ICPT systems. *IEEE Trans. Ind. Electron.* **2011**, *59*, 945–951. [CrossRef]
31. Fotopoulou, K.; Flynn, B.W. Wireless power transfer in loosely coupled links: Coil misalignment model. *IEEE Trans. Magn.* **2010**, *47*, 416–430. [CrossRef]
32. Gao, Y.; Ginart, A.; Farley, K.B.; Tse, Z.T.H. Misalignment effect on efficiency of wireless power transfer for electric vehicles. In Proceedings of the 2016 IEEE applied power electronics conference and exposition (APEC), Long Beach, CA, USA, 20–24 March 2016; IEEE: New York, NY, USA, 2016; pp. 3526–3528.
33. Joy, E.R.; Dalal, A.; Kumar, P. Accurate computation of mutual inductance of two air core square coils with lateral and angular misalignments in a flat planar surface. *IEEE Trans. Magn.* **2013**, *50*, 1–9. [CrossRef]
34. Zhang, B.; Carlson, R.B.; Smart, J.G.; Dufek, E.J.; Liaw, B. Challenges of future high power wireless power transfer for light-duty electric vehicles—technology and risk management. *eTransportation* **2019**, *2*, 100012. [CrossRef]
35. Nanda, N.N.; Yusoff, S.H.; Toha, S.F.; Hasbullah, N.F.; Roszaidie, A.S. A brief review: Basic coil designs for inductive power transfer. *Indones. J. Electr. Eng. Comput. Sci* **2020**, *20*, 1703–1716.
36. Mosammam, B.M.; Rasekh, N.; Mirsalim, M.; Moghani, J.S. Comparative analysis of the conventional magnetic structure pads for the wireless power transfer applications. In Proceedings of the 2019 10th International Power Electronics, Drive Systems and Technologies Conference (PEDSTC), Shiraz, Iran, 12–14 February 2019; IEEE: New York, NY, USA, 2019; pp. 624–628.
37. Mohammad, M.; Choi, S.; Islam, Z.; Kwak, S.; Baek, J. Core design and optimization for better misalignment tolerance and higher range of wireless charging of PHEV. *IEEE Trans. Transp. Electrif.* **2017**, *3*, 445–453. [CrossRef]
38. Barakat, A.; Yoshitomi, K.; Pokharel, R.K. Design approach for efficient wireless power transfer systems during lateral misalignment. *IEEE Trans. Microw. Theory Tech.* **2018**, *66*, 4170–4177. [CrossRef]
39. Mastri, F.; Costanzo, A.; Mongiardo, M. Coupling-independent wireless power transfer. *IEEE Microw. Wirel. Compon. Lett.* **2016**, *26*, 222–224. [CrossRef]
40. Ke, G.; Chen, Q.; Xu, L.; Wong, S.-C.; Chi, K.T. A model for coupling under coil misalignment for DD pads and circular pads of WPT system. In Proceedings of the 2016 IEEE Energy Conversion Congress and Exposition (ECCE), Milwaukee, WI, USA, 18–22 September 2016; IEEE: New York, NY, USA, 2016; pp. 1–6.
41. Li, S.; Mi, C.C. Wireless power transfer for electric vehicle applications. *IEEE J. Emerg. Sel. Top. Power Electron.* **2014**, *3*, 4–17.

42. Deng, J.; Li, W.; Nguyen, T.D.; Li, S.; Mi, C.C. Compact and efficient bipolar coupler for wireless power chargers: Design and analysis. *IEEE Trans. Power Electron.* **2015**, *30*, 6130–6140. [CrossRef]
43. Cabrera, F.L.; de Sousa, F.R. Achieving optimal efficiency in energy transfer to a CMOS fully integrated wireless power receiver. *IEEE Trans. Microw. Theory Tech.* **2016**, *64*, 3703–3713. [CrossRef]
44. Moon, S.; Moon, G.-W. Wireless power transfer system with an asymmetric four-coil resonator for electric vehicle battery chargers. *IEEE Trans. Power Electron.* **2015**, *31*, 6844–6854.
45. Zhang, J.; Yuan, X.; Wang, C.; He, Y. Comparative analysis of two-coil and three-coil structures for wireless power transfer. *IEEE Trans. Power Electron.* **2016**, *32*, 341–352. [CrossRef]
46. Sample, A.P.; Meyer, D.T.; Smith, J.R. Analysis, experimental results, and range adaptation of magnetically coupled resonators for wireless power transfer. *IEEE Trans. Ind. Electron.* **2010**, *58*, 544–554. [CrossRef]
47. Shevchenko, V.; Husev, O.; Strzelecki, R.; Pakhaliuk, B.; Poliakov, N.; Strzelecka, N. Compensation topologies in IPT systems: Standards, requirements, classification, analysis, comparison and application. *IEEE Access* **2019**, *7*, 120559–120580. [CrossRef]
48. Elliott, G.A.; Raabe, S.; Covic, G.A.; Boys, J.T. Multiphase pickups for large lateral tolerance contactless power-transfer systems. *IEEE Trans. Ind. Electron.* **2009**, *57*, 1590–1598. [CrossRef]
49. Budhia, M.; Covic, G.A.; Boys, J.T. Design and optimization of circular magnetic structures for lumped inductive power transfer systems. *IEEE Trans. Power Electron.* **2011**, *26*, 3096–3108. [CrossRef]
50. Bosshard, R.; Kolar, J.W.; Mühlethaler, J.; Stevanović, I.; Wunsch, B.; Canales, F. Modeling and η-α-Pareto Optimization of Inductive Power Transfer Coils for Electric Vehicles. *IEEE J. Emerg. Sel. Top. Power Electron.* **2014**, *3*, 50–64. [CrossRef]
51. Sallán, J.; Villa, J.L.; Llombart, A.; Sanz, J.F. Optimal design of ICPT systems applied to electric vehicle battery charge. *IEEE Trans. Ind. Electron.* **2009**, *56*, 2140–2149. [CrossRef]
52. Ahmad, A.; Alam, M.S.; Chabaan, R. A comprehensive review of wireless charging technologies for electric vehicles. *IEEE Trans. Transp. Electrif.* **2017**, *4*, 38–63. [CrossRef]
53. Miller, J.M.; Daga, A. Elements of wireless power transfer essential to high power charging of heavy duty vehicles. *IEEE Trans. Transp. Electrif.* **2015**, *1*, 26–39. [CrossRef]
54. Bandyopadhyay, S.; Dong, J.; Qin, Z.; Bauer, P. Comparison of Optimized Chargepads for Wireless EV Charging Application. In Proceedings of the 2019 10th International Conference on Power Electronics and ECCE Asia (ICPE 2019-ECCE Asia), Busan, Republic of Korea, 27–30 May 2019; IEEE: New York, NY, USA, 2019; pp. 1–8.
55. Cheng, Y.; Shu, Y. A new analytical calculation of the mutual inductance of the coaxial spiral rectangular coils. *IEEE Trans. Magn.* **2013**, *50*, 1–6. [CrossRef]
56. Wang, J.; Ho, S.L.; Fu, W.; Sun, M. Analytical design study of a novel witricity charger with lateral and angular misalignments for efficient wireless energy transmission. *IEEE Trans. Magn.* **2011**, *47*, 2616–2619. [CrossRef]
57. Chatterjee, S.; Iyer, A.; Bharatiraja, C.; Vaghasia, I.; Rajesh, V. Design optimisation for an efficient wireless power transfer system for electric vehicles. *Energy Procedia* **2017**, *117*, 1015–1023. [CrossRef]
58. Barman, S.D.; Reza, A.W.; Kumar, N.; Karim, M.E.; Munir, A.B. Wireless powering by magnetic resonant coupling: Recent trends in wireless power transfer system and its applications. *Renew. Sustain. Energy Rev.* **2015**, *51*, 1525–1552. [CrossRef]
59. Kalwar, K.A.; Aamir, M.; Mekhilef, S. Inductively coupled power transfer (ICPT) for electric vehicle charging—A review. *Renew. Sustain. Energy Rev.* **2015**, *47*, 462–475. [CrossRef]
60. Jiang, C.; Chau, K.; Liu, C.; Lee, C.H. An overview of resonant circuits for wireless power transfer. *Energies* **2017**, *10*, 894. [CrossRef]
61. Sohn, Y.H.; Choi, B.H.; Lee, E.S.; Lim, G.C.; Cho, G.-H.; Rim, C.T. General unified analyses of two-capacitor inductive power transfer systems: Equivalence of current-source SS and SP compensations. *IEEE Trans. Power Electron.* **2015**, *30*, 6030–6045. [CrossRef]
62. Luo, Z.; Wei, X. Analysis of square and circular planar spiral coils in wireless power transfer system for electric vehicles. *IEEE Trans. Ind. Electron.* **2017**, *65*, 331–341. [CrossRef]
63. Zhang, W.; Mi, C.C. Compensation topologies of high-power wireless power transfer systems. *IEEE Trans. Veh. Technol.* **2015**, *65*, 4768–4778. [CrossRef]
64. Wang, C.-S.; Covic, G.A.; Stielau, O.H. Power transfer capability and bifurcation phenomena of loosely coupled inductive power transfer systems. *IEEE Trans. Ind. Electron.* **2004**, *51*, 148–157. [CrossRef]
65. Khaligh, A.; Dusmez, S. Comprehensive topological analysis of conductive and inductive charging solutions for plug-in electric vehicles. *IEEE Trans. Veh. Technol.* **2012**, *61*, 3475–3489. [CrossRef]
66. ElGhanam, E.; Hassan, M.; Osman, A.; Kabalan, H. Design and performance analysis of misalignment tolerant charging coils for wireless electric vehicle charging systems. *World Electr. Veh. J.* **2021**, *12*, 89. [CrossRef]
67. Wang, C.-S.; Stielau, O.H.; Covic, G.A. Design considerations for a contactless electric vehicle battery charger. *IEEE Trans. Ind. Electron.* **2005**, *52*, 1308–1314. [CrossRef]
68. Chinthavali, M.; Wang, Z.J. Sensitivity analysis of a wireless power transfer (WPT) system for electric vehicle application. In Proceedings of the 2016 IEEE Energy Conversion Congress and Exposition (ECCE), Milwaukee, WI, USA, 18–22 September 2016; IEEE: New York, NY, USA, 2016; pp. 1–8.

69. Spanik, P.; Frivaldsky, M.; Drgona, P.; Jaros, V. Analysis of proper configuration of wireless power transfer system for electric vehicle charging. In Proceedings of the 2016 ELEKTRO, Strbske Pleso, Slovakia, 16–18 May 2016; IEEE: New York, NY, USA, 2016; pp. 231–237.
70. Kalwar, K.A.; Mekhilef, S.; Seyedmahmoudian, M.; Horan, B. Coil design for high misalignment tolerant inductive power transfer system for EV charging. *Energies* **2016**, *9*, 937. [CrossRef]
71. Lam, K.; Ko, K.; Tung, H.; Tung, H.; Tsang, K.; Lai, L. ZigBee electric vehicle charging system. In Proceedings of the 2011 IEEE International Conference on Consumer Electronics (ICCE), Las Vegas, NV, USA, 9–12 January 2011; IEEE: New York, NY, USA, 2011; pp. 507–508.
72. Aditya, K.; Williamson, S.S. Design considerations for loosely coupled inductive power transfer (IPT) system for electric vehicle battery charging-A comprehensive review. In Proceedings of the 2014 IEEE Transportation Electrification Conference and Expo (ITEC), Dearborn, MI, USA, 15–18 June 2014; IEEE: New York, NY, USA, 2014; pp. 1–6.
73. Al-Saadi, M.; Ibrahim, A.; Al-Omari, A.; Al-Gizi, A.; Craciunescu, A. Analysis and comparison of resonance topologies in 6.6 kW inductive wireless charging for electric vehicles batteries. *Procedia Manuf.* **2019**, *32*, 426–433. [CrossRef]
74. Chen, Y.; Yang, B.; Kou, Z.; He, Z.; Cao, G.; Mai, R. Hybrid and reconfigurable IPT systems with high-misalignment tolerance for constant-current and constant-voltage battery charging. *IEEE Trans. Power Electron.* **2018**, *33*, 8259–8269. [CrossRef]

Disclaimer/Publisher's Note: The statements, opinions and data contained in all publications are solely those of the individual author(s) and contributor(s) and not of MDPI and/or the editor(s). MDPI and/or the editor(s) disclaim responsibility for any injury to people or property resulting from any ideas, methods, instructions or products referred to in the content.

Article

Battery Sharing: A Feasibility Analysis through Simulation

Mattia Neroni [1], Erika M. Herrera [2], Angel A. Juan [3,4,*], Javier Panadero [5] and Majsa Ammouriova [2]

1. aHead R&D Team, Spindox SpA, 10149 Torino, Italy; mattia.neroni@ahead-research.com
2. Computer Science Department, Universitat Oberta de Catalunya, 08018 Barcelona, Spain; eherreramac@uoc.edu (E.M.H.); mammouriova@uoc.edu (M.A.)
3. Department of Applied Statistics and Operations Research, Universitat Politècnica de València, 03801 Alcoy, Spain
4. Department of Management, Euncet Business School, 08225 Terrassa, Spain
5. Department of Management, Universitat Politècnica de Catalunya, 08028 Barcelona, Spain; javier.panadero@upc.edu
* Correspondence: ajuanp@upv.es

Abstract: Nowadays, several alternatives to internal combustion engines are being proposed in order to reduce CO_2 emissions in freight transportation and citizen mobility. According to many experts, the use of electric vehicles constitutes one of the most promising alternatives for achieving the desirable reductions in emissions. However, popularization of these vehicles is being slowed by long recharging times and the low availability of recharging stations. One possible solution to this issue is to employ the concept of battery sharing or battery swapping. This concept is supported by important industrial partners, such as Eni in Italy, Ample in the US, and Shell in the UK. This paper supports the introduction of battery swapping practices by analyzing their effects. A discrete-event simulation model is employed for this study. The obtained results show that battery sharing practices are not just a more environmentally and socially friendly solution, but also one that can be highly beneficial for reducing traffic congestion.

Keywords: electric vehicles; mobility; battery sharing; battery swapping; discrete-event simulation

Citation: Neroni, M.; Herrera, E.M.; Juan, A.A.; Panadero, J.; Ammouriova, M. Battery Sharing: A Feasibility Analysis through Simulation. Batteries 2023, 9, 225. https://doi.org/10.3390/batteries9040225

Academic Editor: Quanqing Yu

Received: 31 January 2023
Revised: 30 March 2023
Accepted: 7 April 2023
Published: 11 April 2023

Copyright: © 2023 by the authors. Licensee MDPI, Basel, Switzerland. This article is an open access article distributed under the terms and conditions of the Creative Commons Attribution (CC BY) license (https://creativecommons.org/licenses/by/4.0/).

1. Introduction

Electric vehicles (EVs) are among the most prominent and valid alternatives to internal combustion-based vehicles [1,2]. Even in industrialized and technologically advanced countries, some customers are insecure regarding the transition to EVs [3]. Some studies have assessed the life cycle of EVs and have concluded that greenhouse gas emissions are reduced when these vehicles are employed [4,5]. The life cycle of EVs includes their production, use, and recycling. However, toxicity level increases because of the higher exposure to chemicals and metals during the life cycle of EVs [4]. The initial cost of EVs is expected to be higher than conventional diesel or petrol vehicles. Still, in the long run, the cost of powering them is found to be much cheaper [6,7]. In addition, hydrogen vehicles seem to be an alternative in practice, but unfortunately they are still expensive to refuel. This is due to the fact that hydrogen is expensive to produce and refine [8]. They are similar in design to EVs, since they are also powered by fuel cells, which are characterized by an anode, a cathode, and a catalyst that triggers the separation of protons and electrons. Electrons are removed from the hydrogen, sent to the power motor, and combined with oxygen to form water vapor. The energy generated can be used both to power the electric engine directly or to recharge small lithium-ion batteries, which allows the vehicle to store the energy for later use.

Despite of their huge potential, the large-scale dissemination of EVs is hindered by low efficiency and long charging times. Charging times vary depending on the technology [9] used. These can range from around half an hour to several hours. Waiting for about half an hour (or even more) for vehicles to recharge is not handy, realistic, or compliant with modern commercial transportation requirements, including the car-sharing and ride-sharing mobility modes. In addition, charging at home is not an option for EV users that do not have home charger. Some representatives of the automotive sector and battery manufacturers are working to provide valid alternatives to lithium-ion batteries in terms of duration and charging time. However, this will require big investments and a consistent research effort.

To the best of our knowledge, given the urgency and rapid changes required, a more straightforward solution relies on the concept of battery sharing (BS) [10,11], which is also known as battery swapping. This refers to a situation where drivers of EVs, once arrived at a charging station, remove exhausted batteries and replace them with the charged batteries previously left by somebody else. This approach might be more convenient [12], and it gives rise to a scenario that may be enhanced by previously developed technologies provided by companies such as Ample in the US, Eni in Italy, or NIO in China.

The considered BS scenario has several requirements. First of all, batteries need to be standard or clustered in as few categories as possible. Secondly, they must be redistributed, since there might be more vehicles traveling in one direction rather than the other. Whenever these requirements are correctly fulfilled, BS may lead to an increase in efficiency of commercial travel, a reduction in waiting times at charging stations, and a more handy and user-friendly way to use EVs [13].

Battery sharing is also a more environmentally and socially friendly scenario for the following reasons:

- The service life of batteries is decoupled from the service life of vehicles. In this way, the use of batteries might significantly increase.
- Batteries would be managed by a few easy-to-control companies, thus reducing the risk of illegal disposal.
- The need to redistribute batteries may give rise to many new job positions.

The main goal of this work is to study the performance of some indicators when a BS strategy is employed. For this, we use a discrete-event simulation model that allows us to measure the benefits of employing such a strategy and gain insight into how BS can support the popularization of EVs. The remaining sections are distributed as follows. Section 2 discusses the current state of the art on EVs and battery sharing. Then, Section 3 introduces the main components and the architecture of our simulation model. This simulation model is available at https://github.com/mattianeroni/batteryswap (accessed on 8 April 2023). Section 4 describes the algorithms incorporated into the simulation, which allows the reader to better understand how it works. In Section 5, the BS scenario is validated and compared to a classic approach with no sharing in several road networks across Europe and considering dynamic conditions. Finally, conclusions and future research perspectives are discussed in Section 6.

2. Related Work

Researcher interest in studying energy consumption has increased exponentially during the last decade. According to the Scopus database, 205, 540 documents have been published between 2010 and 2022 concerning energy consumption. This trend is associated with the energy crisis, increased energy demand, and rising energy prices. Many of these documents focused on studying models to estimate energy consumption and investigate solutions to reduce this consumption. One of these solutions is to utilize EVs. These vehicles can substitute traditional fossil fuel vehicles and can also include so-called unmanned aerial vehicles (drones). EVs have a significant role in diminishing environmental pollution and counterbalancing the effects of fossil fuel-based energy [4,14]. However, there are several obstacles that make it difficult to complete the transition to sustainable transportation by utilizing electric power systems. These include the potential increase in electric power demand beyond current generation capabilities and the continued use of fossil fuel-based electric energy sources in the industry [15]. In addition, while EVs are emission-free, the batteries used in them are not environmentally clean in terms of production. Furthermore, an increase in EV penetration can lead to a rise in peak load demand due to unexpected individual EV charging times, which implies that more power plants must be constructed to maintain grid stability. There are several factors that influence consumer perception, such as the limited drive range [14] and extended charging time of EVs [9,16]. International standards have been defined to regulate the charging process of EVs, such as SAE J1772 and IEC 61851-1 [17]. For example, the IEC 61851-1 standard classifies charging systems in Europe and some other countries into four modes. Each mode defines the charging power, protection installation, and socket type.

Sanguesa et al. [2] reviewed batteries and their technology in EVs. These authors highlighted several battery characteristics, such as their capacity, energy density, specific power, life span, internal resistance, and efficacy. These characteristics depend on the battery technology. One of oldest and most widely used types of batteries is the lead-acid battery [2,18]. Because of the low energy density of lead-acid batteries [19], the battery industry faced developments in battery technology, and new types of batteries were developed, such as lithium-ion batteries [2,5] and zinc-nickel batteries [19]. Researchers have compared between the developed battery types and investigated possible improvements in their characteristics [2,19,20].

In recent times, various models for analyzing energy consumption of EVs have been established in the scientific literature. These models generally involve the examination of real-world data and investigating the impact of various factors on the energy consumption of EVs. Li et al. [21] employ a design of experiments to analyze the statistical significance of various factors that affect the energy consumption of EVs. Despite the large number of potential factors, their study focuses on four factors and their interactions, utilizing an empirical approach that has been tested with a real-world case. The results indicate that the use of heating, ventilation, as well as air conditioning and topography has the greatest impact on energy consumption compared to other factors. Yang et al. [22] investigate the impact of the road tilt angle on the energy consumption of EVs. Their study quantifies how the energy consumption increases as the tilt angle of the road increases on uphill sections and decreases on downhill sections. In another work, Mediouni et al. [23] propose a machine learning model for the prediction of the energy consumption based on vehicle speed and acceleration, road tilt angel, road coefficient of rolling resistance, wind speed, and extra weight carried in vehicles. Fiori et al. [14] present the Virginia Tech comprehensive power-based EV energy consumption model (VT-CPEM) as a solution to enhance the limited driving range and harness the energy obtained during braking. The proposed model is an EV energy model that utilizes instantaneous data—such as vehicle speed, acceleration, and roadway grade—to calculate the energy consumption of the vehicle for a specific driving cycle. Chen et al. [24] emphasize the significance of predicting future driving conditions such as vehicle speed, recognizing driving styles and patterns, or anticipating terrain information. They use this information to optimize energy consumption. Neural networks, which are primarily utilized for route and driving pattern recognition, are presented as the most promising equivalent factor adaptation methods for energy consumption management systems in EVs. Moreover, machine learning techniques are shown to have the capability to enhance control decisions through the use of historical and real-time driving data. Wu et al. [25] propose a system for collecting in-use EV data. The system collected and analyzed approximately 5 months of data. It was found that providing the driver with timely information on energy usage could lead to conscious adjustments in driving behavior, which in turn results in a reduction in energy consumption. An analytical model is also presented to estimate the instantaneous power and trip energy consumption in real time. It is based on the principles of vehicle dynamics and the correlation among various essential variables. The authors concluded that driving on urban city streets requires less power compared to driving on freeways—notice that this differs from the typical behavior of internal combustion vehicles. Wager et al. [26] studied energy consumption and driving range of EVs under the influence of various parameters using mathematical modeling. They found that the driving range is significantly reduced under high speed, as well as under scenarios where the overall weight increases. This is particularly true when the vehicle has also to face headwinds (20 km/h or more) and a roof rack. Their study also highlights that the combination of the 80% charge level limit of fast DC charging, battery aging, and battery safety margins result in not utilizing the full battery capacity. In other words, the battery weight is not being used but it significantly contributes to the vehicle's weight. Chang et al. [27] introduce an instantaneous power modeling of EVs to study the effect of velocity, acceleration, road slope, and EV weight on the battery power consumption. In addition, they consider the effect of an onboard charger and regenerative braking on the battery charging power. All the mentioned models found in the literature could potentially be used as an online energy management strategy for on-board control logic [28].

The research directed to batteries is not limited to the estimation of energy consumption and affecting factors, but researchers also considered applying a circular economy regarding the manufacturing of batteries [29,30]. This research trend studies the handling of battery waste and re-manufacturing of batteries, as well as the collection of used batteries. Another research trend is focused on operational decisions concerning the utilization of the EVs that used these batteries. Thus, researchers planned routes that minimize traveled time and distance [31,32], resulting in the green vehicle routing problem [33]. Reducing the travel time and distance could reduce the energy consumption of vehicles and, hence, have a positive impact on the environment. In this context, approaches introducing battery sharing (battery swapping) were introduced [34]. For example, Li et al. [35] studied energy consumption in a vehicle routing problem of EVs where battery swapping is allowed. They aimed to support logistics enterprises so they can reduce energy consumption and gas emissions while delivering goods. This solution is an alternative to battery charging and affects the planning of EVs routes. Utilizing battery sharing might help to overcome one of the EV constraints presented as limiting driving range and long charging times. Thus, this paper investigates this approach using a simulation model.

3. Simulation Components and Architecture

In this paper, the simulation model consists of several components, as illustrated in Figure 1 and described in the following subsections. In addition, our simulation architecture includes algorithmic modules that are described in Section 4.

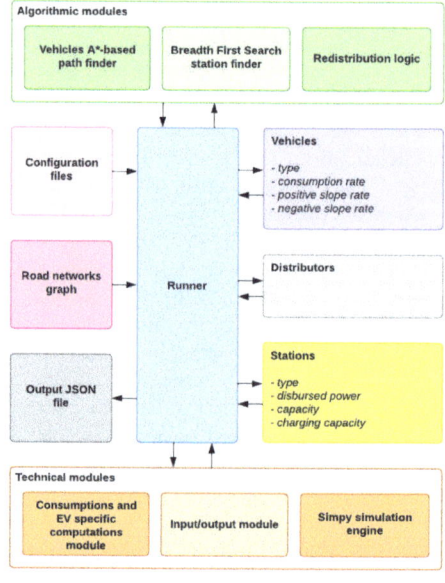

Figure 1. General architecture of the developed simulation library.

3.1. Batteries

Batteries are the lowest-level elements of the simulation. The proposed simulation library offers the possibility of introducing new battery types. Each type is characterized by a certain capacity (expressed in kWh) and can be limited to a specific vehicle, i.e., each vehicle may require a specific type of battery. Each battery instance is characterized by level attributes that describe the remaining energy level. Batteries move around the network carried by vehicles and can be left at charging stations or redistributed among charging stations.

3.2. Vehicles

Vehicles are powered by batteries and move around the road network during a simulation run. Even for vehicles, it is possible to define a set of vehicle types, and each vehicle type will mount a certain number of batteries, all of them belonging to a specific type. It is possible to associate a consumption rate (expressed in kWh per kilometer) with any vehicle type, as well as a positive and negative slope rate, i.e., a percentage that quantifies the effect of street slopes on vehicle consumption. Vehicle consumption is computed as follows:

$$consumption = L \cdot c \cdot (1 + s \cdot \theta)$$

where L is the road length, c is the vehicle consumption rate, θ is the slope of the road, and s refers to the vehicle slope rate, which is positive in the case of positive slope and vice versa. This relationship between road slope and EV consumption is based on the work by Yang et al. [22] and Anselma et al. [28].

Each vehicle type may represent a different class of vehicle characterized by the number of batteries and a specific consumption rate (which is associated with its weight). For instance, we may define three types of vehicles representing trucks, sport utility vehicles (SUVs), and city cars. Each of these vehicle types will have a different consumption rate and will require a different number and type of batteries.

3.3. Distributors

Distributors are particular vehicles in charge of taking unused batteries in a charging station and moving them to another station which is supposed to have a higher demand. For the sake of simplicity, in this work we have assumed that the fleet of distributors is large enough, so we do not need to worry about capacity limitations or recharging operations in vehicles belonging to this fleet.

3.4. Stations

Stations are the network nodes where vehicles can stop to charge or swap batteries. Charging takes place if no sharing is carried out, while swapping takes place only in case of battery sharing. Stations are split into types, and each type is characterized by a certain disbursed power (expressed in kW), a capacity, i.e., the number of vehicles that can be processed at the same time, and a charging capacity, representing the maximum number of batteries of each type the station can charge at the same time; if a station does not handle any battery type, this number can be simply set to 0.

In our experiments, battery charge is assumed to be linear with time, as discussed in Zeng et al. [36] and Cataldo-Díaz et al. [37]. In addition Cataldo-Díaz et al. [37] assumed linear recharge at stations, despite the fact that this assumption might not be fully accurate when using high recharging speeds. Hence, given a battery capacity C, a starting battery power level, and the station disbursed power p, the charging time is computed as follows:

$$time = (C - level)/p$$

3.5. Road Network

The road network is a directed graph including additional details that concern road slope, charging stations, real-time vehicle positions, and distribution of batteries among charging stations.

3.6. Runner

The runner is the main element of the simulation and is in charge of coordinating and running all the simulation processes. Given the number of travelling vehicles in the simulation N, the main process is designed to keep this number constant. Hence, it initially instantiates and executes N processes (one per vehicle). Then, as soon as a process is concluded, a new one is started. It may be represented as in Algorithm 1.

Algorithm 1 Main simulation process

Instantiate N vehicle trip processes
while Simulation is not concluded **do**
    Get the concluded vehicle trip processes
    for each process concluded **do**
        Instantiate a new vehicle trip process
    end for
end while

In the case of sharing, a parallel process is executed; this process is analogous to the one mentioned in Algorithm 1. However, while the main process is in charge of keeping constant the number of traveling EVs, i.e., the traffic intensity, this vehicle handles the redistributors, so that the redistribution of batteries never stops.

4. Incorporated Algorithms

In this section, a deeper description of the algorithmic modules represented in Figure 1 is provided. The algorithms herein described are essential for the simulation to function. Many alternative solutions can be designed and implemented in the simulation model, which allows us to analyze their performance. Notice, however, that the objective of this study is not to find an optimal solution, but rather to reproduce reasonable decisions a standard driver would make in his/her daily activity. The main decision-making processes to consider concern how drivers define the path and how battery redistribution is carried out.

4.1. Driver's Path Definition

The procedure for determining the path of a vehicle when traversing from point a to point b involves the following steps:

1. Using the A^* algorithm [38], find the shortest path from a to b.
2. If a path is not possible, we have a graph error and the trip is considered concluded as well as excluded from the final statistics.
3. If a path can be covered without stopping at the charging stations, the trip is simulated and the process concluded.
4. If b cannot be reached without stops, the algorithm iterates the nodes from a to b looking for a charging station s.
5. If s exists and is on the original path, we simulate a trip from a to s, set $a = s$, and go to step 1. Otherwise, if there are no stations on the path or the station s cannot be reached with residual battery, the algorithm moves to the next step.
6. For each node i from the last reachable node to a, consider i as the root and start a breadth-first search method [39], looking for a station outside the original path. If a station is found, then simulate a trip to s, set $a = s$, and go to step 1.

4.2. Battery Redistribution

The process of redistributing batteries among charging stations is extremely important in periods when most of the vehicles are moving. In these periods, some stations could end up without batteries. As mentioned above, each station has a charging capacity describing the maximum number of batteries that it can charge at the same time. However, vehicles are allowed to leave exhausted batteries in a station even if the station charging capacity is saturated. These batteries will take the name of batteries on-the-side and will remain unused until the station has the opportunity to charge them. Distributors are responsible for identifying the nearest station with a critical number of batteries on-the-side and shipping them to the station with the greatest need for additional batteries. The path traversed by distributors is again determined using the A^* algorithm. As mentioned previously, distributors do not have loading capacity limitations and do not require recharging or refueling operations.

5. Validation and Results

In order to validate the proposed concepts and show the benefits that battery sharing can provide, our simulation model was used to analyze several scenarios. First of all, four different road networks were used (Figure 2): (i) a small section of the Chicago (IL) downtown area, representing a small test network; (ii) the urban area of Modena, a typical Italian medium-sized city; (iii) the urban area of Sassari, an Italian city representing an ancient road network; and (iv) the urban area of Barcelona, representing a modern and large European city. As mentioned, each selected network has some peculiarities that make it a good representative for a specific scenario. The details that concern the road networks were extracted from Open Street Maps (OSM) by using the standard API. Concerning the elevation and altitude, since OSM does not provide this information, the possibility of querying the Open Elevation (https://open-elevation.com/, accessed on 8 April 2023) and National Map (https://www.usgs.gov/the-national-map-data-delivery, accessed on 8 April 2023) services was implemented. For this study, elevation data were taken from Open Elevation. For each road network, we simulated three different situations: (i) a scenario in which no sharing is carried out (i.e., when a vehicle arrives to a charging station, it simply waits to be charged); (ii) a scenario with battery sharing where drivers are allowed to take only fully charged batteries; and (iii) a scenario with battery sharing where even partially charged batteries can be picked up.

The proposed simulation model described in Section 3 offers many possibilities to customize the system's behavior. However, in order to carry out a feasibility analysis and validate the BS approach, we will standardize it using the following configuration:

- The simulation handles three types of batteries—small, medium, and large—with respective capacities 10, 15, and 20 kWh. For experimental purposes, we have considered that the batteries do not reduce their performance with use. Thus, the simulator assumes that the capacity of the batteries does not decrease with the charge cycles.
- In each road network, exactly 2% of nodes are characterized by the presence of a charging station. This results in a different number of charging stations for each network: 4 charging stations in the test network, 69 in Sassari's network, 66 in Modena's network, and 196 in Barcelona's network.
- The simulation handles three types of vehicles. The first two types are more frequent and powered by two batteries each, while the latter is less frequent and powered by three batteries.
- All vehicles have the same consumption rate, 0.3 kWh/km, and their consumption is equally affected by the road slope.
- The simulation handles two types of equally spread charging station, small and large. The first one can process only one vehicle at a time and disburses a power of 10 kWh. The second one can process two vehicles together and disburses a power of 12 kWh. The charging time can be easily estimated using the following equation:

$$ChargingTime = \frac{BatteryCapacity \cdot CurrentBatteryLevel}{DisbursedPower}$$

- The number of traveling vehicles is constant, exactly 1,100 vehicles, during the entire duration of the simulation.
- Battery swapping takes 60 s.
- There are 10 vehicles dedicated to battery redistribution, and a new redistribution is carried out every hour if the previous one is already concluded.
- The simulation runs for 8 h.

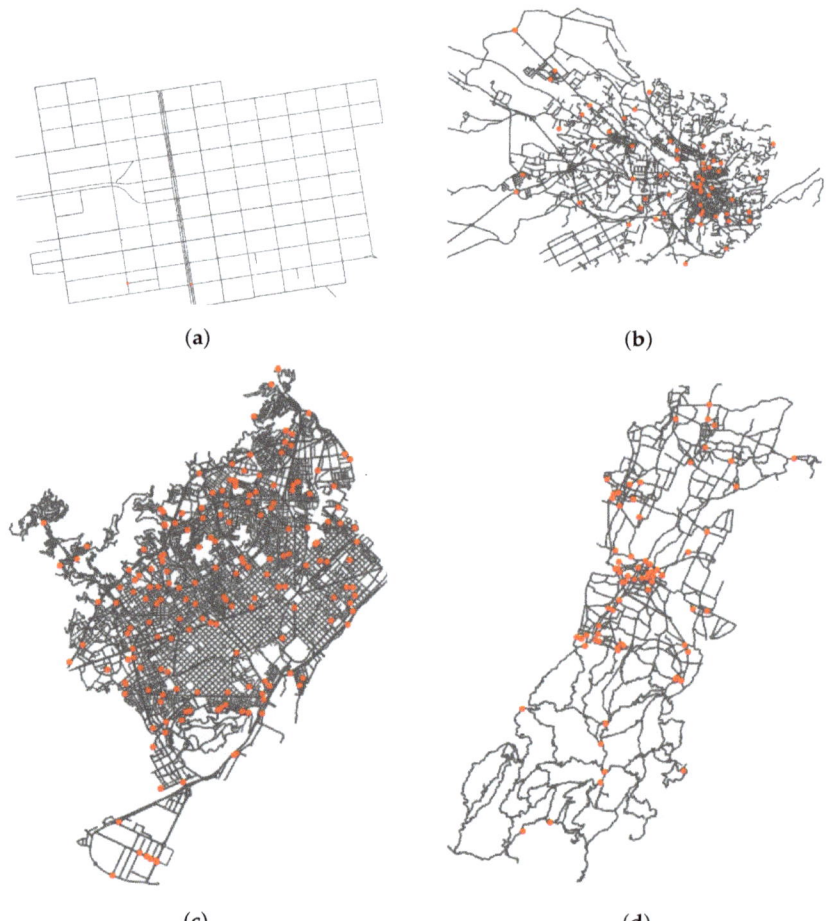

Figure 2. Main streets of considered roads networks. (**a**) Test network (**b**) Sassari (**c**) Barcelona (**d**) Modena.

Notice that all technical and quantitative data we assumed herein are in line with the current technical solutions on the market, as well as well-justified by scientific publications [13,16,30].

Results are presented in terms of relative travel time (i.e., travel time/covered distance), average waiting time at charging stations (i.e., difference between the time at which the vehicle enters the charging station to be processed and the time at which it enters the queue), and average queue at the entrance of the charging station. Results are reported in Tables 1–4, which are also summarized in Figures 3–5.

Table 1. Results concerning the test network.

Scenario	Relative Travel Time [min/km]	Waiting Time [s]	Queue
Sharing of partially charged batteries	1.020	127	3.247
Sharing of fully charged batteries	1.091	3060	77.104
No sharing	1.200	5803	103.570

Table 2. Results concerning Sassari's city network.

Scenario	Relative Travel Time [min/km]	Waiting Time [s]	Queue
Sharing of partially charged batteries	1.020	62	0.154
Sharing of fully charged batteries	1.2	5517	30.004
No sharing	1.38	8917	31.692

Table 3. Results concerning Modena's city network.

Scenario	Relative Travel Time [min/km]	Waiting Time [s]	Queue
Sharing of partially charged batteries	1.020	62	0.126
Sharing of fully charged batteries	1.14	2020	6.679
No sharing	1.200	4091	15.641

Table 4. Results concerning Barcelona's city network.

Scenario	Relative Travel Time [min/km]	Waiting Time [s]	Queue
Sharing of partially charged batteries	1.020	31	0.037
Sharing of fully charged batteries	1.083	1074	2.012
No sharing	1.267	2933	5.553

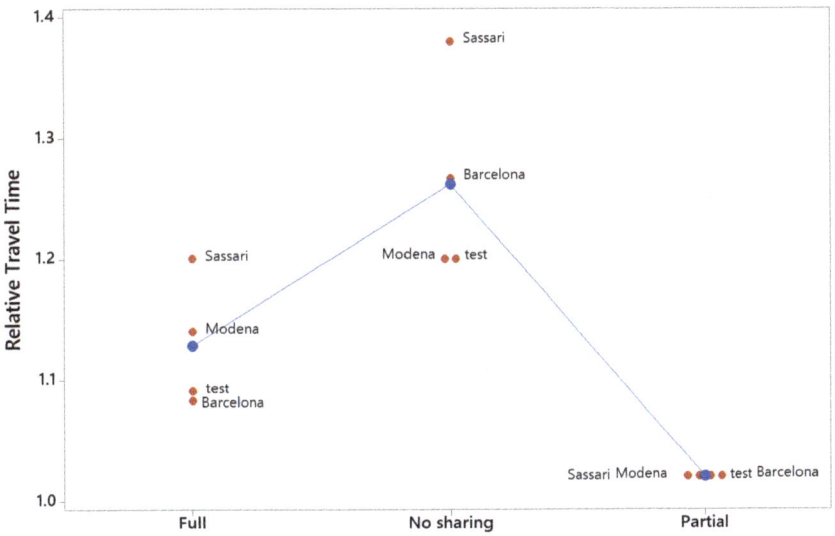

Figure 3. Comparison of relative travel time for each scenario.

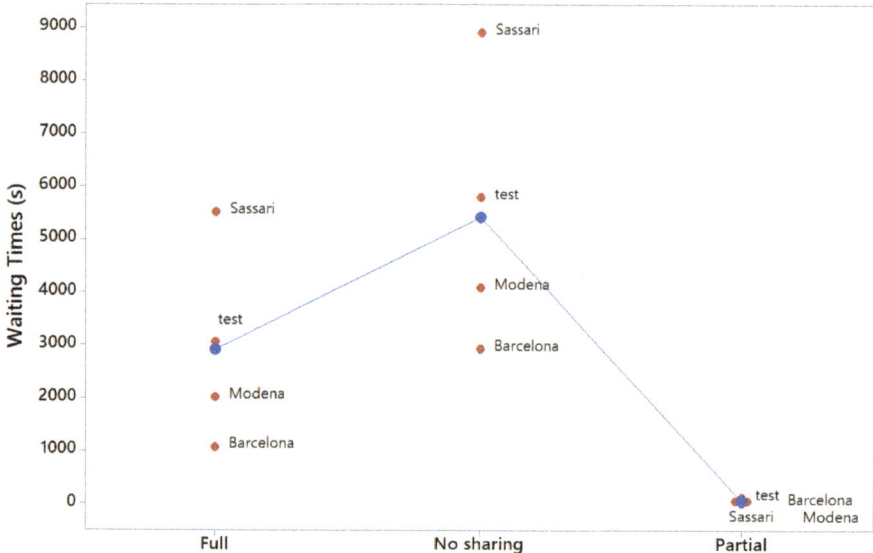

Figure 4. Comparison of waiting times for each scenario.

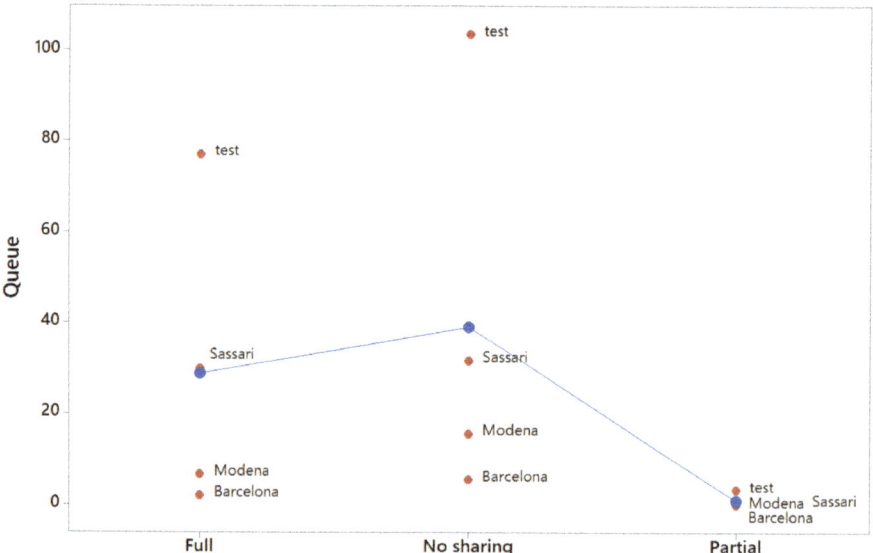

Figure 5. Comparison of queues for each scenario.

As can be seen in these figures, the scenario without battery sharing shows visible traffic congestion that would require the installation of further charging stations. The waiting time at charging stations always exceeds 1 h (and sometimes exceeds 2 h). The relative travel time is higher (because of waiting time at charging stations), and queues exceed 103 vehicles in the test network, 30 vehicles in Sassari city, 15 vehicles in Modena city, and 5 vehicles in Barcelona (since the number of vehicles is constant, the queue length is higher in small networks characterized by fewer charging stations). All conditions being equal, the adoption of partially charged battery sharing leads to a perfectly feasible situation with no need of further charging stations. The relative travel time is reduced, while the waiting time and the queues are almost non-existent. Of course, one should notice that sharing of partially charged batteries generally requires drivers to take more stops during a single trip. Therefore, it may be convenient in terms of the overall network but inefficient from a drivers' perspective. The scenario in which battery sharing is allowed only for fully charged batteries lays in between the other two scenarios. The average waiting time at charging stations is between 30 min and 1 h. This can be considered as an amount of time reasonable according to current standards, even though it can be improved. The queue length at charging stations is often excessive, especially in small networks, and the relative travel time lays in between the results obtained in other scenarios.

As an example, in order to provide an intuitive and visual representation, we can compare in Figures 6 and 7 a representation of the Barcelona city network before and after implementing battery sharing. In the figures, we can see in red, orange, and green the streets with high, medium, and low traffic, respectively. The red points represents the charging stations, and it is easy to see how, in the case of no sharing, the most congested streets are the ones characterized by the presence of a charging station, where vehicles waste a considerable amount of time.

Additional results are obtained by increasing the number of redistributors, i.e., vehicles in charge of moving batteries from one charging station to another, from 10 to 100 vehicles. Table 5 shows how sharing of fully charged batteries may be easily improved with a relatively low budget if compared to the construction of new charging stations required by a scenario with no sharing.

Figure 6. A view of Barcelona city network with no battery sharing.

Figure 7. A view of Barcelona city network after implementing battery sharing.

Table 5. Results obtained increasing the number of redistributors in a scenario of fully charged battery sharing.

Scenario	Relative Travel Time [min/km]	Waiting Time [s]	Queue
Test network	1.158	3010	45.210
Sassari	1.28	2012	5.110
Modena	1.13	1920	3.029
Barcelona	1.083	433	1.512

6. Conclusions and Future Work

This paper has analyzed a novel concept: the impact of using a strategy of battery sharing in electric vehicles as an alternative to a full charging strategy, which is not handy and efficient at the moment mainly due to long charging times. Battery sharing strategies are based on the fact that drivers remove exhausted batteries and replace them with the charged batteries previously left by somebody else (after they have been partially recharged in the station). This allows for significantly reducing the waiting queue at the charging stations. To assess how this approach may affect traffic conditions and how it relates to a classic scenario where no sharing of batteries is used, we have developed a discrete-event simulation model, which validates our approach in four different real road networks and under several environmental conditions. Results show that battery sharing, if feasible from a technological point of view, may be beneficial for traffic networks and easily scalable in case of limited resources, such as the possibility or the budget to install several high-performance charging stations. The current study does not consider imposed costs associated with swapping those batteries and their handling, which are partially caused by the need to hire more staff at the stations. We are aware that research is moving fast in this area and parameters, technologies, and interests may be subject to rapid changes. The objective of this paper was to share a first starting point of a long research project that system would perform in a real-life scenario. Our approach involves training a machine learning model, which can be a slow process in a real-life scenario. Hence, to illustrate how this learning process works, a simulation experiment is proposed. The simulation allows us to emulate how the learning process would occur in a real-life scenario and test the quality of the learning models employed in our approach. Thus, the main contribution of this paper is the combination of reinforcement learning with simulation to illustrate how to make informed decisions in a dynamic OP with electric vehicles and binary random rewards that can be influenced by several factors, including their battery status.

The remainder of the paper is structured as follows: Section 2 gives a more extensive context on the types of EVs and their integration into the ongoing transportation network, current research in the field, battery capacity, materials and charging time, and several factors that might influence the state of the battery. Section 3 provides a brief review of OPs. Section 4 describes the problem in a more formal way. Section 5 introduces the numerical case study that is considered. Section 6 discusses the details of the simulation experiment as well as the reinforcement learning algorithm employed. Section 7 analyzes the results obtained in our computational test. Finally, Section 8 highlights the main conclusions of our work and suggests open research lines.

2. Some Technical Details on EVs

The transportation sector accounted for 23% of energy-related emissions in 2019 [7]. The largest source of transport emissions are road vehicles, predominantly powered by internal combustion engines. Since EVs are a promising mitigation technology to reduce the impact on climate change, air pollutants, and noise [8], researchers have shown significant interest in studying EV energy consumption due to its importance for understanding the efficiency, performance, and environmental impact [9,10]. Several models can be used to calculate energy consumption in EVs, and Qi et al. [11] classified them into three main categories: analytical, statistical, and computational models. While accurate estimation and prediction of EVs' range is crucial for driver confidence and planning, the design and optimization of charging infrastructure networks also play an important role in ensuring efficient charging processes and minimizing energy consumption. There are several types of EVs available on the market [12]. These range from conventional vehicles with no electric components to fully EVs, including hybrid electric vehicles (HEVs) as well as battery electric vehicles (BEVs). Each type has its own advantages and considerations, and the choice depends on factors such as driving habits, range requirements, the availability of charging infrastructure, and personal preferences.

As discussed in [13], BEVs produce zero tailpipe emissions but require charging infrastructure to recharge their batteries. The battery is the core component of an EV, especially for BEVs, and can be categorized by qualities such as its specific energy, specific power, capacity, voltage, and chemistry [14]. Battery modeling is quite complex and many models have been developed: empirical, equivalent circuit, physics-based, and high level stochastic [15]. The initial technology used was a lead–acid battery that was soon replaced because of its drawbacks such as having low energy density and being heavy. Nickel-based batteries were introduced soon after as a mature technology, but they have longer recharging times and poor performance in cold weather. Currently, lithium-based batteries are the most common and widely adopted in the EV industry [16]. Even though some methods have been proposed to predict the remaining useful life of lithium batteries [17], there are still several open issues with their use: in particular, their safety as well as environmental impact concerns.

One of the main challenges raised by BEVs is to obtain acceptable driving ranges. The state of charge (SOC) measurement is particularly important as it indicates the maximum driving range. The capacity of EV batteries can range from around 20 kWh in smaller electric cars to over 100 kWh in larger, high-end electric vehicles. However, accurately estimating driving distance is difficult, and factors such as the vehicle's efficiency, driving style, and weather conditions also influence the actual driving range achieved [18,19]. The optimal SOC range in batteries can vary depending on the specific type of lithium-based battery chemistry and the desired operating parameters [20]. According to Koltermann et al. [21], the capabilities are limited at the border areas of the SOC. The results show that batteries can only safely deliver full power without a detrimental impact on their health and longevity at between 20–80% of the SOC. At both ends, the battery exhibits much higher polarization impedance [22]. The maximum SOC is usually set to ensure safe operation and prevent overcharging, stress on the material, or elevated operating temperatures [23]. Likewise, the minimum SOC tries to avoid an unexpected system shutdown or loss of power that can occur due to low voltage levels and the risk of cell imbalance. In addition, maintaining a low SOC can potentially contribute to driver anxiety. Advancements in charging infrastructure are also reducing concerns about range anxiety by providing faster and more accessible charging options for EV owners [24]. However, the number of charging stations is still relatively limited, so predicting the best place to charge the vehicle in dynamic conditions is important. There are several charging technologies available for EVs depending on the different power levels an EV can be charged to [25]. Level 1 charging (120 V AC) uses a household electrical outlet and provides around 1.4 to 1.9 kW. It provides the slowest charging rate for EVs. Level 2 charging (240 V AC) requires a compatible charging station or wall-mounted charger. It delivers power ranging from 3.3 kW to 22 kW. Level 3 or DC fast charging stations can provide charging powers ranging from 50 kW to over 350 kW. The vehicle's onboard charger and charging port specifications will determine the maximum charging rate it can accept. Among other factors, a vehicle's charge time depends on the level of the charger and the type of EV. According to the U.S. Department of Transportation [26], to reach an 80% battery level from empty can take between 40 and 50 h for a BEV, and between 5 and 6 h for a plug-in HEV (with a Level 1 charger). Likewise, it can take between 4 and 10 h for a BEV and between 1 and 2 h for a plug-in HEV (with a Level 2 charger). Finally, it can take between 20 min and 1 h for a BEV (with a Level 3 charger).

EV penetration causes significant issues on the power distribution grid, such as an increase in power demand, system losses, voltage drops, equipment overloading, and stability impact [27]. At the same time, some exciting opportunities appear with EV deployment on the smart grid, such as grid flexibility through vehicle-to-grid (V2G) technology, demand response, and the integration of renewable energy sources [25]. V2G mode allows EVs to discharge power back to the grid. This can support grid balancing, frequency regulation, and voltage stabilization. The weather conditions directly affect the battery temperature [28], the climate control [29], and the charging efficiency [30]. Extreme weather conditions can affect the temperature of the EV's battery. High temperatures can increase the risk of the battery overheating, which may reduce its performance and lifespan. Likewise, very cold temperatures can decrease the battery's efficiency and capacity temporarily. EVs often rely on climate control systems to maintain a comfortable cabin temperature. This includes features such as air conditioning in hot weather and heating in cold weather. The use of these systems can impact the overall energy consumption and range of the vehicle. Extreme temperatures can also affect the efficiency of charging systems. Congestion can generate an increase in the stop-and-go driving style. This can lead to more energy-intensive acceleration and braking, and thus result in increased energy consumption and a higher load on the battery, potentially reducing its overall range [31]. In addition, when stuck in congestion, EVs may be required to idle for extended periods, especially in situations where traffic is at a standstill. Idling consumes energy from the battery to power auxiliary systems such as climate control and entertainment systems. Prolonged idling can drain the battery charge faster and reduce the available range [32].

3. Related Work on Vehicle Orienteering Problems

The orienteering problem was first introduced by Golden et al. [3], who proved it to be NP hard. Early research examined the deterministic version of the problem within the framework of vehicle routing, where one vehicle chooses the nodes to visit and the order of visits within a defined time frame. In contrast, research on the stochastic OP is relatively recent. The first study to include stochasticity in the OP was conducted by Ilhan et al. [33], which assumed that only node rewards were stochastic. Other researchers, including Campbell et al. [34], Papapanagiotou et al. [35], Verbeeck et al. [36], or Evers et al. [37], focused on cases where service and travel times were stochastic, with service times typically incorporated into travel times. Several approaches have been employed to solve the stochastic orienteering problem, including a combination of branch and bound algorithms with local search [34] and local search simulations [38]. Further refinements were made to these methods, with Varakantham and Kumar [39] utilizing a sample average approximation technique to improve on the results of Lau et al. [38]. Additionally, Zhang et al. [40] expanded the method of Campbell et al. [34] to include time windows for arriving at nodes. Gama and Fernandes [41] introduced a solution to the orienteering problem with time windows, employing Pointer Network models trained through reinforcement learning. A comprehensive review of the orienteering problem and its variants can be found in Gunawan et al. [42]. Still, to the best of our knowledge, our work is the first one that proposes the combined use of RL and simulation to deal with a version of the problem with stochastic rewards that depend upon dynamic conditions. In our view, this dynamic OP with stochastic rewards has relevant applications to scenarios involving EVs that require an efficient management of their batteries.

4. Modeling the Dynamic OP with Binary Random Rewards

Let $G = (V, E)$ be a directed graph with node set V and edge set E. Node $O \in V$ is the origin node, while node $D \in V$ is the destination node. Each node $i \in V \setminus \{O, D\}$ has a reward r_i associated with it, which is a binary random variable that takes the value 1 with probability p_i and 0 with probability $1 - p_i$. Let us denote this by region R_k, with $k \in K = \{1, 2, \ldots, |K|\}$, a subset of $V \setminus \{O, D\}$. The problem is to find a path P that starts at O and ends at D, maximizes the expected total reward collected along the path, and visits at most one node in each region R_k. In our case, the reward is based on extending the battery's lifespan [21], as well as on completing the charging process within a certain time interval. Thus, the probability of obtaining a reward for a recharging node i depends on the type of node as well as on the current status at the region R_k. This status is determined by dynamic context conditions, i.e., $p_i = f(i, R_k)$ for an unknown (black-box) function f.

Let x_i be a binary decision variable that takes the value 1 if node i is visited along the path P, and 0 otherwise. Let e_{ij} be a binary decision variable that takes the value 1 if there is an arc from node i to node $j \neq i$ along the path P, and 0 otherwise. Then, the problem can be formulated as follows:

$$\max \sum_{i \in V \setminus \{O,D\}} r_i p_i x_i \tag{1}$$

subject to:

$$\sum_{i \in V \setminus \{O\}} e_{Oi} = 1 \tag{2}$$

$$\sum_{j \in V \setminus \{i,O\}} e_{ij} - \sum_{j \in V \setminus \{i,D\}} e_{ji} = 0 \quad \forall i \in V \setminus \{O, D\} \tag{3}$$

$$\sum_{j \in V \setminus \{D\}} e_{jD} = 1 \tag{4}$$

$$\sum_{i \in R_k} x_i = 1 \quad \forall k \in K \tag{5}$$

$$x_i \in \{0, 1\} \quad \forall i \in V \tag{6}$$

$$e_{ij} \in \{0, 1\} \quad \forall (i, j) \in E \tag{7}$$

The objective function Equation (1) maximizes the expected total reward collected along the path P. Constraint Equations (2) and (4) ensure that the path starts at the origin node O and ends at the destination node D, respectively. Constraint Equation (3) ensures that the flow of the path P is conserved at each node $i \in V \setminus \{O, D\}$, i.e., the number of incoming arcs is equal to the number of outgoing arcs. Constraint Equation (5) ensures that at most one node in each region R_k is visited. Constraint Equations (6) and (7) enforce the binary variables.

5. A Numerical Case Study

Consider a dynamic OP, with binary random demands $y \in \{0,1\}$, similar to the one represented in Figure 1. Let us assume that a vehicle must now cross six consecutive regions, with each region containing five different types of charging nodes, each type with a different probability of obtaining a reward. In accordance with Section 2, let us assume that such probability will depend on a vector of factors that describe the context conditions at each region–node pair, which are: battery age, congestion, weather, battery status, and time of charge. The logistic sigmoid function $\sigma(z) = \frac{1}{1+e^{-z}}$ maps real-valued numbers to a range between 0 and 1. In particular, for each type of node $i \in \{1, 2, \ldots, 5\}$, let us assume that the real-life probability of obtaining a reward p_i is modeled as a logistic function f, which is defined for each region, k_{1i} to k_{6i} (Table 1), and each of the following factors: (i) battery age (cycles) $ba \in \{10, 1000\}$; (ii) current battery status (in %) $be_i \in \{0, 100\}$; (iii) congestion $c_i \in \{0, 1\}$ (where $c = 1$ represents high congestion and $c = 0$ represents low congestion); (iv) weather conditions $w_k \in \{0, 1\}$ (where $w = 1$ represents good weather and $w = 0$ represents bad weather); and (v) time of charge (in hours) $tc_i \in (0, 30)$. Hence, the real-life probability p_i of obtaining a reward when visiting a node of type i in region R_k under context conditions $(ba, be_i, c_i, w_k, tc_i)$ is given by:

$$p_i = \frac{1}{1 + \exp(-(k_{1i}w_k + k_{2i}c_i + k_{3i}tc_i + k_{4i}ba + k_{5i}(20 - |be_i - 20|) + k_{6i}))} \quad (8)$$

For each of the five considered nodes, Figure 2 shows the real probability of obtaining a reward of 1 when the node is visited under different combinations of congestion, weather conditions, battery age, and battery status. Notice also that charging time has a great influence on this probability for all nodes since the objective is to extend the lifetime of the battery without compromising the time at which the vehicle reaches its destination. Moreover, the node offering a higher probability of reward might vary according to the current weather, congestion, and other conditions in the region.

Table 1. Real-life parameters for each node type (coefficients of the associated logistic function).

Node Type	k_{1i}	k_{2i}	k_{3i}	k_{4i}	k_{5i}	k_{6i}
1	3.0	−0.5	−0.4	−0.0010	−0.00010	3.5
2	1.5	−0.1	−0.41	−0.0004	−0.00030	4.5
3	3.5	−0.1	−0.43	−0.0012	−0.00010	4.5
4	2.0	−0.2	−0.50	−0.0009	−0.00035	4.0
5	4.5	−0.9	−0.42	−0.0008	−0.00030	3.0

Despite the fact that the function f that represents the real probability of obtaining a reward has been properly defined, this will not usually be possible in a real-life application. In effect, the real-life parameters shown in Table 1 will be unknown in many practical applications. Then, given a vector x of five factors associated with a region–node combination, the goal will be to predict y, i.e., $\hat{y} = P(y = 1|x)$, so the next charging node i to be visited can be selected using this estimated probability. Hence, from this point on, it is assumed that the true values of the parameters k_{1i} to k_{6i} are unknown, and a reinforcement learning algorithm is proposed in order to predict the real probability of obtaining a reward. In a real-life application, the algorithm achieves this predictive capacity by making decisions, observing the associated outcomes, and then learning from these interactions with reality. In online RL, the learning agent interacts directly with the environment in real-time. The agent makes decisions, receives feedback (rewards), and updates its policy based on the observed outcomes. The agent explores (tries out different options) and exploits (selects the best-known option) the environment simultaneously, making decisions and adapting its behavior as it interacts with the environment. In order to illustrate this learning process, simulation is employed to emulate this interaction between the algorithm and reality. A total of 10,000 trips were simulated.

At each region R_k, the algorithm must select a node type i to visit. This selection is based on the current estimate of the expected reward for each node type i, and the ϵ parameter controls the balance between exploration and exploitation. Specifically, with probability $1 - \epsilon$, the algorithm selects the node type with the highest expected reward (exploitation), and with probability ϵ, it selects a node type at random (exploration). After selecting a node type, the algorithm receives feedback from reality (the simulation in our case) in the form of a binary reward, and updates its estimate of the expected reward for each node type accordingly.

In the computational experiments a hybrid gradient boosting with decision trees model is utilized to estimate the expected reward for each region–node combination based on the feedback provided by the simulation environment. During the simulation, random values are generated to emulate the context conditions of each region. Then, using the black-box function f, the response that would be obtained in a real-life environment is estimated. The algorithm uses this response to iteratively enhance the model that predicts, for each node and contextual values, the probability of obtaining a reward. Thus, for example, suppose that after several simulation runs, the algorithm learns that nodes of type 1 are the most rewarding under good weather and low congestion conditions, nodes of type 2 are the most rewarding under bad weather and high congestion, nodes of type 3 are the most rewarding under good weather and high congestion, nodes of type 4 are the most rewarding under bad weather and low congestion, and nodes of type 5 are the most rewarding under any other context condition, it then adjusts its policy accordingly and continues to improve over time.

A conceptual schema summarizing the described methodology is provided in Figure 3. The simulation component provides new testing conditions for the RL component to make decisions (step 1). The RL component makes decisions by selecting the next charging node in the routing plan (step 2). Then, the black-box function emulating real life is employed to check the real impact of the decision suggested by the RL agent (step 3) and provide feedback to it (step 4). Finally, the simulation of a new scenario is activated and the process is repeated until a complete solution (a selection of charging nodes connecting the origin depot with the destination depot) is built. At this stage, the entire loop is iterated for a number of runs. As more and more runs are executed, the more trained the RL model becomes and, hence, the better its decisions are in terms of which charging node has to be visited next according to the vector of factors.

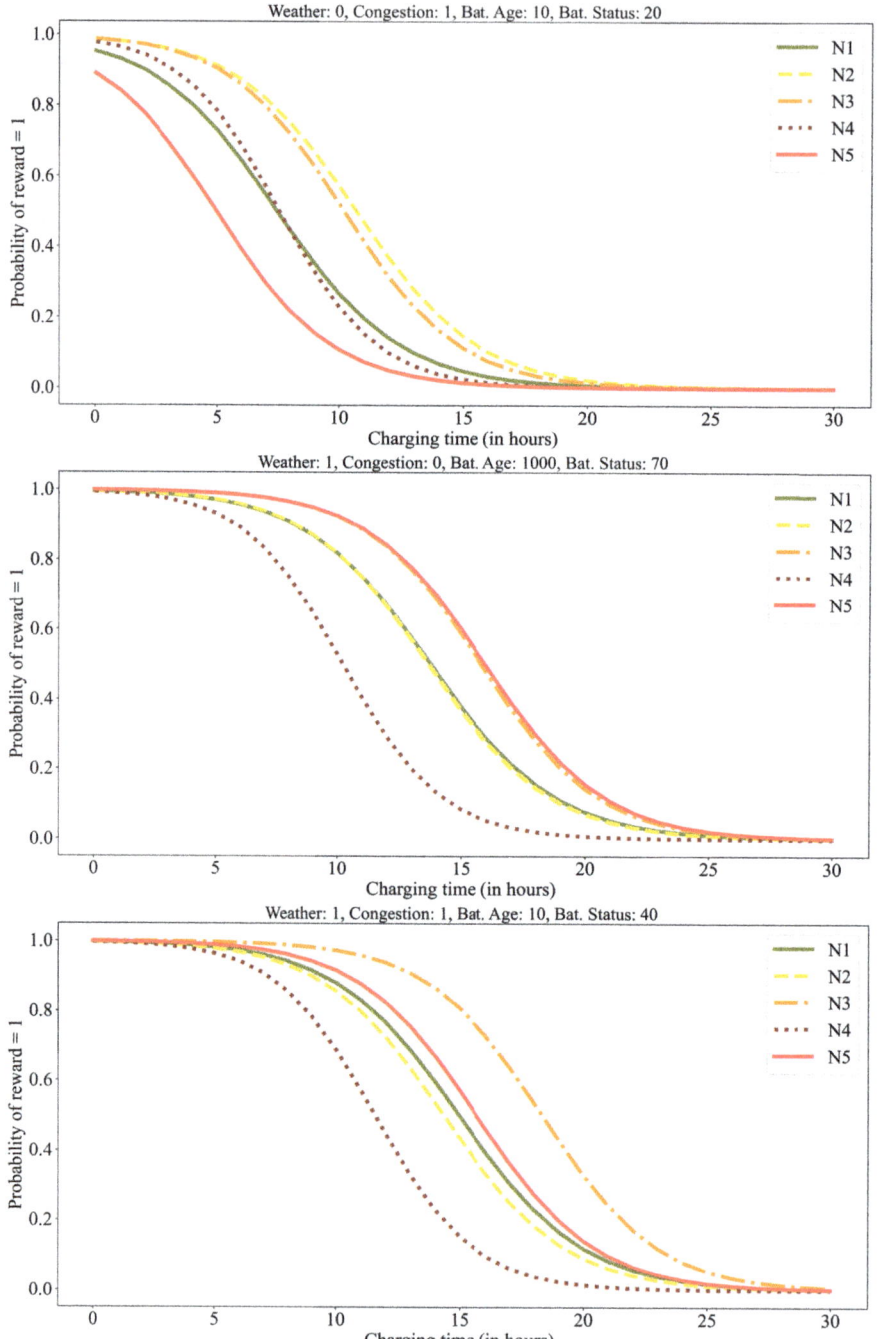

Figure 2. Real probabilities of obtaining a reward of 1 for each node and combination of factors.

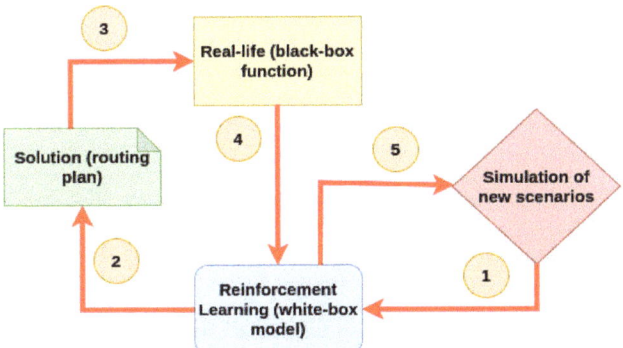

Figure 3. A conceptual schema of the proposed methodology.

6. Simulation and Algorithm Details

Using the Python programming language as a base, this section provides insights on the reinforcement learning algorithm employed to predict the probability of reward for each node and contextual conditions. It also describes the simulation process utilized to illustrate how the learning process works in a practical application. Notice that only the key parts of the code are provided here.

Listing 1 defines a multivariate logistic function, *real_reward_p*, which provides the real probability of obtaining a binary reward of 1 for a given node, based on the inputs weather conditions, congestion, battery age, expected battery, and charging time. This function acts as a black-box model (unknown for the learning algorithm) that emulates reality. This logistic function takes in a node, weather (0 or 1 where 1 is good weather and 0 is bad), congestion (0 or 1 where 1 is high congestion and 0 is low congestion), battery age (an integer between 10 and 1000), expected battery (an integer between 0 and 100), and charging time (an integer between 1 and 40) as parameters. Inside the function, a linear combination of the parameters and some weights stored in the *params* dictionary for that node are computed. The result of the linear combination is passed through the sigmoid function $1/(1 + np.exp(-linear\_comb))$ to obtain a probability of obtaining a reward of 1 for that node given the input parameters. For instance, if the parameters associated with a giving node i are $k_{1i} = 4$, $k_{2i} = -0.5$, $k_{3i} = -0.45$, $k_{4i} = -0.001$, $k_{5i} = -0.0004$, and $k_{6i} = 3.5$, then *real_reward_p(i, 1, 0, 20, 10, 20)* would return the probability of obtaining a reward of 1 for node i when the weather is good, congestion is low, the charging time is 40, the age of the battery is 10 cycles, and the battery status is 20.

In a similar fashion, Listing 2 defines a CatBoost model (gradient boosting with decision trees) [43] to predict the probabilities of obtaining a reward of 1 for a given node–region pair and context inputs. In order to train the model, the data are divided into two sets: training and testing. The training set is utilized to construct the model, while the test data serve the purpose of preventing overfitting by employing an early stopping strategy [44]. The model's performance improves significantly when more data are available. To ensure calibrated probabilities, Platt scaling is applied to fine-tune the model's probability outputs [45]. This model acts as a white-box one that aims at predicting how reality will behave, i.e., it tries to predict the real-life probability provided by the black-box function described in Listing 1. Given that the CatBoost Python library is employed, it is possible to use the inbuilt function *predict_proba* to generate predictions. Notice, however, that a different predictive model—e.g., a neural network—could have been used as a white box.

Listing 1. Python code illustrating the black box used to compute the real probability of reward.

```
1  # Logistic functions (one per node type) defining the probability of obtaining a
       binary reward of 1
2  # weather (1 = good, 0 = bad), congestion (1 = high, 0 = low), battery age (10 to
       1000), time of charge (1 to 40), battery status (0 to 100)
3  # ex: p(reward = 1 / N1) = 1 / (1 + exp(-(4 * weather + -0.5 * congestion + -0.001 *
       battery age + -0.15 * expected battery + -0.0004 * time + 3.5)))
4  def real_reward_p(node, weather, congestion, batery_age, battery_status, time): %
       Attention AE: battery is spelt incorrectly here. The authors should check if this
       is correct.
5      linear_comb = np.dot(params[node], [weather, congestion,batery_age, time, (20-abs(
           battery_status-20)), 1])
6      return 1 / (1 + np.exp(-linear_comb)) # probability of obtaining a reward of 1
```

Listing 2. Python code illustrating the trained model used to predict the probability of reward.

```
1  # Train a CatBoost classifier to predict the probability of reward
2  # Split the data in train and test (80-20) and uses early stopping to avoid
       overfitting
3  # To predict probabilities correctly we apply Platt Scaling to calibrate the model
4  def fit_catboost_model(node, X, y):
5      model = CatBoostClassifier(loss_function="Logloss",max_depth=5,iterations=1000,
           early_stopping_rounds=50, eval_metric="Logloss")
6      node_X = X[X['node'] == node].drop(columns=['node'])
7      node_y = y[X['node'] == node]
8      node_y = [1 if i == True else 0 for i in node_y]
9      X_train, X_val, y_train, y_val = train_test_split(node_X, node_y, test_size=0.2,
           random_state=42)
10     model.fit(X_train, y_train, eval_set=(X_val, y_val), verbose=False)
11     calibrated_model = CalibratedClassifierCV(model, cv='prefit', method='sigmoid')
12     calibrated_model.fit(X_val, y_val)
13     return calibrated_model
14 # Predict the probabilities of obtaining reward = 1 on each node for a new context
15 def predict_reward_p(node, weather, congestion, battery_age, battery_status, time):
16     model = models[node]
17     return model.predict_proba([[weather, congestion, battery_age, battery_status,
           time]])[0][1]
```

The code in Listing 3 defines a Python function called *select_node_eps* that selects a node from a dictionary *predicted_reward_p* based on a slightly modified epsilon-greedy strategy. The function takes in two optional parameters: *eps*, the value of epsilon used in the epsilon-greedy algorithm, and *uniform*, a Boolean flag that specifies whether to choose the node uniformly at random among all nodes with the same probability. The function first sorts the nodes in descending order based on their corresponding probability of reward. It then checks whether a random number between 0 and 1 is greater than *eps*. If it is, the function returns the node with the highest probability of reward (i.e., the first element in the sorted list). If the random number is less than or equal to *eps*, the function randomly selects a node from the sorted list, either uniformly or non-uniformly depending on the value of the uniform flag. If *uniform* is *True*, the function selects a node uniformly at random. If uniform is *False*, the function selects the node with the second highest probability of reward (i.e., the best 'alternative'). This function is typically used in reinforcement learning algorithms to balance the exploration and exploitation of different nodes based on their predicted rewards. Still, other strategies, such as the Thompson sampling [46], could also be used instead.

The code in Listing 4 simulates the generation of new regions in our route and selects the node to visit. The code uses a for loop to iterate over the number of observations specified by the variable *n_obs*. At each iteration, it simulates weather conditions, congestion, battery age, expected battery, and charging time. The code then selects nodes using a round-robin (balanced) selection method for the first *n_update* iterations. From that point on, it uses a trained model and the epsilon-greedy strategy to select the next node. The code then simulates the real probability of reward when this node is selected and updates the accumulated regrets. Since rewards are binary, the expected reward for each node is computed as the estimated probability of success. The code records the new data, increasing the size of observations for X (predictors), y (response), and z (regrets), and updates the models every *n_update* iterations.

Listing 3. Python code illustrating a modified epsilon-greedy strategy.

```python
# Selects a node from nodes based on an epsilon-greedy strategy
def select_node_eps(predicted_reward_p, eps=0.1, uniform=True): # default epsilon
    # sort nodes from higher to lower probability of reward
    sorted_dict = dict(sorted(predicted_reward_p.items(), key=lambda x: x[1], reverse=True))
    sorted_list = list(sorted_dict.keys())
    if random.random() > eps:
        return sorted_list[0] # choose the node with the highest probability
    elif uniform:
        return random.choice(sorted_list)
    else: # choose the node with the second highest probability
        return sorted_list[1]
```

Listing 4. Python code illustrating the core part of the simulation.

```python
# Simulate the generation of new routes and select the node
for i in range(n_obs):
    a_weather = random.randint([0, 1])
    a_congestion = random.randint([0, 1])
    a_battery_age = random.randint(10, 1000)
    a_battery_status = random.randint(0, 100)
    a_time = random.randint(start_time, end_time) # includes end_time
    # Round-robin (balanced) selection of first n_update nodes
    if i+1 < n_update:
        selected_node = list(nodes)[0]
        for node in nodes:
            if n_selected[node] < n_selected[selected_node]:
                selected_node = node
            # Caution!: for this stage, we'll use real p as predicted so we can estimate regrets
            predicted_reward_p[node] = real_reward_p(node, a_weather, a_congestion, a_battery_age, a_battery_status, a_time)
        n_selected[selected_node] = n_selected[selected_node] + 1
    else: # after the first model update, we can predict probabilities and select node
        for node in nodes:
            if i+1 == n_update: # set an initial model for node when n_update is reached
                models[node] = fit_catboost_model(node, X, y)
            predicted_reward_p[node] = predict_reward_p(node, a_weather, a_congestion, a_battery_age, a_battery_status, a_time)
        selected_node = select_node_eps(predicted_reward_p, 0.1, True)
    # Simulate the real probability of reward when this node is selected
    real_reward = random.random() < real_reward_p(selected_node, a_weather, a_congestion, a_battery_age, a_battery_status, a_time)
    # Update accum. regrets. Since rewards are binary, E[reward(node)] = estimated P(success / node)
    regrets = regrets + (max(predicted_reward_p.values()) - predicted_reward_p[selected_node])

    # Record the new data, thus increasing the size of observations for X, y, and z
    new_data = {'weather': a_weather, 'congestion': a_congestion,
                'battery_age': a_battery_age,
                'battery_status': a_battery_status,
                'time': a_time, 'node': selected_node,
                'reward': real_reward, 'regrets': regrets}

    new_df = pd.DataFrame(new_data, index=[0])
    df = pd.concat([df, new_df], ignore_index=True)
    X = df[['weather', 'congestion', 'battery_age', 'battery_status', 'time', 'node']]
    y = df['reward'].astype(bool) # make sure it is a bool data type

    # Update the models every n_update observations
    if (i+1) % n_update == 0 and i+1 > n_update:
        for node in nodes:
            models[node] = fit_catboost_model(node, X, y)
```

7. Computational Experiments

Table 2 shows a comparison between the predicted and the actual (real-life) probability of obtaining a reward of 1 for a randomly selected set of nodes and factor configurations. This table confirms that the trained model is capable of estimating the real-life probability with a relatively low average error.

After training the predictive model, it is possible to apply it to the proposed case study on dynamic OP with binary random rewards. Whenever the EV reaches a new region and receives updated contextual information such as weather, congestion, battery age, expected battery, and time of charge, it will use the predictive model to select one of the nodes with a higher probability of obtaining a reward. Since selecting a node in one region might affect our options in subsequent regions, following a greedy approach in each region—i.e., always selecting the node with the highest expected reward at each step—would not necessarily lead us to an optimal solution for the entire trip. In a situation such as this, it is often convenient to employ diversification strategies based on biased randomization and agile optimization techniques [47].

Table 2. Predicted vs. actual probabilities of reward.

Node Type	Weather	Congest.	Batt. Age (Cycles)	Batt. Status (%)	Charge Time (h)	Predicted	Actual	Error
5	1	0	467	98	18	0.283	0.3974	0.1145
5	0	1	228	82	22	0.032	0.0007	0.0314
1	1	1	69	16	29	0.065	0.0034	0.0616
3	0	1	98	30	17	0.0499	0.0462	0.0038
1	0	0	107	79	25	0.0506	0.0014	0.0493
4	1	1	378	31	22	0.0273	0.0039	0.0234
5	0	1	486	3	15	0.0363	0.0101	0.0263
5	0	1	373	66	21	0.0309	0.0009	0.03
4	0	0	224	92	28	0.0262	0.0	0.0261
5	0	0	235	64	5	0.7532	0.6724	0.0808
1	1	1	354	22	23	0.0544	0.0278	0.0267
5	0	0	402	28	18	0.0312	0.0075	0.0237
3	0	1	196	31	3	0.9524	0.9465	0.0058
3	0	1	464	20	4	0.947	0.8929	0.054
3	1	0	58	44	4	0.9667	0.998	0.0313
4	0	0	234	59	29	0.0261	0.0	0.0261
4	1	1	314	59	7	0.7149	0.8833	0.1684
2	1	1	56	26	19	0.0691	0.1282	0.0591
5	1	1	159	52	28	0.0341	0.005	0.029
5	1	1	349	6	3	0.9603	0.9937	0.0334
Average:								0.0452

A total of 10,000 trips for the case study were simulated, which considered six regions and five types of nodes per region. Figure 4 shows boxplots with the expected accumulated reward for solutions generated by our approach (in which the trip is guided by the predictive model) and a non-guided approach (in which a node is randomly chosen in each region). Hence, while solutions proposed by our methodology show an average value of 2.83 and a standard deviation of 1.02 for the expected accumulated reward, solutions provided by a non-guided approach show an average value of 2.37 and a standard deviation of 0.99 for the expected accumulated reward. As expected, a t-test provides a t-statistic = 32.44, with an associated p-value = $4.82 * 10^{-225}$, which clearly indicates that the solutions generated by our reinforcement learning approach are significantly better than the ones constructed without taking into account the dynamic contextual conditions.

Figure 5 illustrates a comparative analysis between a non-guided solution and the route provided by our approach. The blue nodes (N1, N2, N3, N4, N5) in the figure represent the chargers in each region, and the weight of the edges in the graph represents the real probabilities of reward based on the original solution. Notably, the selections made by our approach consistently demonstrate higher probabilities of obtaining rewards.

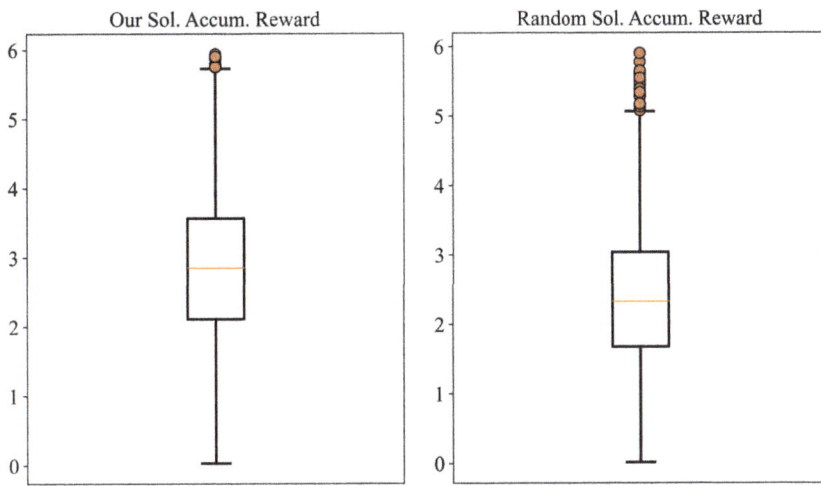

Figure 4. Boxplots for comparing our guided routing solutions with non-guided ones.

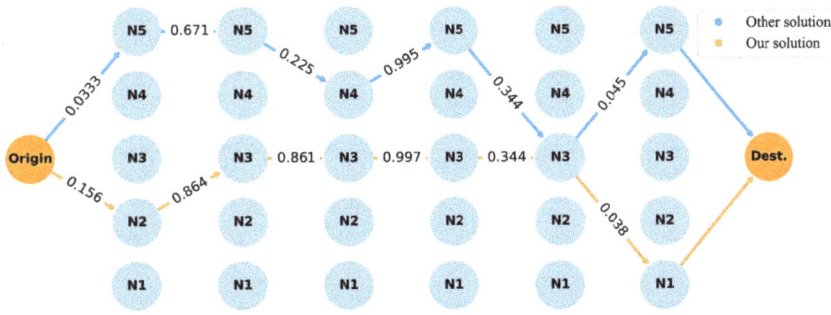

Figure 5. Visual representation of our guided routing solution and a non-guided one.

8. Conclusions

This paper proposes a hybrid methodology combining simulation and reinforcement learning to explore a vehicle orienteering problem with binary random rewards and dynamic conditions. This problem is discussed in the context of electric vehicles that, while covering a long trip, have to cross different regions and choose, in each of these region, a charging station with a high probability of reward. The challenge lies in the fact that the reward for visiting each charging node is a binary random variable and depends on dynamic context conditions, such as: weather conditions, road or on-site congestion, current battery status, etc. In order to emulate the learning process in real-life conditions under uncertainty, a simulation is employed. Thus, a black-box model is used to estimate the real-life probability of obtaining a reward for each node based on the dynamic context conditions at a given time. A reinforcement learning mechanism is then employed to make informed decisions at each stage of the problem, and a logistic regression model is used to predict the aforementioned probabilities. Through the simulation results, it is shown that the proposed reinforcement learning approach can effectively learn to make informed decisions based on the dynamic context conditions. The computational experiments show that a statistically significant improvement is obtained when the proposed approach is utilized. All in all, the proposed approach can be useful in a range of real-life scenarios such as transportation logistics, delivery services, and resource allocation.

One potential direction is to explore more complex models for the probability of obtaining a reward, such as neural networks or models that account for additional context conditions. To mitigate some of the downsides of the strategy chosen in this paper (ε greedy), various enhancements could be carried out, such as using decaying exploration rates or dynamically adjusting the exploration rate based on the agent's learning progress. Another potential area for future work is to investigate the impact of different reinforcement learning algorithms and strategies on the performance of the model. For example, analyzing different exploration strategies (beyond the epsilon-greedy strategy used in this study) may lead to more efficient learning and better decision making. Furthermore, it may be valuable to consider the use of more advanced optimization techniques, such as metaheuristics and simheuristics [48], that can be combined with the reinforcement learning approach introduced in this paper. Finally, this study focused on a single vehicle traveling through a fixed set of regions. Future work could explore more complex scenarios, such as the involvement of multiple vehicles, i.e., a dynamic team orienteering problem with stochastic rewards.

Author Contributions: Conceptualization, A.A.J.; methodology, A.A.J., Y.A., and J.P.; software, Y.A., C.A.M. and X.A.M.; validation, J.P. and X.A.M.; writing—original draft preparation, A.A.J., C.A.M. and R.F.; writing—review and editing, C.A.M., Y.A. and R.F.; supervision, J.P., X.A.M. and A.A.J. All authors have read and agreed to the published version of the manuscript.

Funding: This work was partially funded by the European Commission projects SUN (HORIZON-CL4-2022-HUMAN-01-14-101092612), and AIDEAS (HORIZON-CL4-2021-TWIN-TRANSITION-01-07-101057294).

Institutional Review Board Statement: Not applicable.

Informed Consent Statement: Not applicable.

Data Availability Statement: Not applicable.

Conflicts of Interest: The authors declare no conflict of interest.

Abbreviations

The following abbreviations are used in this manuscript:

EV	Electric vehicle
BMS	Battery management system
OP	Orienteering problem
AC	Alternating current
DC	Direct current
RL	Reinforcement learning
HEV	Hybrid electric vehicle
BEV	Battery electric vehicle
SOC	State of charge
V2G	Vehicle-to-grid

References

1. Almouhanna, A.; Quintero-Araujo, C.L.; Panadero, J.; Juan, A.A.; Khosravi, B.; Ouelhadj, D. The location routing problem using electric vehicles with constrained distance. *Comput. Oper. Res.* **2020**, *115*, 104864. [CrossRef]
2. Abdollahi, A.; Han, X.; Avvari, G.V.; Raghunathan, N.; Balasingam, B.; Pattipati, K.K.; Bar-Shalom, Y. Optimal battery charging, Part I: Minimizing time-to-charge, energy loss, and temperature rise for OCV-resistance battery model. *J. Power Sources* **2015**, *303*, 388–398. [CrossRef]
3. Golden, B.; Levy, L.; Vohra, R. The Orienteering Problem. *Nav. Res. Logist.* **1987**, *34*, 307–318. [CrossRef]

4. Panadero, J.; Currie, C.; Juan, A.A.; Bayliss, C. Maximizing Reward from a Team of Surveillance Drones under Uncertainty Conditions: A Simheuristic Approach. *Eur. J. Ind. Eng.* **2020**, *14*, 1–23. [CrossRef]
5. Bayliss, C.; Juan, A.A.; Currie, C.S.; Panadero, J. A learnheuristic approach for the team orienteering problem with aerial drone motion constraints. *Appl. Soft Comput.* **2020**, *92*, 106280. [CrossRef]
6. Bilgin, E. *Mastering Reinforcement Learning with Python: Build Next-Generation, Self-Learning Models using Reinforcement Learning Techniques and Best Practices*; Packt Publishing Ltd.: Birmingham, UK, 2020.
7. IPCC. *Climate Change 2022: Mitigation of Climate Change*; Contribution of Working Group III to the Sixth Assessment Report of the Intergovernmental Panel on Climate Change; IPCC: Geneva, Switzerland, 2022.
8. IEA. *Global EV Outlook 2023*; International Energy Agency: Paris, France, 2023.
9. Dixit, M.; Muralidharan, N.; Parejiya, A.; Essehli, R.; Belharouak, I.; Amin, R. Electrochemical energy storage systems. In *Emerging Trends in Energy Storage Systems and Industrial Applications*; Elsevier: Amsterdam, The Netherlands, 2023; pp. 259–282.
10. Liang, J.; Wu, T.; Wang, Z.; Yu, Y.; Hu, L.; Li, H.; Zhang, X.; Zhu, X.; Zhao, Y. Accelerating perovskite materials discovery and correlated energy applications through artificial intelligence. *Energy Mater* **2022**, *2*, 200016. [CrossRef]
11. Qi, X.; Wu, G.; Boriboonsomsin, K.; Barth, M.J. Data-driven decomposition analysis and estimation of link-level electric vehicle energy consumption under real-world traffic conditions. *Transp. Res. Part D Transp. Environ.* **2018**, *64*, 36–52. [CrossRef]
12. Kwang, H.N. *AC Motor Control and Electric Vehicle Applications*; CRC Press, Taylor & Francis Group: Boca Raton, FL, USA, 2010.
13. Alanazi, F. Electric Vehicles: Benefits, Challenges, and Potential Solutions for Widespread Adaptation. *Appl. Sci.* **2023**, *13*, 6016. [CrossRef]
14. Böhme, T.J.; Frank, B. *Hybrid Systems, Optimal Control and Hybrid Vehicles. Theory, Methods and Applications*; Springer International Publishing: Cham, Switzerland, 2017.
15. Townsend, A.; Gouws, R. A Comparative Review of Lead-Acid, Lithium-Ion and Ultra-Capacitor Technologies and Their Degradation Mechanisms. *Energies* **2022**, *15*, 4930. [CrossRef]
16. Chau, K. Pure electric vehicles. In *Alternative Fuels and Advanced Vehicle Technologies for Improved Environmental Performance*; Woodhead Publishing: Cambridge, UK, 2014; Volume 8176, pp. 655–684.
17. Guo, F.; Wu, X.; Liu, L.; Ye, J.; Wang, T.; Fu, L.; Wu, Y. Prediction of remaining useful life and state of health of lithium batteries based on time series feature and Savitzky-Golay filter combined with gated recurrent unit neural network. *Energy* **2023**, *270*, 126880. [CrossRef]
18. Li, W.; Stanula, P.; Egede, P.; Kara, S.; Herrmann, C. Determining the main factors influencing the energy consumption of electric vehicles in the usage phase. *Procedia CIRP* **2016**, *48*, 352–357. [CrossRef]
19. Bi, J.; Wang, Y.; Zhang, J. A data-based model for driving distance estimation of battery electric logistics vehicles. *EURASIP J. Wirel. Commun. Netw.* **2018**, *251*, 1–13. [CrossRef]
20. Preger, Y.; Barkholtz, H.M.; Fresquez, A.; Campbell, D.L.; Juba, B.W.; Romàn-Kustas, J.; Ferreira, S.R.; Chalamala, B. Degradation of Commercial Lithium-Ion Cells as a Function of Chemistry and Cycling Conditions. *J. Electrochem. Soc.* **2020**, *167*, 1–9.
21. Koltermann, L.; Cortés, M.C.; Figgener, J.; Zurmühlen, S.; Sauer, D.U. Power Curves of Megawatt-Scale Battery Storage Technologies for Frequency Regulation and Energy Trading. *Appl. Energy* **2023**, *347*, 121428. [CrossRef]
22. Jiang, J.; Shi, W.; Zheng, J.; Zuo, P.; Xiao, J.; Chen, X.; Xu, W.; Zhang, J.G. Optimized Operating Range for Large-Format LiFePO4/Graphite Batteries. *J. Electrochem. Soc.* **2013**, *161*, 336–341. [CrossRef]
23. Ameli, M.T.; Ameli, A. Electric vehicles as means of energy storage: Participation in ancillary services markets. In *Energy Storage in Energy Markets. Uncertainties, Modelling, Analysis and Optimization*; Academic Press: Cambridge, MA, USA, 2021; pp. 235–249.
24. Mastoi, M.S.; Zhuang, S.; Hafiz Mudassir, M.; Haris, M.; Hassan, M.; Usman, M.; Bukhari, S.S.H.; Ro, J.S. An in-depth analysis of electric vehicle charging station infrastructure, policy implications, and future trends. *Energy Rep.* **2022**, *8*, 11504–11529. [CrossRef]
25. Yong, J.Y.; Ramachandaramurthy, V.K.; Tan, K.M.; Mithulananthan, N. A Review on the State-of-the-Art Technologies of Electric Vehicle, Its Impacts and Prospects. *Renew. Sustain. Energy Rev.* **2015**, *49*, 365–385. [CrossRef]
26. U.S. Department of Transportation. Charging Speeds. Rural Electric Vehicle Toolkit. Available online: https://www.transportation.gov/rural/ev/toolkit/ev-basics/charging-speeds (accessed on 1 August 2023).
27. Brenna, M.; Foiadelli, F.; Leone, C.; Longo, M. Electric Vehicles Charging Technology Review and Optimal Size Estimation. *J. Electr. Eng. Technol.* **2020**, *15*, 2539–2552. [CrossRef]
28. Solntsev, A.; Asoyan, A.; Nikitin, D.; Bagrin, V.; Fediushkina, O.; Evtykov, S.; Marusin, A. Influence of temperature on the performance and life cycle of storage batteries. *Transp. Res. Procedia* **2021**, *57*, 652–659. [CrossRef]
29. Zhang, Z.; Li, W.; Zhang, C.; Chen, J. Climate control loads prediction of electric vehicles. *Appl. Therm. Eng.* **2017**, *110*, 1183–1188. [CrossRef]
30. Yang, X.G.; Zhang, G.; Ge, S.; Wang, C.Y. Fast charging of lithium-ion batteries at all temperatures. *Appl. Therm. Eng.* **2018**, *115*, 7266–7271. [CrossRef] [PubMed]
31. Jonas, T.; Hunter, C.D.; Macht, G.A. Quantifying the Impact of Traffic on Electric Vehicle Efficiency. *World Electr. Veh. J.* **2022**, *13*, 15. [CrossRef]
32. Hu, K.; Wu, J.; Schwanen, T. Differences in Energy Consumption in Electric Vehicles: An Exploratory Real-World Study in Beijing. *J. Adv. Transp.* **2017**, *2017*, 1–17.
33. Ilhan, T.; Iravani, S.; Daskin, M. The Orienteering Problem with Stochastic Profits. *IIE Trans.* **2008**, *40*, 406–421. [CrossRef]

34. Campbell, A.; Gendreau, M.; Thomas, B. The Orienteering Problem with Stochastic Travel and Service Times. *Ann. Oper. Res.* **2011**, *186*, 61–81. [CrossRef]
35. Papapanagiotou, V.; Montemanni, R.; Gambardella, L. Objective Function Evaluation Methods for the Orienteering Problem with Stochastic Travel and Service Times. *J. Appl. Oper. Res.* **2014**, *6*, 16–29.
36. Verbeeck, C.; Vansteenwegen, P.; Aghezzaf, E.H. Solving the Stochastic Time-Dependent Orienteering Problem with Time Windows. *Eur. J. Oper. Res.* **2016**, *255*, 699–718.
37. Evers, L.; Glorie, K.; van der Ster, S.; Barros, A.; Monsuur, H. A Two-Stage Approach to the Orienteering Problem with Stochastic Weights. *Comput. Oper. Res.* **2014**, *43*, 248–260. [CrossRef]
38. Lau, H.C.; Yeoh, W.; Varakantham, P.; Nguyen, D.T.; Chen, H. Dynamic Stochastic Orienteering Problems for Risk-Aware Applications. *arXiv* **2012**, arXiv:1210.4874.
39. Varakantham, P.; Kumar, A. Optimization Approaches for Solving Chance-Constrained Stochastic Orienteering Problems. In *Algorithmic Decision Theory*; Springer: Cham, Switzerland, 2013; Volume 8176, pp. 387–398.
40. Zhang, S.; Ohlmann, J.; Thomas, B. A Priori Orienteering with Time Windows and Stochastic Wait Times at Customers. *Eur. J. Oper. Res.* **2014**, *239*, 70–79. [CrossRef]
41. Gama, R.; Fernandes, H.L. A reinforcement learning approach to the orienteering problem with time windows. *Comput. Oper. Res.* **2021**, *133*, 105357. [CrossRef]
42. Gunawan, A.; Lau, H.; Vansteenwegen, P. Orienteering Problem: A Survey of Recent Variants, Solution Approaches and Applications. *Eur. J. Oper. Res.* **2016**, *255*, 315–332. [CrossRef]
43. Prokhorenkova, L.; Gusev, G.; Vorobev, A.; Dorogush, A.V.; Gulin, A. CatBoost: Unbiased boosting with categorical features. *Adv. Neural Inf. Process. Syst.* **2018**, *31*, 1–11.
44. Yao, Y.; Rosasco, L.; Caponnetto, A. On early stopping in gradient descent learning. *Constr. Approx.* **2007**, *26*, 289–315. [CrossRef]
45. Platt, J. Probabilistic Outputs for Support Vector Machines and Comparisons to Regularized Likelihood Methods. *Adv. Large Margin Classif.* **2000**, *10*.
46. Umami, I.; Rahmawati, L. Comparing Epsilon Greedy and Thompson Sampling Model for Multi-Armed Bandit Algorithm on Marketing Dataset. *J. Appl. Data Sci.* **2021**, *2*, 14–26. [CrossRef]
47. Do C. Martins, L.; Hirsch, P.; Juan, A.A. Agile optimization of a two-echelon vehicle routing problem with pickup and delivery. *Int. Trans. Oper. Res.* **2021**, *28*, 201–221.
48. Chica, M.; Juan, A.A.; Bayliss, C.; Cordón, O.; Kelton, W.D. Why simheuristics? Benefits, limitations, and best practices when combining metaheuristics with simulation. *SORT-Stat. Oper. Res. Trans.* **2020**, *44*, 311–334. [CrossRef]

Disclaimer/Publisher's Note: The statements, opinions and data contained in all publications are solely those of the individual author(s) and contributor(s) and not of MDPI and/or the editor(s). MDPI and/or the editor(s) disclaim responsibility for any injury to people or property resulting from any ideas, methods, instructions or products referred to in the content.

Article

Artificial Neural Network Modeling to Predict Thermal and Electrical Performances of Batteries with Direct Oil Cooling

Kunal Sandip Garud [1], Jeong-Woo Han [2], Seong-Guk Hwang [1] and Moo-Yeon Lee [1,*]

[1] Department of Mechanical Engineering, Dong-A University, 37 Nakdong-Daero 550, Saha-gu, Busan 49315, Republic of Korea; 1876936@donga.ac.kr (K.S.G.); 2178735@donga.ac.kr (S.-G.H.)
[2] Thermal Management R&D Center, Korean Automotive Technology Institute, 303 Pungse-ro, Pungse-Myun, Cheonan 31214, Republic of Korea; jwhan1@katech.re.kr
* Correspondence: mylee@dau.ac.kr; Tel.: +82-51-200-7642

Abstract: The limitations of existing commercial indirect liquid cooling have drawn attention to direct liquid cooling for battery thermal management in next-generation electric vehicles. To commercialize direct liquid cooling for battery thermal management, an extensive database reflecting performance and operating parameters needs to be established. The development of prediction models could generate this reference database to design an effective cooling system with the least experimental effort. In the present work, artificial neural network (ANN) modeling is demonstrated to predict the thermal and electrical performances of batteries with direct oil cooling based on various operating conditions. The experiments are conducted on an 18650 battery module with direct oil cooling to generate the learning data for the development of neural network models. The neural network models are developed considering oil temperature, oil flow rate, and discharge rate as the input operating conditions and maximum temperature, temperature difference, heat transfer coefficient, and voltage as the output thermal and electrical performances. The proposed neural network models comprise two algorithms, the Levenberg–Marquardt (LM) training variant with the Tangential-Sigmoidal (Tan-Sig) transfer function and that with the Logarithmic-Sigmoidal (Log-Sig) transfer function. The ANN_LM-Tan algorithm with a structure of 3-10-10-4 shows accurate prediction of thermal and electrical performances under all operating conditions compared to the ANN_LM-Log algorithm with the same structure. The maximum prediction errors for the ANN_LM-Tan and ANN_LM-Log algorithms are restricted within ±0.97% and ±4.81%, respectively, considering all input and output parameters. The ANN_LM-Tan algorithm is suggested to accurately predict the thermal and electrical performances of batteries with direct oil cooling based on a maximum determination coefficient (R^2) and variance coefficient (COV) of 0.99 and 1.65, respectively.

Keywords: artificial neural network; battery; direct oil cooling; electrical performance; electric vehicle; thermal performance

Citation: Garud, K.S.; Han, J.-W.; Hwang, S.-G.; Lee, M.-Y. Artificial Neural Network Modeling to Predict Thermal and Electrical Performances of Batteries with Direct Oil Cooling. Batteries 2023, 9, 559. https:// doi.org/10.3390/batteries9110559

Academic Editor: Carlos Ziebert

Received: 11 October 2023
Revised: 13 November 2023
Accepted: 14 November 2023
Published: 16 November 2023

Copyright: © 2023 by the authors. Licensee MDPI, Basel, Switzerland. This article is an open access article distributed under the terms and conditions of the Creative Commons Attribution (CC BY) license (https:// creativecommons.org/licenses/by/ 4.0/).

1. Introduction

The excessive consumption of fossil fuels by internal combustion engine vehicles is causing a rapid increase in greenhouse gas emissions, including environmental contaminates of carbon dioxide (CO_2), carbon monoxide (CO), nitrogen oxides (NOx), sulfur dioxide (SO_2), and particulate matter (PM) [1]. To assure a low carbon future, the governments of various countries have passed regulations on emissions, for example, the United States has issued the "1990 Clean Air Act", European countries have implemented the "Low Emission Zone Program", and Japan has recirculated the "2007 NOx and PM Law" [2]. Furthermore, the European Commission has introduced the package "Fit for 55" to reduce greenhouse gas emissions up to 55% by 2030 [3]. In global greenhouse gas emissions, the transportation sector is the second largest contributor [4]. Therefore, the current transportation sector is undergoing a drastic change of replacing internal combustion engine vehicles with electric vehicles to achieve a carbon-free and energy-sustainable future [5,6]. Electric vehicles offer several benefits, such as a safe and clean environment and improved safety and human health [7,8]. In addition, the ecological benefits of electric vehicles could be extended by charging their batteries from renewable energy sources [9].

Electric vehicles are provided with batteries as the main energy storage system on board the vehicle, the energy densities of which are continuously increasing to improve the performance of electric vehicles [10,11]. The battery temperature should be maintained within a range of 20 °C to 45 °C to ensure safe and efficient operations [12,13]. However, the increasing energy densities result in higher heat generation, thus degrading the performance and operational life of batteries [14]. Furthermore, the excessive heat generation during high charging/discharging operations results in thermal runaway and explosion in batteries [15,16]. Therefore, an advanced cooling technique should enable effective thermal management of batteries, which could improve their efficiency and life span and thus the safety and performance of electric vehicles.

Air cooling and indirect liquid cooling are commercially adopted for the thermal management of electric vehicle batteries [17]. The cooling performance of air cooling is poor and indirect liquid cooling imposes high thermal resistance owing to the existence of cooling channels/plates, which reduces its cooling performance for high power density batteries [18,19]. To overcome the limitations of the existing cooling strategies, research has been initiated to find next-generation thermal management techniques for batteries. Direct liquid cooling diminishes the thermal resistance by enabling direct contact between batteries and the dielectric cooling fluid, hence improving the cooling performance [20,21]. In the last few years, numerous research studies have reported on direct liquid cooling as an emerging battery thermal management technique. Li et al. restricted the temperature of 18650 batteries within 34 °C and 34.5 °C at 4C and 7C discharge rates, respectively, using SF33 coolant-based direct liquid cooling [22]. Patil et al. employed immersion cooling to maintain the maximum temperature of a battery pack within 28 °C at a 3C discharge rate and 10 L/min flow rate [23]. Sundin et al. proposed single-phase immersion cooling, which maintained the battery temperature within 30 °C at a 2C discharge rate, and Zhou et al. further demonstrated that the thermal runaway of batteries was suppressed using two-phase immersion cooling [24,25]. Dubey et al. showed improvements in maximum temperature and pumping power for 21700 batteries with Novec 7500-based direct cooling compared to that with water/glycol-based indirect cooling [26].

In recent years, machine learning is gaining popularity to predict and optimize the performance of physical systems based on various influential variables compared to other prediction approaches. Furthermore, ANN models are widely adopted to replicate the behaviors of specific systems/devices under several conditions owing to the benefits of faster response, minimal error, and least complex mathematical manipulation [27,28]. Numerous studies have confirmed the potential of neural network models to accurately predict the performance of batteries. Panchal et al. proposed a neural network model to predict the thermal and electrical characteristics of batteries under real driving conditions [29]. Furthermore, Wang et al. also predicted the thermal and electrical characteristics of lithium-ion batteries using a coupled thermal-equivalent circuit model integrated with a neural network [30]. Feng et al. predicted the voltage and temperature of batteries using a neural network model under several conditions of current and temperature [31]. Xie et al. developed a back-propagation-based neural network model to estimate battery internal resistance and battery temperature [32]. Arora et al. proposed a neural network model with battery heat generation as the output parameter and battery nominal capacity as the input parameter [33]. Liu et al. developed an ANN model with the structure of 1-30-1 to predict the surface temperature of batteries and pressure drop [34]. Jaliliantabar et al. predicted the battery temperature for input conditions of phase change material with and without paraffin/graphene composite, phase change material thickness, time, and discharge rate using a neural network model with an accuracy of 0.99 and mean square error of 0.0173 [35].

The open literature reveals that significant research has reported on direct liquid cooling for next-generation battery thermal management. However, the reported research is not sufficient to commercialize direct liquid cooling and there is scope to extend the research to prove its reliability for battery thermal management. The development of a final-stage direct liquid cooling system for batteries needs extensive efforts in terms of prototype fabrication, testing, and optimization, which are time consuming and expensive. In this scenario, accurately trained neural network models could effectively replicate the performance behavior of a cooling system under actual operating conditions with comparatively less effort. However, there is no concrete research that elaborates neural network modeling for batteries with a direct liquid cooling system. Therefore, in the present study, ANN models based on experimental data are developed to predict the thermal and electrical performances of batteries with direct oil cooling. Two combinations of an algorithm, namely ANN_LM-Tan and ANN_LM-Log, are developed to predict the maximum temperature, temperature difference, heat transfer coefficient, and voltage, considering operating conditions of oil temperature, oil flow rate, and discharge rate. The ANN model with the best algorithm is suggested to accurately replicate the performance data of batteries with direct oil cooling under several operating conditions.

2. Experimental Method

2.1. Experimental Set-Up Description

The experimental set-up of the 18650 battery module with direct oil cooling is depicted in Figure 1. The considered batteries are INR18650 MJ1 3.5 Ah, comprising silicone-graphite as the anode material and NMC-811 as the cathode material and outsourced from LG Chem Ltd. (Seoul, South Korea). The battery module is composed of 16 cylindrical cells with a 4-series and 4-parallel configuration. The specifications of the selected battery cells are presented in Table 1. The battery module is contained within an acryl box, which is filled with dielectric thermal oil manufactured by Shell company. The thermophysical properties of the selected dielectric thermal oil are presented in Table 2 [20]. The oil is distributed around the battery cells through inlet and outlet ports, located at the center on opposite faces of the acryl box. Sixteen T-type thermocouples with a range of $-200\ °C$ to $400\ °C$ are used to measure the temperature of the battery module. A thermocouple is attached at the center of each battery cell, considering a negligible temperature difference over the battery cell surface. Li et al. observed that there was no significant difference in battery temperature over its various surface locations [22]. Pt-100 temperature sensors with a range of $0\ °C$ to $850\ °C$ are provided at the inlet and outlet of the acryl box to measure the temperature of the oil. A peristaltic pump with a range of 0.4 mL/min to 2.2 L/min is used to circulate the oil in the acryl box. The heated oil from the battery module is cooled using a 30 L chiller with range of $-25\ °C$ to $80\ °C$. The terminals of the battery module are connected to the 1.2 kW DC electronic loader (TOYOTECH TLF1200, Incheon, South Korea) with a voltage range of 1 V to 150 V and current range of 0 A to 240 A. The electronic loader is used to discharge the battery module at constant current mode, considering different discharge rates ranging from 1C to 4C. The battery module is discharged until the voltage is cut off and rested for 2 h. The fully discharged battery module is charged using a DC power supply with a voltage range of 0 V to 30 V and current range of 0 A to 10 A, considering a charging rate of 0.5C. The battery module is charged in constant current mode with 7 A until it reaches 16.8 V, and then the battery module is maintained at constant voltage mode until the current approaches to 0.2 A. The fully charged battery is rested for 2 h. The battery management unit with passive cell balancing is employed. Furthermore, the rest time after full discharging and charging operations ensures cell balancing of the battery module. All measuring devices are connected to a GL840 data logger to monitor and record the data. The temperature and voltage data of the battery module with direct oil cooling are recorded at each second over the discharge period of battery. The temperature difference and heat transfer coefficient are calculated corresponding to the measured data. The measured and calculated data of the battery module with direct oil cooling are evaluated considering four oil flow rate conditions of 0.4 L/min, 0.6 L/min, 0.8 L/min, and 1.0 L/min, four oil temperature conditions of 15 °C, 20 °C, 30 °C, and 35 °C, and four discharge rates of 1C, 2C, 3C, and 4C. These measured and calculated data under several operating conditions are employed to develop the ANN models. The sample of the experimental data used for neural network modeling is depicted in Figure 2. The experimental set-up is housed within a chamber with temperature and humidity ranges of −30 °C to 60 °C and 30% to 95%, respectively; therefore, the ambient temperature during all experiments is controlled at 25 °C.

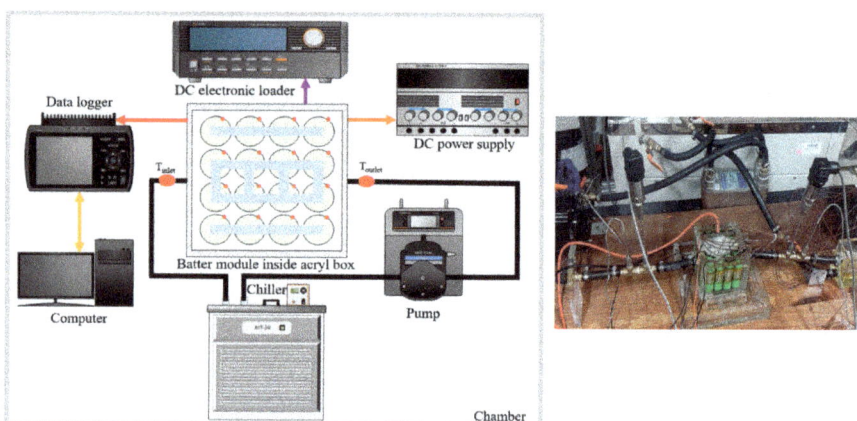

Figure 1. Experimental set-up of 18650 battery module with direct oil cooling.

Table 1. Specifications of the selected battery cells.

Specification	Value
Nominal capacity (Ah)	3.5
Nominal voltage (V)	3.653
Max voltage (V)	4.2
Discharge cut-off voltage (V)	2.5
Standard charge current (A)	1.7
Standard charge cut-off current (A)	0.050

Table 2. Thermophysical properties of dielectric thermal oil.

Property	Value
Density (kg/m^3)	810
Thermal conductivity (W/m-K)	0.14
Specific heat (J/kg-K)	2100
Viscosity (cSt)	19.4

2.2. Experimental Parameters and Uncertainty Analysis

Uncertainty analysis was conducted on the experimental parameters to consider the accuracies of the experimental devices and errors owing to probe position, calibration, and measurement [36,37]. The accuracies labeled with the T-type thermocouples, Pt-100 temperature sensors, DC loader, data logger, and pump were ±0.5%, ±0.25%, ±0.1%, ±0.1%, and ±0.2%, respectively. The uncertainties in various experimental parameters were evaluated using Equation (1) [38]:

$$U_R = \left[\left(\frac{\partial R}{\partial X_1}U_1\right)^2 + \left(\frac{\partial R}{\partial X_2}U_2\right)^2 + \cdots + \left(\frac{\partial R}{\partial X_n}U_n\right)^2\right]^{\frac{1}{2}} \quad (1)$$

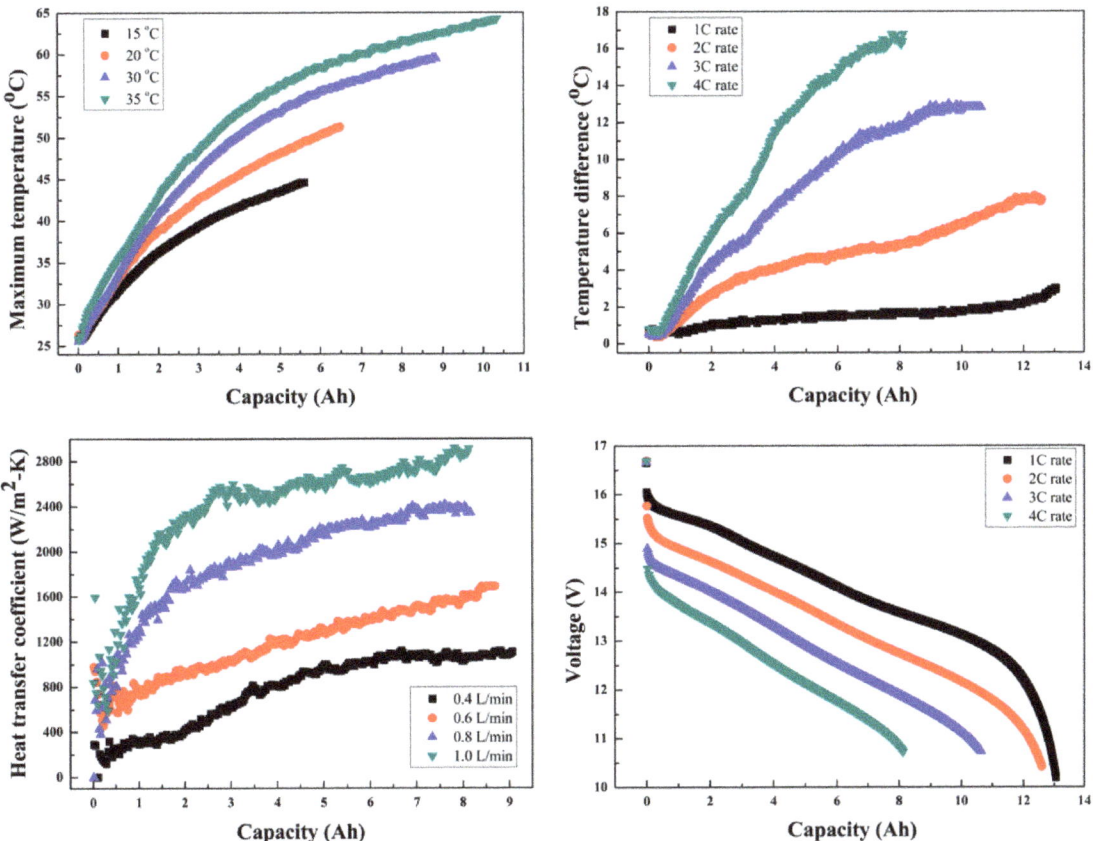

Figure 2. Sample of experimental data used for neural network modeling.

Here, R is the dependent experimental parameter and U_R is the uncertainty in the dependent experimental parameter, whereas $X_1, X_2, \ldots X_n$ are independent experimental parameters, and $U_1, U_2, \ldots U_n$ are the uncertainties in the independent experimental parameters. The uncertainties in measured temperature, voltage, and heat transfer coefficient were evaluated as 3.13%, 1.29%, and 6.45%, respectively.

The temperature difference of the battery module was calculated as follows [39]:

$$\Delta T = T_{max,\ battery} - T_{min, battery} \tag{2}$$

Here, $T_{max,\ battery}$ and $T_{min,battery}$ are the maximum and minimum temperatures of the battery module, respectively.

The heat transfer coefficient for the battery module with direct oil cooling was calculated as follows [40]:

$$h = \frac{Q_{convection,\ oil}}{A_{battery}\left(T_{mean,battery} - T_{mean,oil}\right)} \tag{3}$$

The convective heat transfer from the battery module to the oil was calculated as follows [40]:

$$Q_{convection,oil} = \dot{m}_{oil} C_{p,oil}(T_{outlet,\ oil} - T_{inlet,oil}) \tag{4}$$

Here, $A_{battery}$ is the surface area of the battery module, $T_{mean,battery}$ is the mean temperature of the battery module, $T_{mean,oil}$ is the mean temperature of the oil, $\dot{m}_{oil}$ is the mass flow rate of the oil, $C_{p,oil}$ is the specific heat of the oil, and $T_{outlet,\,oil}$ and $T_{inlet,oil}$ are outlet and inlet temperatures of the oil, respectively.

3. Artificial Neural Network Modeling

The non-linear and complex relationship between various performances and influential factors can be effectively mapped with the least computational time using neural network models. The ANN mimics the biological neural structure, which relates larger datasets of various parameters for any physical system [41]. The neural network involves the integration of the neurons in respective layers of input, output and hidden [42]. The input and output parameters decide the number of neurons in respective layers, whereas the numbers of hidden layers and neurons in hidden layers are considered based on the optimized training error [43]. The weights are the connection link between each layer of neurons. Various combinations of algorithms with transfer functions and training variants are employed to train the neural network [44]. The effective mapping pattern between input and output datasets is established by adjusting the weights while neural network training [45].

In the present work, ANN was modeled to predict the thermal and electrical performances of batteries for direct oil cooling considering different operating conditions. The developed ANN model comprised oil temperature, oil flow rate, and battery discharge rate as input neurons in the input layer. And the discharge voltage, maximum temperature, temperature difference, and heat transfer coefficient were included as output neurons in the output layer. The input conditions of oil temperature, oil flow rate, and discharge rate were varied in the ranges of 15 °C to 35 °C, 0.4 L/min to 1.0 L/min, and 1C to 4C, respectively, for neural network modeling. After trying several combinations, the number of hidden layers was adjusted to 2, each with 10 hidden neurons to achieve the minimum error and computational time. Hence, the structure of the proposed ANN model was presented by 3-10-10-4, as depicted in Figure 3. The neural network was trained using a back-propagation algorithm comprising of two combinations of the LM training variant, that with transfer functions of the Tan-Sig and the Log-Sig. MATLAB R2018a software was used for neural network modeling considering the aforementioned parameters and algorithms.

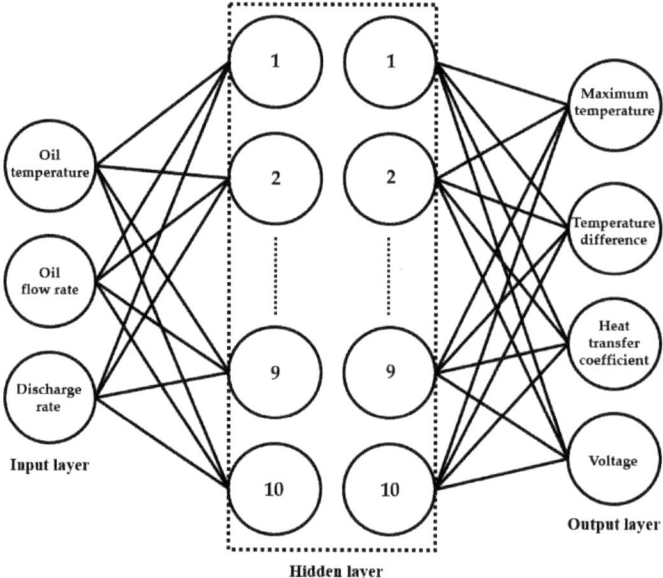

Figure 3. Structure of proposed ANN model.

The predicted output from the developed ANN model was expressed as follows [46]:

$$Y^i = G\left(A^i + U^T Y^{i-1}\right) \tag{5}$$

Here, Y is the predicted output, G indicates the transfer function, A and U present the connection and weight matrixes, respectively, and i stands for the layer number.

The mathematical expressions for the considered Tan-Sig and Log-Sig transfer functions are as follows [41,46]:

$$f(Y) = \frac{1 - e^{-Y}}{1 + e^{-Y}} \tag{6}$$

$$f(Y) = \frac{1}{1 + e^{-Y}} \tag{7}$$

To evaluate the accuracy of the predicted results from the ANN model, three statistical parameters, the coefficient of determination (R^2), variance coefficient (COV), and mean square error (MSE), were calculated using Equations (8)–(10) [47]:

$$R^2 = 1 - \frac{\sum_{i=1}^{n}(Y_{pre,i} - Y_{mea,i})^2}{\sum_{i=1}^{n}(Y_{mea,i})^2} \tag{8}$$

$$COV = \frac{\sqrt{\frac{\sum_{i=1}^{n}(Y_{pre,i} - Y_{mea,i})^2}{n}}}{|\overline{Y}_{mea}|} \times 100 \tag{9}$$

$$MSE = \frac{1}{n}\sum_{i=1}^{n}\left(Y_i - Y\right)^2 \tag{10}$$

Here, $Y_{pre,i}$ is the predicted value at ith data point, $Y_{mea,i}$ is the measured value at ith data point, $\overline{Y}_{mea}$ is the average value of measured data points, and n indicates the maximum data points.

The experimental data of the considered input and output parameters were evaluated over the discharge period of the batteries at each second. A time series of experimental data was considered to develop the ANN models. The 1000 data points comprising four performances (output parameters) were evaluated corresponding to three operating conditions (input parameters). Thus, the neural network models were developed for input and output parameters comprising 1000 experimental data points. The total dataset of 1000 data points was divided into three subgroup proportions of 60%, 20%, and 20%, corresponding to training, validation, and testing, respectively. Based on the considered data, the deducted training, validation, and testing errors for the developed ANN model with two algorithms are presented in Figure 4. It should be noted that the input parameter values were fixed; however, the output parameter values were predicted in form of time series over the discharge period of the batteries. And from the predicted results in form of time series, the thermal performance was compared at the end of discharge and electrical performance was compared at the same discharge capacity, considering the variations in various influential factors.

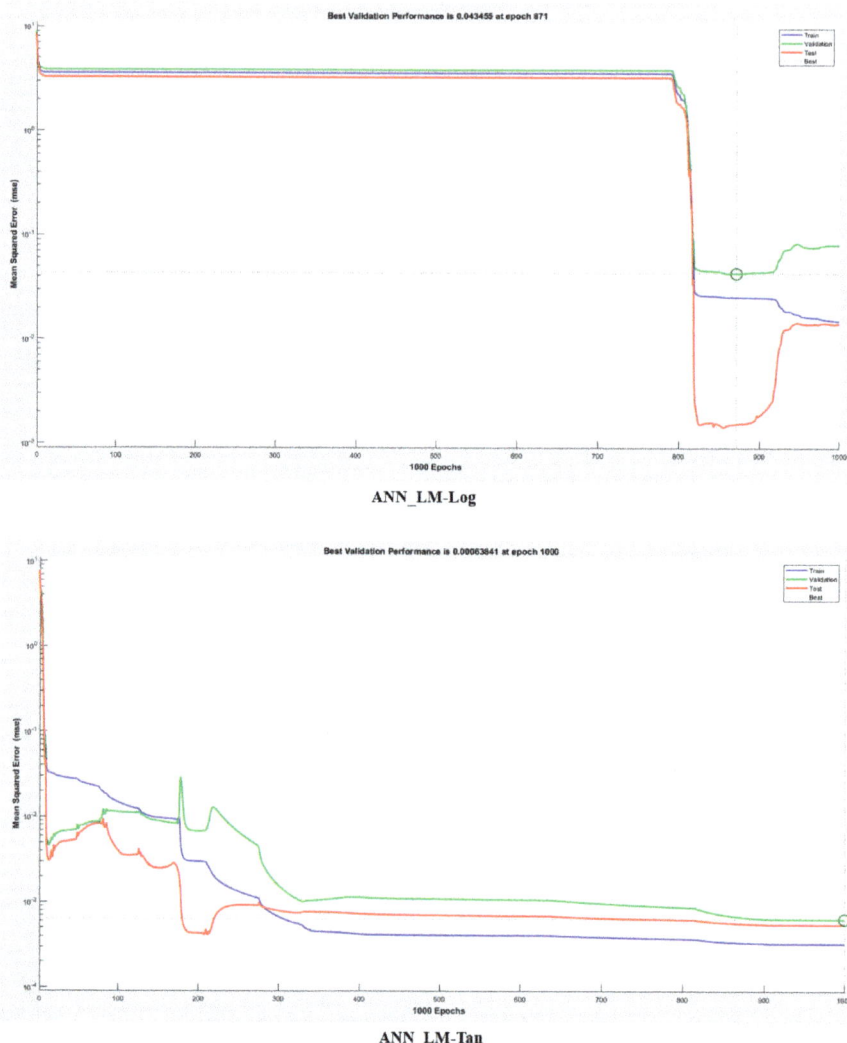

Figure 4. Deducted training, validation, and testing errors for developed ANN model with two algorithms.

4. Results and Discussion

The thermal and electrical performances of batteries with direct oil cooling are evaluated and discussed in this section for various operating conditions. Furthermore, the ANN models with two combinations of the algorithm are compared to predict the battery performance under similar operating conditions. The thermal performance under various conditions is elaborated in Section 4.1, followed by Section 4.2 with a discussion of the electrical performance under various conditions, and Section 4.3 presents a replication of the thermal and electric performances by the best ANN algorithm with the experimental results under various discharge capacities.

4.1. Thermal Performance

The maximum temperature, temperature difference, and heat transfer coefficient were evaluated and predicted as the thermal performance of the batteries under various oil temperatures, oil flow rates, and discharge rates. The thermal performance of the batteries in terms of maximum temperature, temperature difference, and heat transfer coefficient was predicted in form of time series over the discharge period of the batteries, considering variations in oil temperature, oil flow rate, and discharge rate. However, for comparison, the predicted and experimental results of thermal performance at the end of discharge are presented.

4.1.1. Maximum Temperature

The variation in experimental and predicted maximum temperatures of the batteries with change in oil temperature is presented in Figure 5. The difference in temperature between battery and oil decides the rate of dissipated heat generated in the battery. The heat transfer rate is superior when the difference in temperatures between the two sources is high. Therefore, the heat transfer rate from battery module to oil was maximum when oil at a lower temperature contacted the battery surface. The maximum temperature of the batteries increased from 44.6 °C to 64.3 °C when the oil temperature rose from 15 °C to 35 °C. Accurate training with lower prediction error in the case of the ANN_LM-Tan algorithm resulted in closer agreement between predicted and experimental maximum temperatures. The maximum temperature increased from 44.32 °C to 63.76 °C and 43.23 °C to 61.69 °C for the ANN_LM-Tan and ANN_LM-Log algorithms, respectively, with an increase in the oil temperature from 15 °C to 35 °C, which indicated corresponding maximum prediction errors of 0.94% and 4.05%.

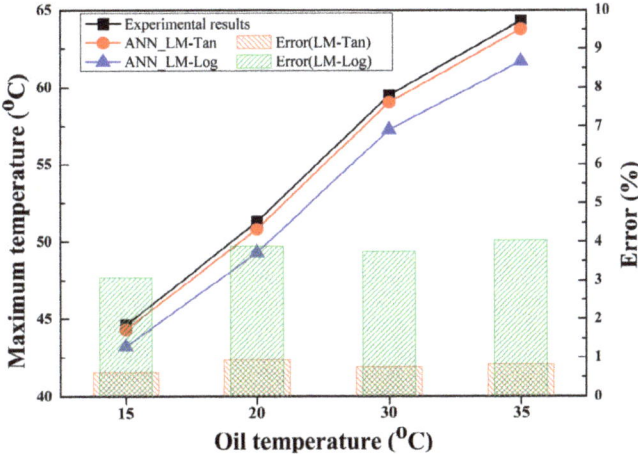

Figure 5. Variation in experimental and predicted maximum temperatures of batteries with change in oil temperature.

The experimental and predicted maximum temperatures of the batteries decreased with an increase in oil flow rate, as shown in Figure 6. The convective heat transfer from the battery module to the oil improved as the oil flow rate increased owing to the increase in local obstruction of flowing oil around the battery cells. The predicted maximum temperatures by the ANN_LM-Tan and ANN_LM-Log algorithms followed the same trend as the experimental maximum temperature variation with oil flow rate, corresponding to maximum errors of 0.97% and 4.30%. The experimental results and the ANN_LM-Tan and ANN_LM-Log algorithms showed decreases in maximum temperature from 63.40 °C to 45.20 °C, 63.91 °C to 45.46 °C, and 65.73 °C to 46.91 °C, respectively, with an increase in oil flow rate from 0.4 L/min to 1.0 L/min.

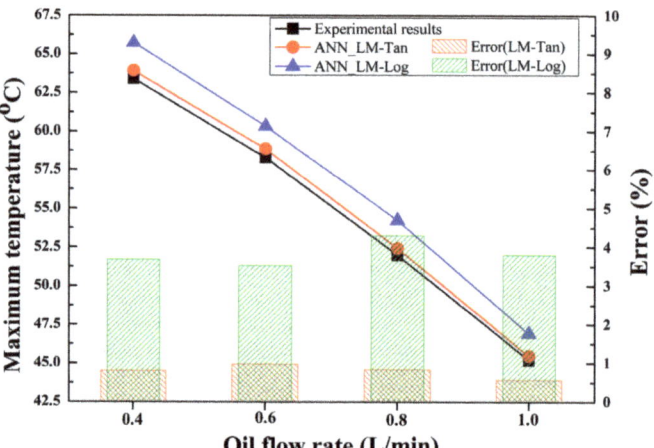

Figure 6. Variation in experimental and predicted maximum temperatures of batteries with change in oil flow rate.

The maximum temperature was predicted for the two algorithms and compared with experimental results under various discharge rates, as shown in Figure 7. The batteries generated a larger amount of heat during the high discharge rates; hence, for the same direct oil cooling conditions, the maximum temperature of the batteries increased with an increase in discharge rate. With an increase in discharge rate from 1C to 4C, the experimental results and the ANN_LM-Tan- and ANN_LM-Log-predicted maximum temperatures showed increases from 32.40 °C to 60.20 °C, 32.69 °C to 60.68 °C, and 33.51 °C to 62.18 °C, respectively. It can be observed that the predicted maximum temperature by the ANN_LM-Tan algorithm showed closer agreement with the experimental maximum temperature compared to the ANN_LM-Log algorithm, with corresponding maximum errors of 0.89% and 4.20%.

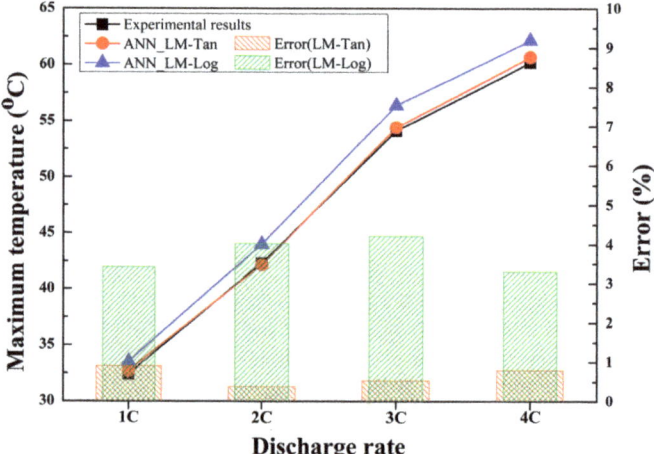

Figure 7. Variation in experimental and predicted maximum temperatures of batteries with change in discharge rate.

4.1.2. Temperature Difference

The difference between the maximum and minimum temperatures of the battery module improved with a decrease in oil temperature and increase in oil flow rate. As explained, the lower oil temperature and high oil flow rate improved the heat transfer rate between battery module and oil, which resulted in a lower temperature difference. In addition, similar to the maximum temperature, the temperature difference also increased as the discharge rate increased owing to an increase in battery heat generation under the same cooling conditions. Figures 8–10 show the variations in experimental and predicted temperature differences with changes in oil temperature, oil flow rate, and discharge rate, respectively. The temperature difference increased from 6.9 °C to 17.3 °C and 3 °C to 16.8 °C with an increase in oil temperature from 15 °C to 35 °C and increase in discharge rate from 1C to 4C, respectively. However, the temperature difference dropped from 16.7 °C to 7.3 °C with an increase in flow rate from 0.4 L/min and 1.0 L/min. In the case of temperature difference, the ANN_LM-Tan algorithm also had high prediction accuracy with the experimental results compared to the ANN_LM-Log algorithm for all oil temperatures, oil flow rates, and discharge rates. The maximum errors between the predicted and experimental temperature differences were 0.95% and 4.29% in the case of oil temperature, 0.96% and 4.86% in the case of oil flow rate, and 0.81% and 3.52% in the case of discharge rate, corresponding to the ANN_LM-Tan and ANN_LM-Log algorithms, respectively.

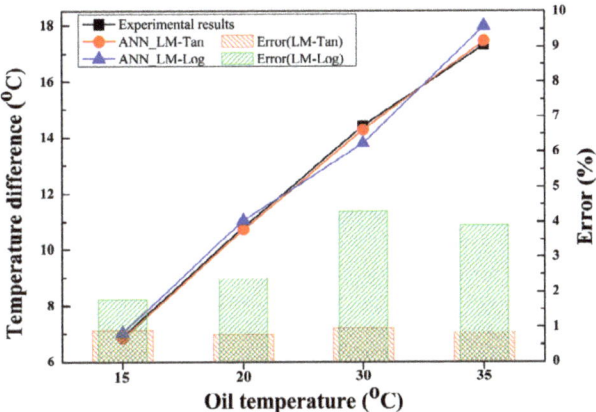

Figure 8. Variation in experimental and predicted temperature difference of batteries with change in oil temperature.

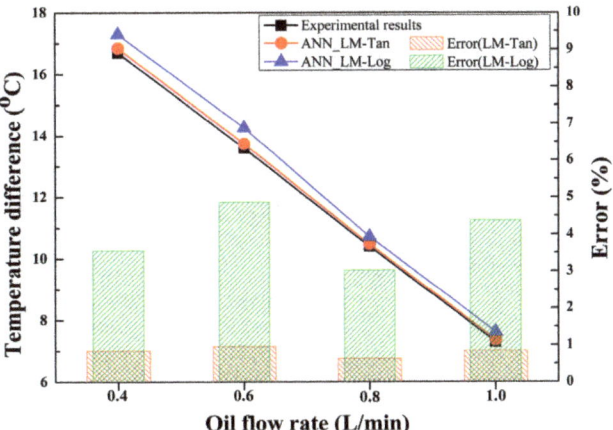

Figure 9. Variation in experimental and predicted temperature difference of batteries with change in oil flow rate.

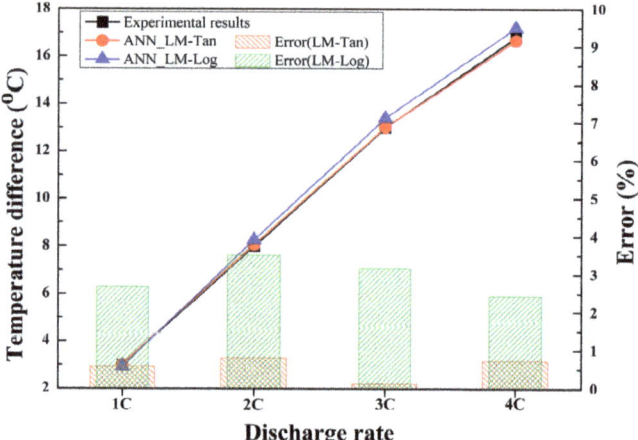

Figure 10. Variation in experimental and predicted temperature difference of batteries with change in discharge rate.

4.1.3. Heat Transfer Coefficient

To assess the effectiveness of direct oil cooling for the battery module, the heat transfer coefficient was evaluated under various conditions of oil temperature, oil flow rate, and discharge rate, as shown in Figures 11–13. The convective heat transfer between batteries and oil improved when the oil temperature decreased and oil flow rate increased. Therefore, the lower battery temperature at the lower oil temperature and higher oil flow rate indicated a maximum heat transfer coefficient. The maximum heat transfer coefficients of 3908.83 W/m^2-K and 2374.92 W/m^2-K were evaluated corresponding to 15 °C oil temperature and 1.0 L/min oil flow rate, respectively. There was no significant difference in heat transfer coefficient with change in discharge rate; however, the higher discharge rate enabled the opportunity for increased convective heat transfer from battery to oil owing to higher heat generation compared to the lower discharge rate. The maximum heat transfer coefficient of 2741.22 W/m^2-K was observed at a discharge rate of 4C. Closer agreement between the actual and predicted heat transfer coefficients was observed for the ANN_LM-Tan algorithm under conditions of oil temperature, oil flow rate, and discharge rate with corresponding lowest maximum errors of 0.93%, 0.94%, and 0.79%, respectively. The highest maximum errors of 4.14%, 4.73%, and 4.17% were observed between ANN_LM-Log-predicted and actual heat transfer coefficients in the case of oil temperature, oil flow rate, and discharge rate, respectively.

4.2. Electrical Performance

The discharge voltage was evaluated and predicted using the ANN model as the electrical performance of the batteries under various conditions of oil temperature, oil flow rate, and discharge rate. The voltage results were predicted in the form of time series over the discharge period of the batteries for changes in influential factors. However, it should be noted that the voltage at the end of discharge was the same; hence, to compare the experimental and predicted results of voltage under variations of oil temperature, oil flow rate, and discharge rate, the voltage results were considered at the same discharge capacity.

Voltage

The effect of oil temperature on the voltage of batteries is depicted in Figure 14. The presented voltage results are compared at the same discharge capacity. The operating temperature of a battery affects the electrochemical characteristics of the battery; therefore, a change in oil temperature has a significant impact on the voltage of a battery during the discharge condition. The lower oil temperature showed a decreased voltage value, which increased as the oil temperature increased because the lower oil temperature had a higher heat transfer rate from the batteries, which raised the internal resistance of the batteries. A drop in surrounding temperature results in the enhancement of ohmic resistance, which degrades the voltage of a battery [48]. In addition, Lu et al. claimed that the ionic conductivities of the SEI layer, electrode, and electrolyte were minimum at low temperature, which generates a decreasing voltage trend for the battery [49]. The voltage of the batteries dropped from 12.181 V to 10.996 V when the oil temperature decreased from 35 °C to 15 °C. Furthermore, the predicted voltages from the ANN model with two algorithms are compared to the experimental results under various oil temperatures in Figure 14. The overall error combining training, validation, and testing was higher in the case of the ANN_LM-Log algorithm compared to the ANN_LM-Tan algorithm, as presented in Figure 4. Therefore, the voltage predicted by the ANN_LM-Tan algorithm showed closer agreement with the experimental voltage at all oil temperatures compared to that by the ANN_LM-Log algorithm. The maximum errors between the experimental and predicted voltages by the ANN_LM-Tan and ANN_LM-Log algorithms were 0.88% and 4.81%, respectively.

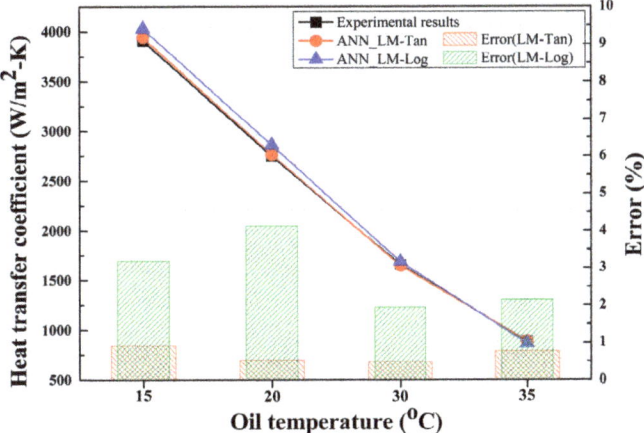

Figure 11. Variation in experimental and predicted heat transfer coefficient with change in oil temperature.

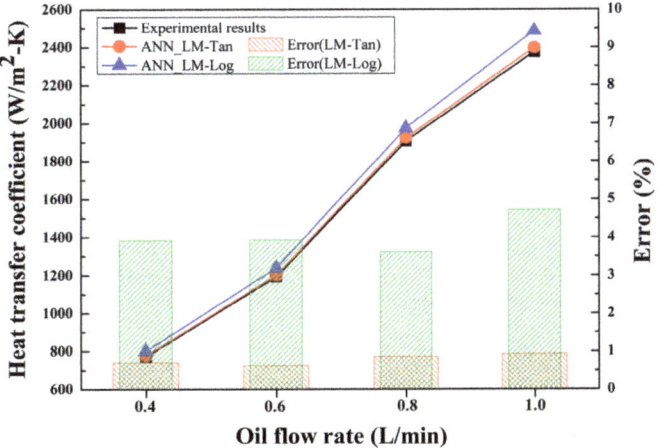

Figure 12. Variation in experimental and predicted heat transfer coefficient with change in oil flow rate.

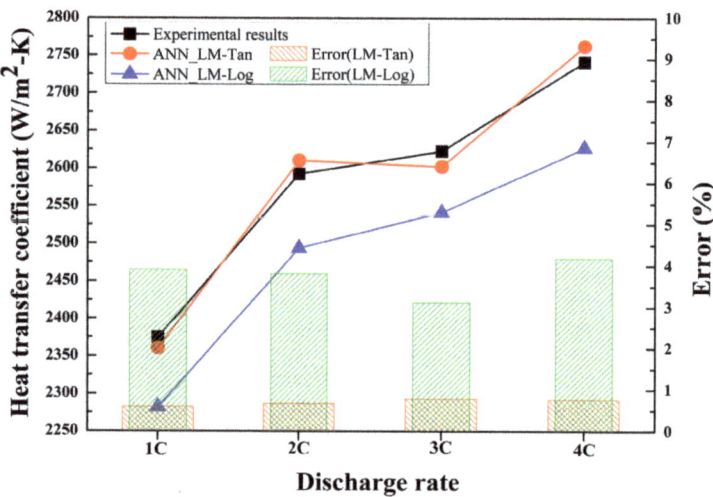

Figure 13. Variation in experimental and predicted heat transfer coefficient with change in discharge rate.

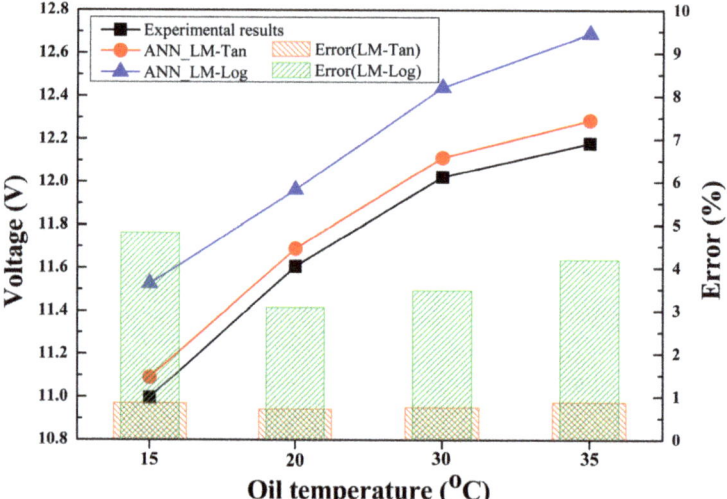

Figure 14. Variation in experimental and predicted voltage of batteries with change in oil temperature.

The effect of oil flow rate on experimental voltage and predicted voltage using both algorithms is depicted in Figure 15. An increase in oil flow rate indicates an improvement in battery cooling performance, which means the internal resistance of the battery increases with an increase in oil flow rate and thus a decrease in battery voltage during the discharge condition. Tong et al. also observed that the voltage of batteries dropped owing to a rise in the internal resistance of batteries when the battery cooling rate improved [50]. Therefore, the experimental and predicted voltage results showed a decreasing trend with an increase in oil flow rate. The voltage dropped from 11.113 V to 10.725 V with an increase in oil flow rate from 0.4 L/min to 1.0 L/min. The prediction accuracy for the ANN_LM-Tan algorithm was higher compared to the ANN_LM-Log algorithm with experimental voltages at each oil flow rate. The maximum prediction errors for the ANN_LM-Tan and ANN_LM-Log algorithms were 0.62% and 3.43%, respectively.

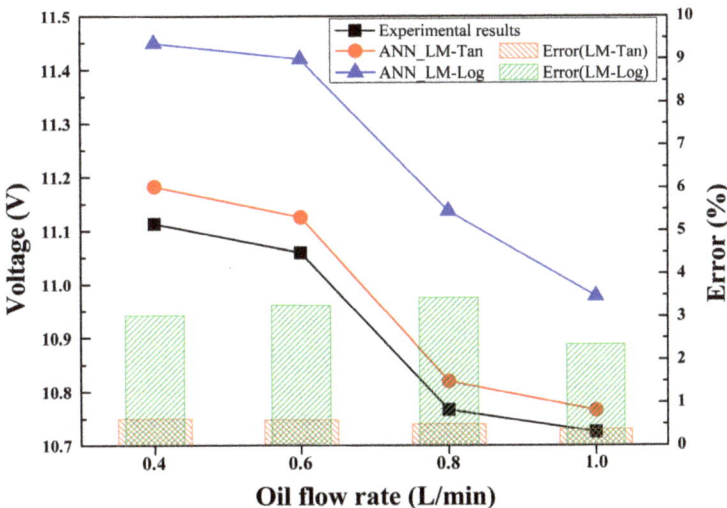

Figure 15. Variation in experimental and predicted voltage of batteries with change in oil flow rate.

The comparison of experimental and predicted voltages for different discharge conditions is shown in Figure 16. The voltage dropped rapidly as the discharge rate increased; therefore, at the same discharge capacity, the lower and higher discharge rates showed maximum and minimum voltages. The voltage dropped from 13.553 V to 10.725 V with an increase in discharge rate from 1C to 4C. The predicted voltages for both algorithms showed the same decreasing trend as the experimental voltage with the rise in discharge rate. However, the ANN_LM-Tan algorithm was found to be an accurate model to predict closer voltages with corresponding experimental values compared to the ANN_LM-Log algorithm. Considering all discharge rates, the maximum errors between the predicted voltages by the ANN_LM-Tan and ANN_LM-Log algorithms with the experimental voltage were 0.57% and 3.79%, respectively. As explained, the experimental and predicted results of the voltage were compared at the same discharge capacity of 8.136 Ah.

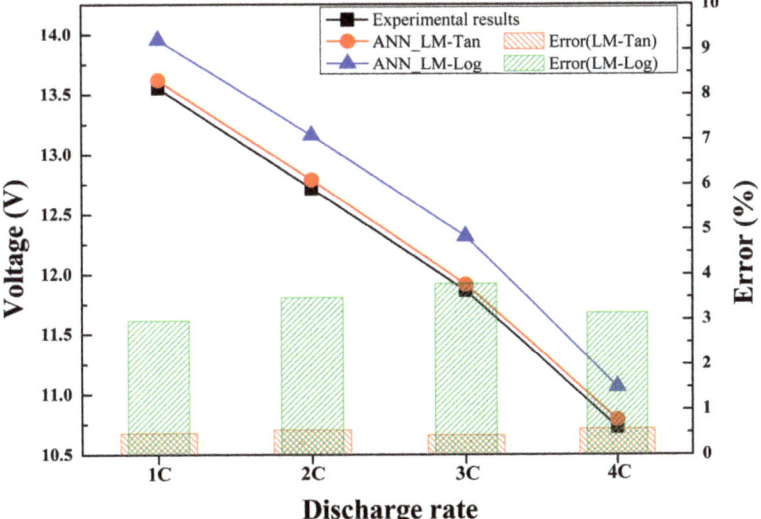

Figure 16. Variation in experimental and predicted voltage of batteries with change in discharge rate.

4.3. Accuracy of Proposed ANN Model

The ANN_LM-Tan algorithm depicted accurate predictions of all thermal and electrical performances under several conditions of oil temperature, oil flow rate, and discharge rate compared to the ANN_LM-Log algorithm. Therefore, the ANN_LM-Tan algorithm is suggested to replicate the various performances of battery modules with direct oil cooling under real operating conditions. Furthermore, to assure the accuracy and reliability of the suggested ANN model, the maximum temperature and voltage were predicted as thermal and electrical performances with change in discharge capacity and compared with the corresponding experimental results. The variations in experimental and ANN_LM-Tan-predicted maximum temperature and voltage with discharge capacity are presented in Figure 17. This comparison is presented for an oil temperature of 30 °C, oil flow rate of 1.0 L/min, and discharge rate of 4C. For each condition of discharge capacity, the proposed ANN model depicted accurate replications of maximum temperature and voltage compared to the experimental data. The statistical parameters were calculated for the comparison, as presented in Figure 17, to quantify the accuracy of the predicted thermal and electrical performances. The calculated R^2 and COV were 0.9998 and 1.55, respectively, in the case of maximum temperature, and 0.9997 and 1.66, respectively, in the case of voltage, indicating the reliability of the proposed ANN model to accurately mimic the actual condition data.

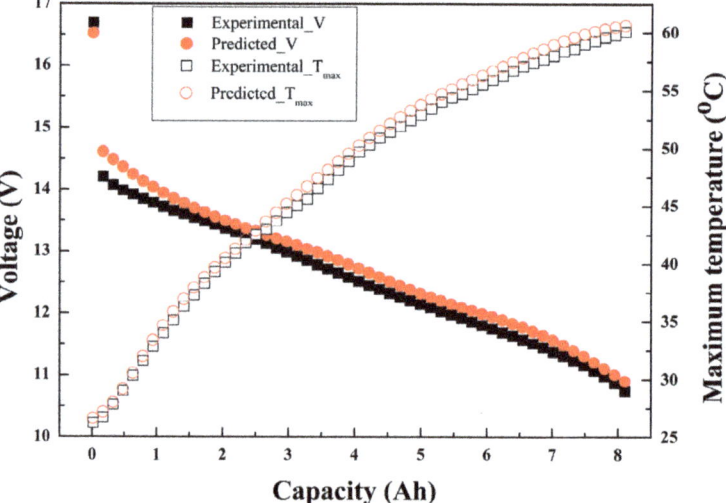

Figure 17. Variations in experimental and ANN_LM-Tan-predicted maximum temperature and voltage with discharge capacity.

The development of an accurate neural network model enables replication of the performance of a battery with direct oil cooling under realistic operating conditions with minimal errors. The proposed neural network model could be used to generate a database relating the influential parameters and performance of batteries with direct oil cooling. Thus, several efforts in the development of prototypes could be minimized to fabricate a final-stage direct oil cooling system using the generated comprehensive reference database.

5. Conclusions

The thermal and electrical performances of a battery module with direct oil cooling were experimentally evaluated and predicted using neural network models under several operating conditions. The following key findings are listed from the conducted present work.

(a) The thermal performance in terms of maximum temperature, temperature difference, and heat transfer coefficient improves with a decrease in oil temperature. The lower maximum temperature and temperature difference of 44.6 °C and 6.9 °C, respectively, and higher heat transfer coefficient of 3908.83 W/m^2-K were evaluated at a lower oil inlet temperature of 15 °C. The electrical performance in terms of voltage drops with a decrease in oil temperature, such that oil temperatures of 15 °C and 35 °C showed voltages of 10.996 V and 12.181 V, respectively.

(b) An increase in oil flow rate reduces the maximum temperature, temperature difference, and voltage, whereas the heat transfer coefficient is enhanced. With an increase in oil flow rate from 0.4 L/min to 1.0 L/min, drops of 18.2 °C, 9.4 °C, and 0.388 V and an improvement of 1602.78 W/m^2-K were observed in the maximum temperature, temperature difference, voltage, and heat transfer coefficient, respectively.

(c) The maximum temperature and temperature difference increased by 27.8 °C and 13.8 °C, respectively, and the voltage dropped by 2.828 V with an increment in discharge rate from 1C to 4C. The maximum heat transfer coefficient of 2741.22 W/m^2-K was evaluated at a higher discharge rate of 4C.

(d) The ANN_LM-Tan and ANN_LM-Log algorithms showed maximum errors of 0.97% and 4.30% in the case of maximum temperature, 0.96% and 4.86% in the case of temperature difference, 0.94% and 4.73% in the case of heat transfer coefficient, and 0.88% and 4.81% in the case of voltage, respectively, considering all conditions of oil temperature, oil flow rate, and discharge rate. The prediction accuracy of the ANN_LM-Tan algorithm was superior compared to the ANN_LM-Log algorithm for all thermal and electrical performances under the considered operating conditions.

(e) The ANN_LM-Tan algorithm is recommended as the best neural network model to generate data of thermal and electrical performances under influential conditions for batteries with direct oil cooling. The reliability of the best neural network model was further established by predicting the maximum temperature and voltage for various discharge capacities, reflecting a maximum R^2 and COV of 0.99 and 1.66, respectively.

(f) The proposed prediction model and prediction database could guide mapping the relationship between operating conditions and performance, which could be utilized to design and fabricate a direct liquid cooling system for high energy density batteries in electric vehicles. In future, tests will be conducted to develop prediction models for a battery module with direct oil cooling under fast charging and discharging conditions to assure the safety and reliability of the proposed next-generation battery thermal management technique.

Author Contributions: Conceptualization, K.S.G. and M.-Y.L.; methodology, K.S.G.; software, K.S.G.; formal analysis, K.S.G. and J.-W.H.; investigation, K.S.G. and S.-G.H.; resources, K.S.G. and M.-Y.L.; data curation, K.S.G. and S.-G.H.; writing—original draft preparation, K.S.G.; writing—review and editing, K.S.G. and M.-Y.L.; visualization, K.S.G., J.-W.H. and S.-G.H.; supervision, M.-Y.L.; project administration, M.-Y.L.; funding acquisition, M.-Y.L. All authors have read and agreed to the published version of the manuscript.

Funding: This work was supported by the Dong-A University research fund.

Data Availability Statement: The data presented in this study are available upon request to the corresponding author. The data are not publicly available due to privacy.

Conflicts of Interest: The authors declare no conflict of interest.

References

1. Alanazi, F. Electric Vehicles: Benefits, Challenges, and Potential Solutions for Widespread Adaptation. *Appl. Sci.* **2023**, *13*, 6016. [CrossRef]
2. Stoma, M.; Dudziak, A. Future Challenges of the Electric Vehicle Market Perceived by Individual Drivers from Eastern Poland. *Energies* **2023**, *16*, 7212. [CrossRef]
3. Fresia, M.; Bracco, S. Electric Vehicle Fleet Management for a Prosumer Building with Renewable Generation. *Energies* **2023**, *16*, 7213. [CrossRef]

4. Ramraj, R.; Pashajavid, E.; Alahakoon, S.; Jayasinghe, S. Quality of Service and Associated Communication Infrastructure for Electric Vehicles. *Energies* **2023**, *16*, 7170. [CrossRef]
5. Liu, H.; Wei, Z.; He, W.; Zhao, J. Thermal issues about Li-ion batteries and recent progress in battery thermal management systems: A review. *Energy Convers. Manag.* **2017**, *150*, 304–330. [CrossRef]
6. Wilberforce, T.; El-Hassan, Z.; Khatib, F.N.; Al Makky, A.; Baroutaji, A.; Carton, J.G.; Olabi, A.G. Developments of electric cars and fuel cell hydrogen electric cars. *Int. J. Hydrog. Energy* **2017**, *42*, 25695–25734. [CrossRef]
7. Kumar, M.; Panda, K.P.; Naayagi, R.T.; Thakur, R.; Panda, G. Comprehensive Review of Electric Vehicle Technology and Its Impacts: Detailed Investigation of Charging Infrastructure, Power Management, and Control Techniques. *Appl. Sci.* **2023**, *13*, 8919. [CrossRef]
8. Dan, D.; Zhao, Y.; Wei, M.; Wang, X. Review of Thermal Management Technology for Electric Vehicles. *Energies* **2023**, *16*, 4693. [CrossRef]
9. Irfan, M.; Deilami, S.; Huang, S.; Veettil, B.P. Rooftop Solar and Electric Vehicle Integration for Smart, Sustainable Homes: A Comprehensive Review. *Energies* **2023**, *16*, 7248. [CrossRef]
10. Liu, H.; Xiao, Q.; Jin, Y.; Mu, Y.; Meng, J.; Zhang, T.; Jia, H.; Teodorescu, R. Improved LightGBM-Based Framework for Electric Vehicle Lithium-Ion Battery Remaining Useful Life Prediction Using Multi Health Indicators. *Symmetry* **2022**, *14*, 1584. [CrossRef]
11. Lu, L.; Han, X.; Li, J.; Hua, J.; Ouyang, M. A review on the key issues for lithium-ion battery management in electric vehicles. *J. Power Sources* **2013**, *226*, 272–288. [CrossRef]
12. Liu, Z.; Huang, J.; Cao, M.; Jiang, G.; Yan, Q.; Hu, J. Experimental study on the thermal management of batteries based on the coupling of composite phase change materials and liquid cooling. *Appl. Therm. Eng.* **2021**, *185*, 116415. [CrossRef]
13. Huang, Y.; Wei, C.; Fang, Y. Numerical investigation on optimal design of battery cooling plate for uneven heat generation conditions in electric vehicles. *Appl. Therm. Eng.* **2022**, *211*, 118476. [CrossRef]
14. Behi, H.; Karimi, D.; Behi, M.; Ghanbarpour, M.; Jaguemont, J.; Sokkeh, M.A.; Gandoman, F.H.; Berecibar, M.; Van Mierlo, J. A new concept of thermal management system in Li-ion battery using air cooling and heat pipe for electric vehicles. *Appl. Therm. Eng.* **2020**, *174*, 115280. [CrossRef]
15. Zhang, X.; Li, Z.; Luo, L.; Fan, Y.; Du, Z. A review on thermal management of lithium-ion batteries for electric vehicles. *Energy* **2022**, *238*, 121652. [CrossRef]
16. Jiaqiang, E.; Yi, F.; Li, W.; Zhang, B.; Zuo, H.; Wei, K.; Chen, J.; Zhu, H.; Zhu, H.; Deng, Y. Effect analysis on heat dissipation performance enhancement of a lithium-ion-battery pack with heat pipe for central and southern regions in China. *Energy* **2021**, *226*, 120336.
17. Panchal, S.; Khasow, R.; Dincer, I.; Agelin-Chaab, M.; Fraser, R.; Fowler, M. Thermal design and simulation of mini-channel cold plate for water cooled large sized prismatic lithium-ion battery. *Appl. Therm. Eng.* **2017**, *122*, 80–90. [CrossRef]
18. Akbarzadeh, M.; Kalogiannis, T.; Jaguemont, J.; Jin, L.; Behi, H.; Karimi, D.; Beheshti, H.; Van Mierlo, J.; Berecibar, M. A comparative study between air cooling and liquid cooling thermal management systems for a high-energy lithium-ion battery module. *Appl. Therm. Eng.* **2021**, *198*, 117503. [CrossRef]
19. Tan, X.; Lyu, P.; Fan, Y.; Rao, J.; Ouyang, K. Numerical investigation of the direct liquid cooling of a fast-charging lithium-ion battery pack in hydrofluoroether. *Appl. Therm. Eng.* **2021**, *196*, 117279. [CrossRef]
20. Roe, C.; Feng, X.; White, G.; Li, R.; Wang, H.; Rui, X.; Li, C.; Zhang, F.; Null, V.; Parkes, M.; et al. Immersion cooling for lithium-ion batteries—A review. *J. Power Sources* **2022**, *525*, 231094. [CrossRef]
21. Wu, S.; Lao, L.; Wu, L.; Liu, L.; Lin, C.; Zhang, Q. Effect analysis on integration efficiency and safety performance of a battery thermal management system based on direct contact liquid cooling. *Appl. Therm. Eng.* **2022**, *201*, 117788. [CrossRef]
22. Li, Y.; Zhou, Z.; Hu, L.; Bai, M.; Gao, L.; Li, Y.; Liu, X.; Li, Y.; Song, Y. Experimental studies of liquid immersion cooling for 18650 lithium-ion battery under different discharging conditions. *Case Stud. Therm. Eng.* **2022**, *34*, 102034. [CrossRef]
23. Patil, M.S.; Seo, J.H.; Lee, M.Y. A novel dielectric fluid immersion cooling technology for Li-ion battery thermal management. *Energy Convers. Manag.* **2021**, *229*, 113715. [CrossRef]
24. Sundin, D.W.; Sponholtz, S. Thermal management of Li-ion batteries with single-phase liquid immersion cooling. *IEEE Open J. Veh. Technol.* **2020**, *1*, 82–92. [CrossRef]
25. Zhou, H.; Dai, C.; Liu, Y.; Fu, X.; Du, Y. Experimental investigation of battery thermal management and safety with heat pipe and immersion phase change liquid. *J. Power Sources* **2020**, *473*, 228545. [CrossRef]
26. Dubey, P.; Pulugundla, G.; Srouji, A.K. Direct comparison of immersion and cold-plate based cooling for au-tomotive Li-ion battery modules. *Energies* **2021**, *14*, 1259. [CrossRef]
27. Mazzeo, D.; Herdem, M.S.; Matera, N.; Bonini, M.; Wen, J.Z.; Nathwani, J.; Oliveti, G. Artificial intelligence application for the performance prediction of a clean energy community. *Energy* **2021**, *232*, 120999. [CrossRef]
28. Pang, Z.; Niu, F.; O'Neill, Z. Solar radiation prediction using recurrent neural network and artificial neural network: A case study with comparisons. *Renew. Energy* **2020**, *156*, 279–289. [CrossRef]
29. Panchal, S.; Dincer, I.; Agelin-Chaab, M.; Fraser, R.; Fowler, M. Design and simulation of a lithium-ion battery at large C-rates and varying boundary conditions through heat flux distributions. *Measurement* **2018**, *116*, 382–390. [CrossRef]
30. Wang, Q.K.; He, Y.J.; Shen, J.N.; Ma, Z.F.; Zhong, G.B. A unified modeling framework for lithium-ion batteries: An artificial neural network based thermal coupled equivalent circuit model approach. *Energy* **2017**, *138*, 118–132. [CrossRef]

31. Feng, F.; Teng, S.; Liu, K.; Xie, J.; Xie, Y.; Liu, B.; Li, K. Co-estimation of lithium-ion battery state of charge and state of temperature based on a hybrid electrochemical-thermal-neural-network model. *J. Power Sources* **2020**, *455*, 227935. [CrossRef]
32. Xie, Y.; He, X.J.; Hu, X.S.; Li, W.; Zhang, Y.J.; Liu, B.; Sun, Y.T. An improved resistance-based thermal model for a pouch lithium-ion battery considering heat generation of posts. *Appl. Therm. Eng.* **2020**, *164*, 114455. [CrossRef]
33. Arora, S.; Shen, W.; Kapoor, A. Neural network based computational model for estimation of heat generation in LiFePO4 pouch cells of different nominal capacities. *Comput. Chem. Eng.* **2017**, *101*, 81–94. [CrossRef]
34. Liu, J.; Tavakoli, F.; Sajadi, S.M.; Mahmoud, M.Z.; Heidarshenas, B.; Aybar, H.Ş. Numerical evaluation and artificial neural network modeling of the effect of oval PCM compartment dimensions around a triple lithium-ion battery pack despite forced airflow. *Eng. Anal. Bound. Elem.* **2022**, *142*, 71–92. [CrossRef]
35. Jaliliantabar, F.; Mamat, R.; Kumarasamy, S. Prediction of lithium-ion battery temperature in different operating conditions equipped with passive battery thermal management system by artificial neural networks. *Mater. Today Proc.* **2022**, *48*, 1796–1804. [CrossRef]
36. James, A.; Srinivas, M.; Mohanraj, M.; Raj, A.K.; Jayaraj, S. Experimental studies on photovoltaic-thermal heat pump water heaters using variable frequency drive compressors. *Sustain. Energy Technol. Assess.* **2021**, *45*, 101152. [CrossRef]
37. Holman, J.P. *Experimental Methods for Engineers*, 8th ed.; McGraw Hill Publisher: New York, NY, USA, 2021.
38. Raj, A.K.; Srinivas, M.; Jayaraj, S. A cost-effective method to improve the performance of solar air heaters using discrete macro-encapsulated PCM capsules for drying applications. *Appl. Therm. Eng.* **2019**, *146*, 910–920. [CrossRef]
39. Han, J.W.; Garud, K.S.; Kang, E.H.; Lee, M.Y. Numerical Study on Heat Transfer Characteristics of Dielectric Fluid Immersion Cooling with Fin Structures for Lithium-Ion Batteries. *Symmetry* **2022**, *15*, 92. [CrossRef]
40. Han, J.W.; Garud, K.S.; Hwang, S.G.; Lee, M.Y. Experimental Study on Dielectric Fluid Immersion Cooling for Thermal Management of Lithium-Ion Battery. *Symmetry* **2022**, *14*, 2126. [CrossRef]
41. Maduabuchi, C. Thermo-mechanical optimization of thermoelectric generators using deep learning artificial intelligence algorithms fed with verified finite element simulation data. *Appl. Energy* **2022**, *315*, 118943. [CrossRef]
42. Mohanraj, M.; Jayaraj, S.; Muraleedharan, C. Performance prediction of a direct expansion solar assisted heat pump using artificial neural networks. *Appl. Energy* **2009**, *86*, 1442–1449. [CrossRef]
43. Islam, K.T.; Raj, R.G.; Mujtaba, G. Recognition of traffic sign based on bag-of-words and artificial neural network. *Symmetry* **2017**, *9*, 138. [CrossRef]
44. Ullah, I.; Fayaz, M.; Kim, D. Improving accuracy of the Kalman filter algorithm in dynamic conditions using ANN-based learning module. *Symmetry* **2019**, *11*, 94. [CrossRef]
45. Moya-Rico, J.D.; Molina, A.E.; Belmonte, J.F.; Tendero, J.C.; Almendros-Ibanez, J.A. Characterization of a triple concentric-tube heat exchanger with corrugated tubes using Artificial Neural Networks (ANN). *Appl. Therm. Eng.* **2019**, *147*, 1036–1046. [CrossRef]
46. Kishore, R.A.; Mahajan, R.L.; Priya, S. Combinatory finite element and artificial neural network model for predicting performance of thermoelectric generator. *Energies* **2018**, *11*, 2216. [CrossRef]
47. Gunasekar, N.; Mohanraj, M.; Velmurugan, V. Artificial neural network modeling of a photovoltaic-thermal evaporator of solar assisted heat pumps. *Energy* **2015**, *93*, 908–922. [CrossRef]
48. Wu, H.; Zhang, X.; Cao, R.; Yang, C. An investigation on electrical and thermal characteristics of cylindrical lithium-ion batteries at low temperatures. *Energy* **2021**, *225*, 120223. [CrossRef]
49. Lu, Z.; Yu, X.L.; Wei, L.C.; Cao, F.; Zhang, L.Y.; Meng, X.Z.; Jin, L.W. A comprehensive experimental study on temperature-dependent performance of lithium-ion battery. *Appl. Therm. Eng.* **2019**, *158*, 113800. [CrossRef]
50. Tong, W.; Somasundaram, K.; Birgersson, E.; Mujumdar, A.S.; Yap, C. Numerical investigation of water cooling for a lithium-ion bipolar battery pack. *Int. J. Therm. Sci.* **2015**, *94*, 259–269. [CrossRef]

Disclaimer/Publisher's Note: The statements, opinions and data contained in all publications are solely those of the individual author(s) and contributor(s) and not of MDPI and/or the editor(s). MDPI and/or the editor(s) disclaim responsibility for any injury to people or property resulting from any ideas, methods, instructions or products referred to in the content.

Article

A Novel Method for State of Health Estimation of Lithium-Ion Batteries Based on Deep Learning Neural Network and Transfer Learning

Zhong Ren [1,2,3], Changqing Du [1,2,3,*] and Yifang Zhao [4]

[1] Hubei Key Laboratory of Advanced Technology for Automotive Components, Wuhan University of Technology, Wuhan 430070, China; renzhong@whut.edu.cn
[2] Foshan Xianhu Laboratory of the Advanced Energy Science and Technology Guangdong Laboratory, Foshan 528200, China
[3] Hubei Research Center for New Energy & Intelligent Connected Vehicle, Wuhan University of Technology, Wuhan 430070, China
[4] SAIC-GM-Wuling Automobile, Liuzhou 545007, China; yifan.zhao@sgmw.com.cn
* Correspondence: cq_du@whut.edu.cn

Abstract: Accurate state of health (SOH) estimation of lithium-ion batteries is critical for maintaining reliable and safe working conditions for electric vehicles (EVs). The machine learning-based method with health features (HFs) is encouraging for health prognostics. However, the machine learning method assumes that the training and testing data have the same distribution, which restricts its application for different types of batteries. Thus, in this paper, a deep learning neural network and fine-tuning-based transfer learning strategy are proposed for accurate and robust SOH estimation toward different types of batteries. First, a universal HF extraction strategy is proposed to obtain four highly related HFs. Second, a deep learning neural network consisting of long short-term memory (LSTM) and fully connected layers is established to model the relationship between the HFs and SOH. Third, the fine-tuning-based transfer learning strategy is exploited for SOH estimation of various types of batteries. The proposed methods are comprehensively verified using three open-source datasets. Experimental results show that the proposed deep learning neural network with the HFs can estimate the SOH accurately in a single dataset without using the transfer learning strategy where the mean absolute error (MAE) and root mean square error (RMSE) are constrained to 1.21% and 1.83%. For the transfer learning between different aging datasets, the overall MAE and RMSE are limited to 1.09% and 1.41%, demonstrating the reliability of the fine-tuning strategy.

Keywords: lithium-ion battery; state of health estimation; machine learning; transfer learning

1. Introduction

After years of development in the electric vehicle (EV) industry, lithium-ion batteries (LiBs) have become the main source of EVs [1,2]. Although EVs' market share has steadily increased in recent years, their safety and lifespan have always been essential issues restricting their long-term use. Therefore, the battery management system (BMS) is implemented to monitor the health state of the battery system and to ensure safety and reliable operation. State of health (SOH) is an important indicator to evaluate the health state of the battery, and it can be expressed as follows:

$$SOH_c = \frac{C_t}{C_0} \times 100\% \quad (1)$$

where C_t and C_0 are the actual and rated capacity, respectively.

SOH estimation is a key function of the BMS [3]. In the past decades, the existing SOH estimation approaches can be classified as direct measurement methods and indirect analytical methods [4]. Typical direct measurement methods include capacity measurement based on the coulomb counting method, internal resistance measurement based on specific tests, and the impedance measurement method, which relies on electrochemical impedance spectroscopy (EIS). The direct measurement methods suit laboratory conditions but not onboard applications. Indirect analytical methods can be further classified as model-based and data-driven methods. A high-fidelity battery model, such as the electrochemical model [5], equivalent circuit model (ECM) [6], empirical model, and stochastic degradation model, is first built for the model-based methods. Then, the well-parameterized battery model is integrated with filter algorithms to estimate the SOH [7,8].

Owing to advanced techniques, such as 5G, cloud computing, and the Internet of Things (IoT), the data-driven approaches have found successful applications across different domains of LiBs, such as the production of LiBs [9,10], material design [11], fast charging strategy [12], safety control [13], as well the state estimation, especially the SOH estimation. Compared with the model-based method, the superiority of the data-driven method is that it considers the battery as a black box, and the pre-determined battery model is not required anymore. Machine learning (ML) methods can mine the hidden degradation information from the aging data [14]. The typical flowchart to develop an ML-based SOH estimation method is shown in Figure 1. Data processing and model training are two critical steps that determine the performance of the SOH estimation [15]. For data preprocessing, the so-called health features (HFs) [16] are extracted from the raw data, which is the foundation and key for ML methods. The extraction methods include direct extraction and indirect extraction methods [17]. The direct extraction method obtains HFs based on the measured raw data directly. For example, the charging time of constant current (CC) and constant voltage (CV) processes [18,19], the slope of the curve during the end of the CC charging process [20], and voltage discrepancy at uniform time intervals [21] were used as HFs in recently published papers. The indirect extraction method obtains HFs based on the reconstructed curves. Commonly used reconstructing methods are incremental capacity (IC) analysis [22], differential voltage (DV) analysis [23], and differential temperature (DT) analysis [16]. For example, multiple peaks can be seen in the IC curve, and each peak reflects the phase transition when the battery is working. With the aging of LiBs, the IC curve shows a specific trend of change, especially the peaks. Therefore, the value, position, width, slope, and area under each peak are often used as HFs. For these kinds of reconstructed curves, filtering methods are required to eliminate the influence of noise. Other indirect HFs include open circuit voltage and ohmic resistance within the ECM or polarization capacitance and resistance derived from electrochemical impedance [24,25]. The main drawback is that additional algorithms are required, increasing its difficulty. Other variables, namely, the sample entropy [26,27] and Kullback–Leibler distance [28], are also used as HFs for SOH estimation.

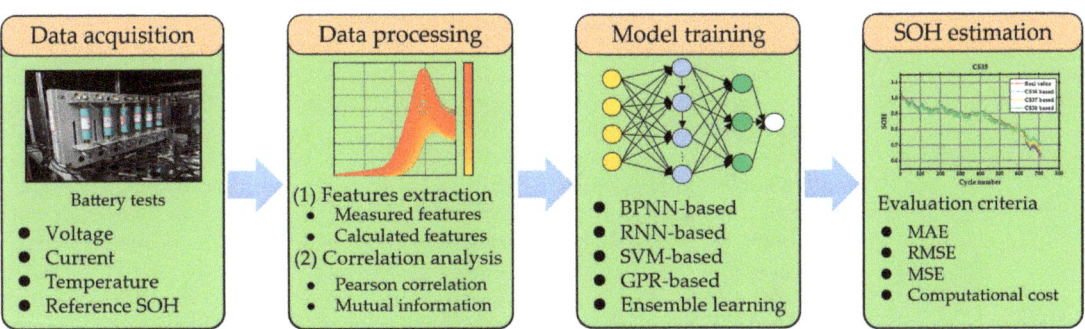

Figure 1. The procedures to build an ML-based SOH estimation method.

Based on the pre-processed HFs and SOH values, different ML approaches are employed to train the SOH estimation model, including shallow neural networks, deep learning methods, Gaussian process regression (GPR), support vector machine (SVM), and ensemble learning methods. Although developing the data-driven-based SOH estimation method requires numerous aging data, more and more public aging datasets are available online thanks to the joint efforts of all researchers, which is beneficial for developing advanced health prognostic methods. Some popular public aging datasets are given in Table 1.

Table 1. Summary of the publicly available battery datasets.

Sources	Refs.
Research & development data repository from Sandia National Labs	[29]
Oxford battery degradation dataset	[30]
Center for Advanced Life Cycle Engineering (CALCE) at the University of Maryland	[31,32]
NASA Ames Prognostic Data Repository (NASA)	[33]
Massachusetts Institute of Technology (MIT) dataset	[34]
The Tongji University dataset	[35]
The Huazhong University of Science and Technology dataset	[36]

Currently, the HF-based ML methods for SOH estimation are usually developed based on a single battery dataset, which means that the effectiveness and applicability of the HF extraction strategy cannot be guaranteed toward other battery datasets. For example, Fan et al. [37] used different combinations of HFs to train the SOH estimation model toward the NASA and Oxford aging datasets, respectively. When choosing appropriate HFs, in addition to the principles of easy acquisition, suitability for practical conditions, and a highly relevant degree, the adaptability and universality of different types of batteries are significant. Therefore, it is essential to explore a robust HF extraction strategy that is especially applicable to diverse material types and varied working conditions [38].

On the other hand, when using a single battery dataset to train the SOH estimation model, it usually assumes that the training and testing data have the same distribution. However, such an assumption does not work for different types of batteries or working conditions. Therefore, the well-trained SOH estimation model cannot guarantee an acceptable estimation performance toward other battery datasets. In this case, transfer learning (TL) technology is utilized for such problems [39]. Two strategies, namely, the model parameter fine-tuning and domain adaption methods, are used in recent works for SOH estimation [40].

For the fine-tuning strategy, the data from the source domain is used to train the base model first, and then the specific layer in the base model is re-trained using the data from the target domain while other layers keep frozen. Huang et al. [41] proposed a deep learning model, and the principles of either fine-tuning or rebuilding were applied based on whether the target domain shared the same type of LiBs. However, the HFs used in the source domain and target domain were different according to different source and target domain selections. In another work [42], according to whether the feature expression score (FES) was greater than a threshold, the last fully connected (FC) layer of the base model was fine-tuned or reconstructed. However, the proposed method was only validated inside the same aging datasets (e.g., the NASA aging dataset). In addition, the need for individual fine-tuning for each battery in the target domain significantly increases computational expenses. Zhu et al. [35] randomly selected a battery from the target domain and used its complete cyclic data for TL. They also compared different data selection strategies and concluded that using data from a randomly selected battery can achieve much better results than that from the time-series-based data. In addition, the fine-tuning strategy was applied for the SOH estimation of the battery pack [43].

Domain adaptation methods aim to reduce the disparity in feature distributions between the source and target domains, ultimately enhancing the generalization and accuracy of data-driven models. [44]. Li et al. [45] used transfer component analysis (TCA) to minimize data differences and eliminate redundancy across various datasets. In another work [46], joint distribution adaptation (JDA) was employed to achieve simultaneous adaptation of both the marginal probability and the conditional probability distribution. Fu et al. [38] proposed a feature mapping strategy that first identified the reference cell (RC) and optimal matching cell (OMC) in the source domain and target domain, respectively. Then, a linear matching approach was developed. The cyclic data of the OMC in the target domain was used to re-train the base model. However, for different source and target domains, it is necessary to identify the RC and OMC first, increasing the complexity of the algorithm.

Based on the analysis above, some challenges need to be considered: (1) a universal and effective HF extraction strategy is imperative, especially for different types of batteries and different working conditions; (2) how accurate and robust SOH estimation toward different types of batteries using TL can be ensured. To address these challenges, a universal HF extraction strategy and a deep learning neural network-based transfer learning method are established for SOH estimation in this paper. Three open-source aging datasets, namely the Oxford, CALCE, and NASA aging datasets, are used to validate the effectiveness of the proposed algorithms. The main contributions are as follows:

(1) To comprehensively reflect the aging characteristics of LiBs and apply them to different battery types and working conditions, a universal HF extraction strategy is proposed. Only partial voltage and current information is required to extract four straightforward and highly relevant HFs.
(2) To learn the long-term dependency between the HFs and capacity and fulfill TL, a deep learning neural network consisting of the long short-term memory (LSTM) and FC layers is proposed in this paper. For the same battery type (or in the same aging dataset), the proposed neural network is trained using a random-selected battery, and other batteries are used to test the model directly without using the TL strategy.
(3) To achieve accurate and robust SOH estimation of different types of batteries and different working conditions, the fine-tuning-based TL strategy is used in this paper. The basic principle is to use a random-selected battery in the source domain to train the base model, and then use a random-selected battery in the target domain to re-train the base model, where only the last layer is re-trained and the other layers are frozen.
(4) Comprehensive verifications are conducted using three popular open-source aging datasets, namely Oxford, CALCE, and NASA aging datasets. Sixteen batteries, featuring two distinct cathode material types, are subjected to cycling under five different operating conditions to assess the efficacy of the proposed methods.

The remainder of the paper is organized as follows: Section 2 introduces the experimental datasets and feature extraction strategy. Section 3 explains the used algorithms. Section 4 gives the results and discussions. Finally, conclusions are summarized in Section 5.

2. Experimental Datasets

Referring to recent research, three open-source aging datasets provided by the University of Oxford [30], the CALCE at the University of Maryland [31,32], and NASA Ames Prognostics Center of Excellence [33] are used in this paper. Note that only the most used data from each dataset is used in this paper, e.g., Cell 1 to 8 from the Oxford aging dataset. Details are explained as follows.

2.1. Oxford Aging Dataset

Eight LiCoO2 (LCO) pouch batteries with 0.74 Ah nominal capacity were selected from the Oxford aging dataset, labeled Cell 1 to Cell 8, respectively. The aging experimental procedures were as follows: (1) The batteries were discharged using a driving cycle obtained from the urban Artemis profile (average current = 1.36 A), and the termination voltage was 2.7 V. (2) Then, the CC charging process was conducted with the batteries with a constant current rate of 2 C until the terminal voltage reached 4.2 V. (3) The above charging–

discharging cycle was repeated 100 times, and then a characterization test was conducted to measure the capacity of the batteries. The aging experiment was conducted under 40 °C. The SOH curves of the Oxford aging dataset are shown in Figure 2a. Note that the cycle number in the x-axis represents the number of characterization tests. For example, 45 cycles mean the 45th characterization test after 4500 charging–discharging cycles.

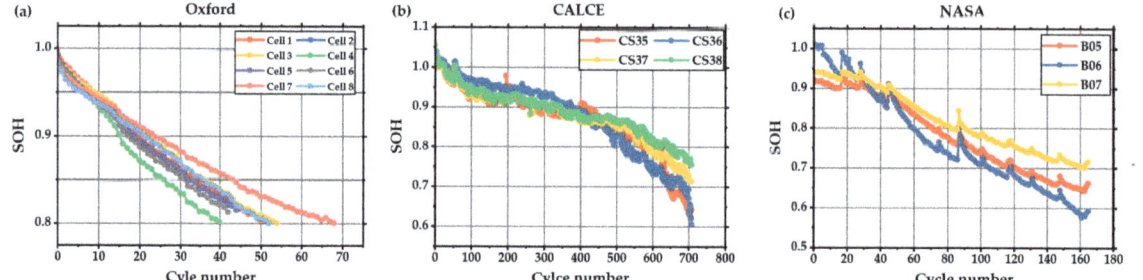

Figure 2. Degradation paths of three aging datasets: (**a**) Oxford dataset, (**b**) CALCE dataset, and (**c**) NASA dataset.

2.2. CALCE Aging Dataset

Four LCO prismatic batteries with 1.14 Ah nominal capacity were selected from the CALCE aging, labeled CS35, CS36, CS37, and CS38, respectively. The aging experimental procedures were as follows: (1) The batteries were discharged using a constant current rate of 1 C, and the termination voltage was 2.7 V. (2) Then, the batteries underwent the same constant current–constant voltage (CC–CV) protocol, where the CC charging current was 0.5 C, and the CV process ended when the current dropped to 0.05 A. (3) The above charging–discharging cycle was repeated until the end-of-experiment criteria were met. The aging experiments were conducted at room temperature (around 25 °C). The SOH curves of the CALCE aging dataset are shown in Figure 2b.

2.3. NASA Aging Dataset

Three 18650 LiNiCoAlO2 (NCA) batteries with 2 Ah nominal capacity were selected from the NASA aging dataset, labeled as B05, B06, and B07, respectively. The aging experimental procedures were as follows: (1) The batteries were discharged with a constant current of 2 A, and the termination voltages were 2.7 V, 2.5 V, and 2.2 V, respectively. (2) The batteries were then fully charged with the CC–CV protocols, where the CC charging current was 1.5 A, and the CV process ended when the current dropped to 0.02 A. (3) The above charging–discharging cycle was repeated until the end-of-experiment criteria were met. The aging experiments were conducted at room temperature (around 24 °C). The SOH curves of the NASA aging dataset are shown in Figure 2c.

According to Figure 2, it is clear that significant differences between the degradation paths of the three aging datasets due to different materials, structures, and working conditions can be observed. In addition, as for the Oxford or CALCE aging datasets, even though the batteries have the same cathode materials and experience the same test protocols, they have different aging paths because of the initial difference. For example, the first charged capacity of the four batteries from CALCE are 1.135 Ah, 1.142 Ah, 1.130 Ah, and 1.137 Ah, respectively, indicating the aging inconsistency. The initial difference could result in more aggressive internal variations as the battery ages. The characteristics of the three aging datasets are given in Table 2.

Table 2. Characteristics of three open-source aging datasets.

Aging Dataset	Battery Label	Cell Types	Charge C Rate (C)	Charging Cut-Off Voltage (V)	Discharge C Rate (C)	Discharging Cut-Off Voltage (V)	Temperature (°C)	Nominal Capacity (Ah)
Oxford	Cell 1 to Cell 8	LCO	2	4.2	2	2.7	40	0.74
CALCE	CS35 to CS38	LCO	0.5	4.2	1	2.7	25	1.1
NASA	B05	NCA	0.75	4.2	1	2.7	24	2
	B06					2.5		
	B07					2.2		

2.4. Aging Analysis and Feature Extraction

The HF extraction is the foundation and key for building the ML method for SOH estimation. When choosing appropriate HFs, in addition to the principles of easy acquisition, suitability for practical conditions, and high relevant degree, the adaptability and universality toward different types of batteries are significant. Based on this principle, although many valuable HFs are summarized in the Introduction, we want to choose as few neural network inputs as possible but maintain accuracy and robustness simultaneously. Therefore, a universal HF extraction strategy is proposed in this section based on the analysis of three open-source aging datasets.

Note that the CC–CV charging protocol was used in the experiments of the CALCE and NASA, while the CC charging protocol was used in the experiments of Oxford. Hence, only the data from CC charging duration is used to obtain HFs. Figure 3 shows the terminal voltage curves during the CC charging process under different aging states. Taking the Oxford aging dataset as an example, it can be observed that the time for LiBs to reach the upper cut-off voltage (4.2 V) decreases as the battery ages. This phenomenon directly reflects the reduction in usable capacity. Therefore, the charged time during the CC charging process is selected as an HF in many existing research. However, to fulfill the HF-selection principle stated at the beginning of this section, the whole CC charging process cannot be used to obtain the time-related HFs because LiBs are hardly charged from 0% to 100% in practical applications. Hence, the charging duration derived from the specific segment (from 3.8 V to 4.1 V) of the CC charging curves is selected as an HF to represent the battery degradation, denoted as T1. Correspondingly, the charged capacity in this voltage range can be calculated easily since the current remains constant. Therefore, the charged capacity is considered as another HF, denoted as Q1.

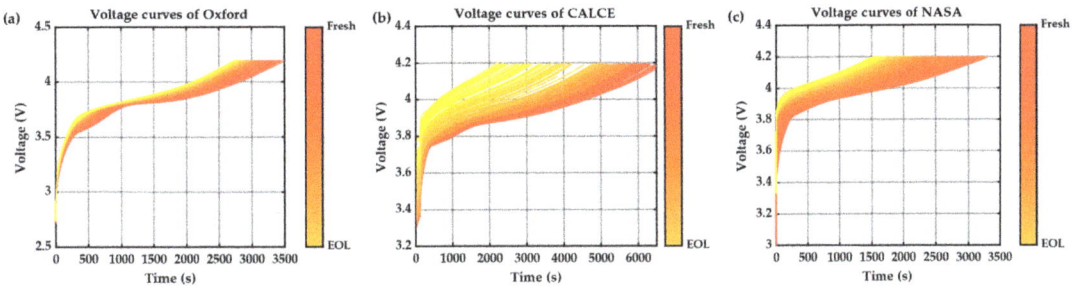

Figure 3. The terminal voltage curves under different aging states: (a) Oxford dataset, (b) CALCE dataset, and (c) NASA dataset. (EOL means end of life).

The incremental curve analysis (ICA) is a popular method to analyze the aging mechanism of LiBs and is used to extract high-related HFs. The IC curves can be calculated from the CC charging process using a differential equation as follows:

$$\text{IC} = \frac{dQ}{dV} = I \cdot \frac{dt}{dV} \qquad (2)$$

where Q and V represent the capacity and voltage, respectively, I represents the current, and t is the sampling time.

Figure 4 shows the IC curves under different aging states. We note that the IC curves are further smoothed by the Gaussian filter. Taking the IC curves of the Oxford aging dataset as an example, two peaks can be observed in the middle range of about 3.6–4.1 V. Each peak represents the phase transition process during active material insertion and delamination. With the aging of LiBs, the first peak gradually disappears, and the second peak decreases with a clear trend. In addition, the region beneath the peaks diminishes as the age of the LiBs decreases, indicating a loss of active material and a loss of lithium [47]. As for the other two aging datasets, the IC curves of the CALCE aging dataset also have two peaks and have the same trend as the Oxford aging dataset, while there is only one obvious peak in the IC curves of the NASA aging dataset. To guarantee the universality of the HF extraction strategy, the second peak value and the region beneath the second peak (between 3.8 V and 4.1 V) are selected as HFs, denoted as P1 and A1, respectively.

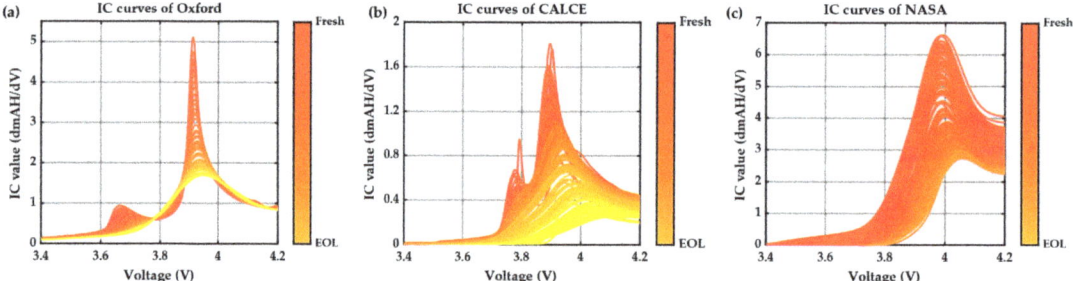

Figure 4. The IC curves under different aging states: (**a**) Oxford dataset, (**b**) CALCE dataset, and (**c**) NASA dataset.

In many existing research and our previous work [48], temperature information is used to extract valuable HFs. However, to build a uniform feature-extraction strategy considering the commonality of different datasets where the CALCE aging dataset does not provide the temperature information, the temperature information is ignored in this paper. Then, the Pearson correlation coefficient, an efficient method to evaluate the correlation degree between two data sets, is employed in this paper. The expression is as follows:

$$\rho = \frac{\sum_{i=1}^{n}\left(HF_i - \overline{HF}\right)\left(C_i - \overline{C}\right)}{\sqrt{\sum_{i=1}^{n}\left(HF_i - \overline{HF}\right)^2 \sum_{i=1}^{n}\left(C_i - \overline{C}\right)^2}} \qquad (3)$$

where HF_i and C_i represent HF and SOH, respectively, $\overline{HF}$ and $\overline{C}$ are their mean values, and n is the number of samples. Typically, the Pearson correlation coefficient spans from -1 to 1, with a closer absolute value to 1 indicating a stronger degree of correlation. Figure 5 gives the heat map of the Pearson correlation coefficients of each battery in three open-source aging datasets. It is evident that every Pearson correlation coefficient surpasses 0.95, signifying a robust correlation. In summary, four straightforward and high-related HFs are obtained from the specific segments of the voltage and current curves based on analyzing three aging datasets' degradation characteristics and commonalities.

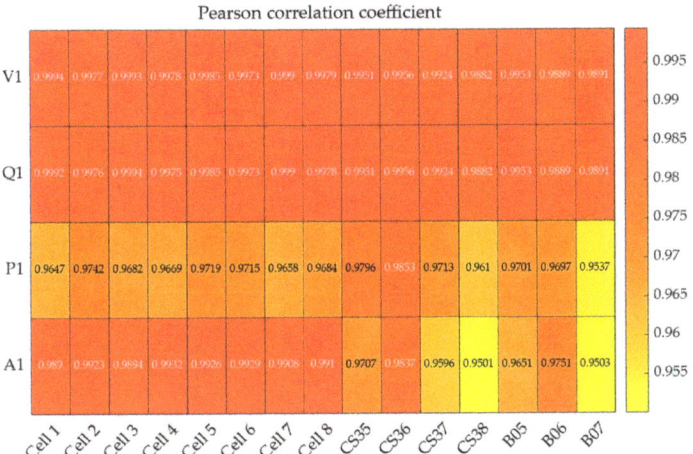

Figure 5. Heat map of the Pearson correlation coefficient of each battery.

3. Related Algorithms

3.1. The Deep Learning Neural Network

Referring to [41,42,46], a deep learning neural network consisting of one LSTM layer, one dropout layer, and three FC layers is established to achieve accurate SOH estimation toward a single aging dataset and fulfill a TL strategy toward different aging datasets. For simplicity, the proposed deep learning neural network is abbreviated as LSTM-FC-NN.

The LSTM layer is used because of its superiority in dealing with time-series data. The LSTM was first created by Hochreiter et al. [49] to solve the information decay during the algorithm back-propagation through time (BPTT). Most gradient-based learning algorithms, such as the Elman neural network [50], suffer either gradient vanishing or gradient explosion when the backflow transfers the information over a long period. To overcome the time dependence problem, the LSTM is proposed with a cell state (memory cell), a hidden state, a forget gate, an input gate, and an output gate, as shown in Figure 6. The expressions are as follows:

$$\begin{aligned} f_t &= \sigma\left(W_f x_t + U_f H_{t-1} + b_f\right) \\ i_t &= \sigma(W_i x_t + U_i H_{t-1} + b_i) \\ o_t &= \sigma(W_o x_t + U_o H_{t-1} + b_o) \\ c_t &= f_t \odot c_{t-1} + i_t \odot tanh(W_c x_t + U_C H_{t-1} + b_c) \\ H_t &= o_t \odot tanh(c_t) \end{aligned} \quad (4)$$

where x_t represents the input of the LSTM, H_t represents the hidden state, W and U represent the weight matrices, b represents the bias, i_t, f_t, o_t, and c_t represent the input, forget, output gates, and memory cell, respectively, and σ is the sigmoid activation function. The significant feature of LSTM is the use of different gates to control the information flows. In particular, the input gate determines the new information allowed to pass into the memory cell, while the forget gate identifies the information from the previous memory cell (c_{t-1}) that should be disregarded. The output gate is used to calculate the outputs. More details can be found in [51].

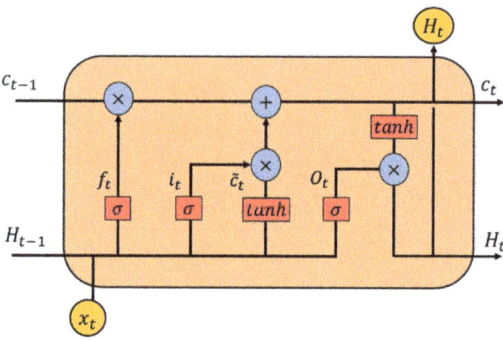

Figure 6. The structure of LSTM.

Although different batteries have similar declining degradation curves, the specific characteristics are different. As referred to in [52], the FC layers in the base model can serve as the "firewall" to achieve outstanding transfer learning performance. Therefore, three FC layers are used in this paper. In addition, to prevent overfitting, especially when there is limited data, a dropout layer is added to the base model. In the end, the structure of the proposed deep learning neural network is shown in Figure 7. For reproducibility, important hyperparameters are listed in Table 3. Note that all algorithms and computations are executed using MATLAB 2022b.

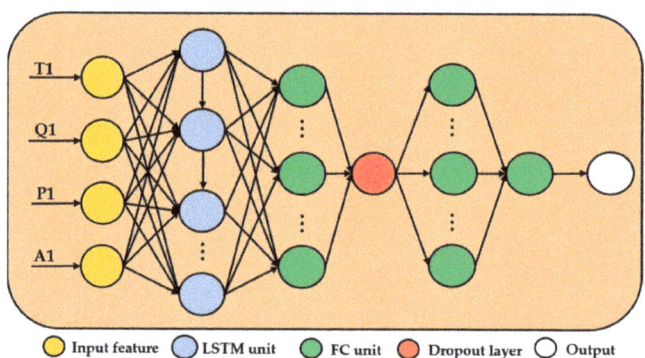

Figure 7. The structure of the proposed deep learning neural network.

Table 3. Hyperparameters of the deep learning neural network.

Hyperparameters	Values and Settings
Optimization method	Adam
Number of units in LSTM layer	100
Number of units in FC layers	10/10/1
Dropout rate	0.5
Data split ratio for training base model	80%/10%/10%
Data split ratio for re-training	80%/20%

3.2. Transfer Learning Strategy

Parameter fine-tuning and domain adaptation methods are two commonly used strategies in TL. The parameter fine-tuning method is more straightforward than the domain adaption method. Thus, the fine-tuning-based TL strategy is used in this paper. The principle of the fine-tuning-based TL is shown in Figure 8:

(1) First, three open-source aging datasets, the Oxford, CALCE, and NASA aging datasets, are represented using different shapes, respectively. Note that different color shades represent the inconsistency between the batteries in the same aging dataset, caused by different initial states, working conditions, and internal variations in material properties from battery manufacturing.
(2) Second, one of the three aging datasets is selected as the source domain, and the other two are considered the target domains. Then, the cyclic data of a randomly selected battery from the source domain is used to train the base model. And other batteries in the source domain are used to test the base model.
(3) Third, the fine-tuning-based TL strategy is applied when the target domain is available. The data-selection principle is used to randomly select a battery in the target domain and re-train the base model with its cyclic data using the fine-tuning method. The fine-tuning method in this paper means that only the last FC layer is updated during re-training, while other layers are frozen.
(4) Finally, other batteries in the target domain are used to test the re-trained model.

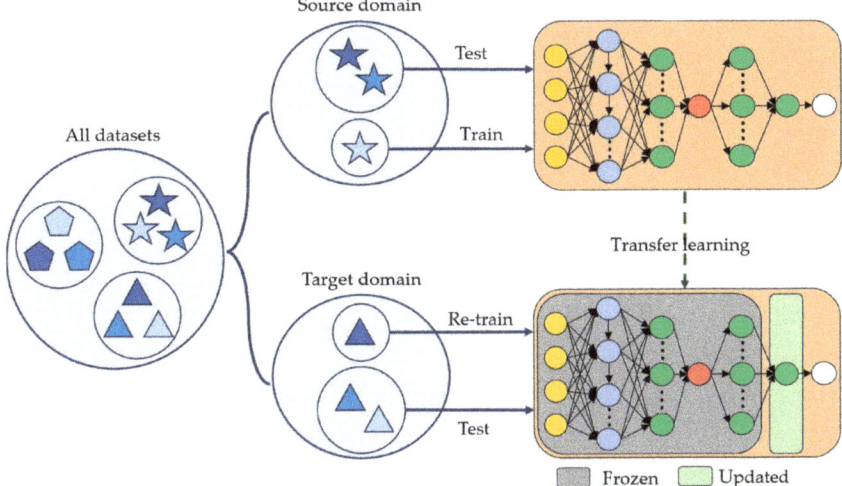

Figure 8. The principle of the fine-tuning-based TL strategy.

By using the random-selected strategy, the accuracy and robustness of the proposed deep learning neural network and fine-tuning-based TL strategy can be thoroughly evaluated.

4. Results and Discussion

This section gives the results of the proposed universal HF extraction strategy and LSTM-FC-NN for SOH estimation on the single aging dataset, as well as the effectiveness of the proposed fine-tuning-based TL strategy toward different types of batteries in detail. Commonly used statistical metrics, namely, the mean absolute error (MAE), root mean square error (RMSE), and the maximum error (MaxE), are used to evaluate the performance quantitively. The MAE gives equal weight to all errors, which can be used to evaluate the accuracy of a regression model. At the same time, the RMSE is more sensitive to outliers and can be used to evaluate the robustness of a model. The MaxE is useful for understanding the magnitude of the largest error. The expressions are as follows:

$$\text{MAE} = \frac{1}{N}\sum_{i=1}^{N}\left|\widehat{SOH}_i - SOH_i\right| \qquad (5)$$

$$\text{RMSE} = \sqrt{\frac{1}{N}\sum_{i=1}^{N}\left(\widehat{SOH}_i - SOH_i\right)^2} \tag{6}$$

$$\text{MaxE} = max\left|\widehat{SOH}_i - SOH_i\right| \tag{7}$$

where $\widehat{SOH}_i$ represents the predicted value, SOH_i represents the reference value, and N is the number of samples.

4.1. Results of the Single Aging Dataset

This section separately evaluates the proposed deep learning neural network for SOH estimation using three open-source aging datasets. It first verifies the effectiveness of the proposed universal HFs extraction strategy and proves that the proposed LSTM-FC-NN model can achieve accurate and robust SOH estimation toward different battery types. Note that the training and evaluation principle for the same type of batteries is that a random-selected battery is used to train the base model, and other batteries are used to test the base model directly without using the TL strategy.

4.1.1. Oxford Aging Dataset

Table 4 gives the statistical metrics of the case that Cell 1 in the Oxford aging dataset is randomly selected to train the base model, and other batteries are used to test the base model directly without using the TL strategy. Table 5 gives the results where another battery is randomly selected to train the model, e.g., Cell 2 to Cell 7, and the MAE, RMSE, and MaxE in Table 5 are the mean values. For example, the second column in Table 5 represents that Cell 2 is used to train the base model, and the other seven cells are used to test the base model. Then, the MAE, RMSE, and MaxE are the mean values of the seven testing results. Correspondingly, the eight curves in Figure 9 represent which battery is used to train the base model.

Table 4. Estimation results of the Oxford aging dataset.

Train	Test	MAE (%)	RMSE (%)	MaxE (%)
Cell 1	Cell 2	0.22	0.28	0.59
	Cell 3	0.21	0.27	0.52
	Cell 4	0.29	0.36	0.89
	Cell 5	0.49	0.56	1.05
	Cell 6	0.56	0.61	1.02
	Cell 7	0.45	0.53	1.07
	Cell 8	0.56	0.62	1.09

Table 5. Estimation results of all cases in the Oxford aging dataset.

Unit: %	Cell 1	Cell 2	Cell 3	Cell 4	Cell 5	Cell 6	Cell 7	Cell 8
MAE *	0.37	0.39	0.32	0.34	0.32	0.34	0.29	0.32
RMSE *	0.43	0.48	0.40	0.42	0.39	0.39	0.35	0.37
MaxE *	0.84	0.95	0.83	0.86	0.82	0.75	0.71	0.70

* The MAE, RMSE, and MaxE are the mean values.

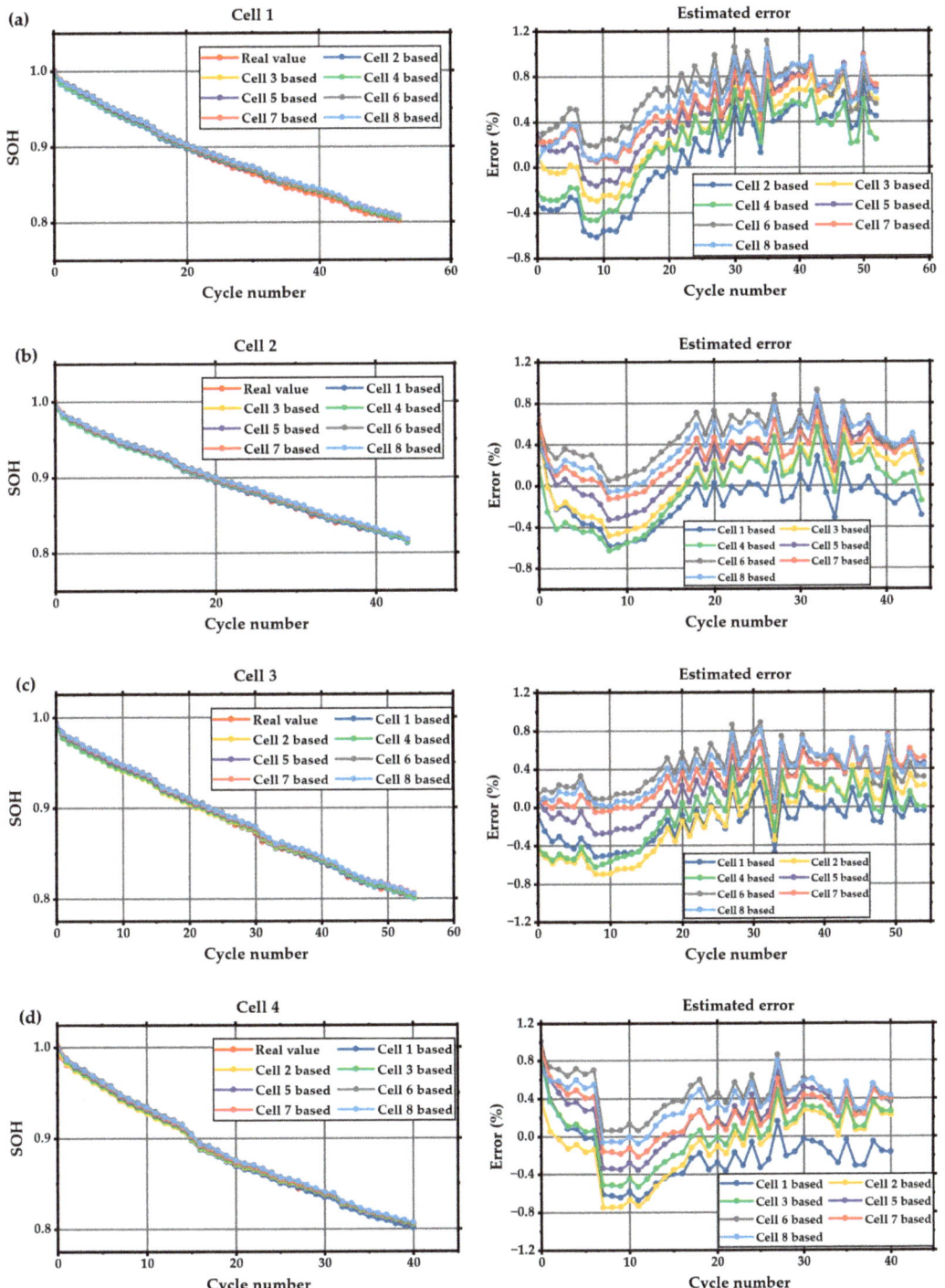

Figure 9. Estimation results of the Oxford aging dataset: (**a**) Cell 1; (**b**) Cell 2; (**c**) Cell 3; (**d**) Cell 4.

Overall, it can be seen from Figure 9 that no matter which battery is randomly selected to train the base model, the base model can achieve accurate SOH estimation results toward other batteries without using the TL strategy. Specifically, as an example, Figure 9a gives the results where Cell 2 to Cell 7 are used to train the base model in turn, and Cell 1 is used to test the base model. It can be observed that all seven estimated SOH curves can follow the actual SOH curves quite well. The MaxE is merely 1.09% when Cell 8 is used to train the base model, according to Table 4. In addition, the mean values of MAE and RMSE are only 0.37% and 0.43%, respectively, according to Table 5. As for other evaluation cases in Table 5, the estimation results demonstrate that the proposed universal HF strategy and LSTM-FC-NN model can achieve accurate and robust SOH estimation toward the Oxford aging dataset.

4.1.2. CALCE Aging Dataset

Table 6 gives the results that a random-selected battery from the CALCE aging dataset is used to train the base model, and other batteries are used to test the base model without using the TL strategy. Correspondingly, Figure 10 illustrates the estimation results.

Table 6. Estimation results of the CALCE aging dataset.

Train	Test	MAE (%)	RMSE (%)	MaxE (%)
CS35	CS36	0.51	0.70	2.55
	CS37	0.48	0.59	3.06
	CS38	0.63	0.75	3.32
CS36	CS35	0.51	0.61	2.54
	CS37	0.57	0.70	3.38
	CS38	0.80	0.94	3.70
CS37	CS35	0.63	0.94	3.87
	CS36	0.74	1.04	5.08
	CS38	0.48	0.60	3.04
CS38	CS35	1.11	1.87	8.30
	CS36	1.21	1.83	9.71
	CS37	0.66	0.90	3.60
Mean values	-	0.69	0.96	4.35

Similar to the Oxford aging dataset results, on which a battery is used to train the base model, the model can achieve accurate and robust SOH estimation toward other batteries without the TL strategy. For example, Figure 10d shows the estimation results of CS38 using different battery-based base models. It can be observed that no matter which battery is used to train the base model, the base model can obtain accurate estimation results toward CS38. The estimated SOH curves can always follow the actual aging path, even in some slight range fluctuation. In addition, the estimation errors rarely exceed absolute 2%, demonstrating the proposed method's robustness. However, for CS35 and CS36, when evaluated using the CS38-based base model, the results are slightly worse than other battery-based models. This is because the aging path of CS38 is quite different from that of CS35 and CS36, especially when the SOH drops below 80%, as shown in Figure 2b. When the SOH drops below 80%, the internal electrochemical reaction becomes more uncontrollable, resulting in quicker and more different aging paths. Therefore, significant estimation errors and MaxE occur when the SOH drops below 80%. At the same time, there is consistency between the estimated and actual SOH curves when the SOH is higher than 80%, as shown in Figure 10a,b. Overall, the last row in Table 6 represents the mean values of the corresponding columns. It can be concluded that the mean values of MAE and RMSE are 0.69% and 0.96%, respectively, demonstrating the accuracy and robustness of the proposed model toward the CALCE aging dataset.

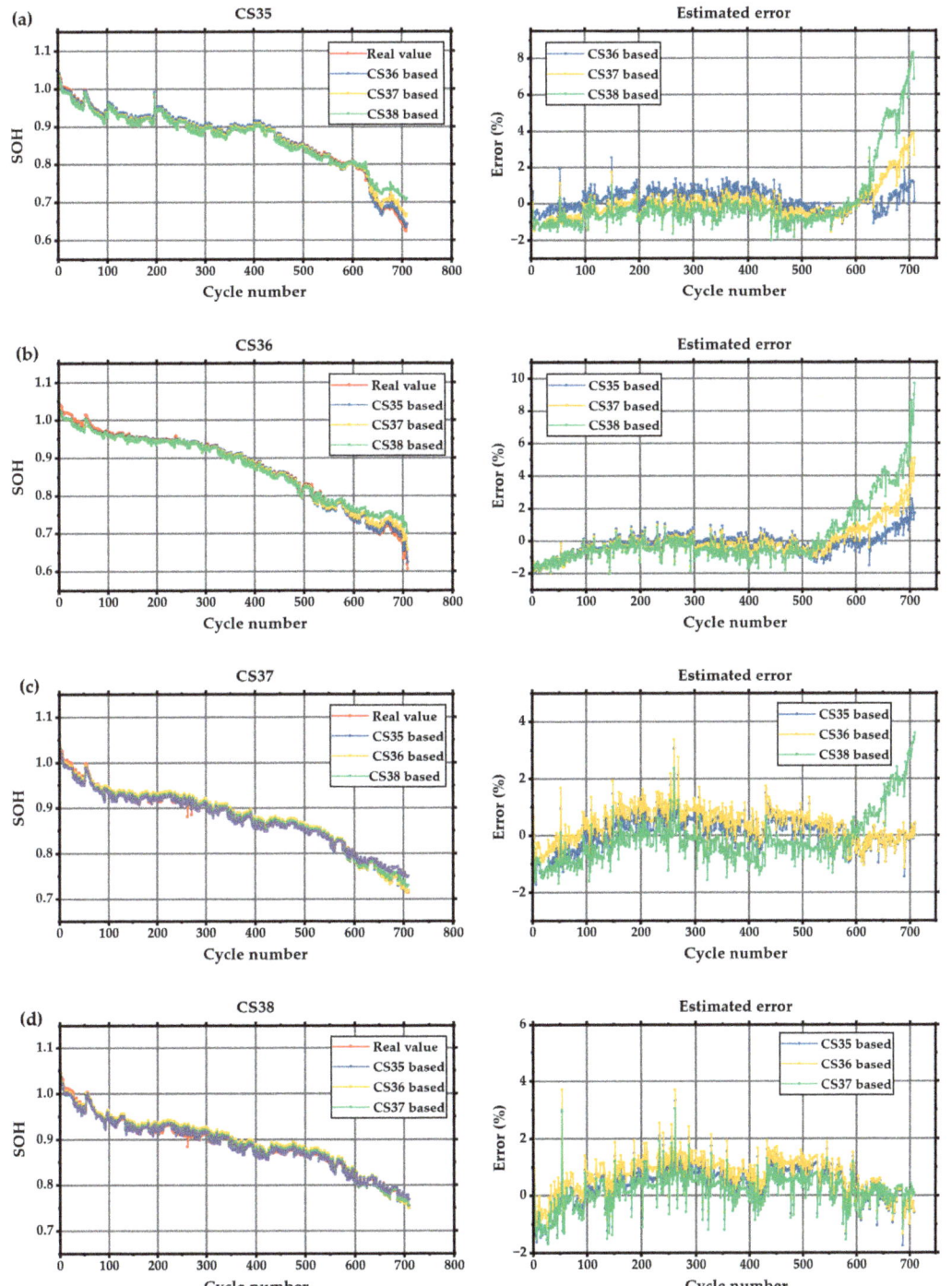

Figure 10. Estimation results of the CALCE aging dataset: (**a**) CS35; (**b**) CS36; (**c**) CS37; (**d**) CS38.

4.1.3. NASA Aging Dataset

Table 7 gives the results that a random-selected battery from the NASA aging dataset is used to train the base model, and other batteries are used to test the model directly without using the TL strategy. Figure 11 compares the estimated SOH curves based on different base models and actual SOH curves and the estimated errors.

Table 7. Estimation results of the NASA aging dataset.

Train	Test	MAE (%)	RMSE (%)	MaxE (%)
	B06	3.40	3.88	12.43
B05	B06_retrain	1.39	1.86	8.21
	B07	0.63	0.97	5.68
B06	B05	2.94	3.12	4.85
	B07	1.99	2.22	5.00
	B05	0.72	0.95	5.10
B07	B06	3.40	3.91	12.72
	B06_retrain	1.37	1.93	8.66

According to Table 7, it can be concluded that when the B05 is used to train the base model, the base model can achieve accurate SOH estimation toward B07 but not B06. The MAE and RMSE of B07 are 0.63% and 0.97%, respectively, while the MAE and RMSE of B06 are 3.40% and 3.88%, respectively. As shown in Figure 11c, the estimated SOH curve based on B05 can follow the actual aging path quite well, but in Figure 11b, there is a significant difference between the estimated SOH curve based on B05 and the actual SOH curve. The same results can be observed when B07 is used to train the base model. As for the case when B06 is used to train the base model, it can be seen from Figure 11a,c that the estimated curves based on B06 cannot follow the real SOH curves well, e.g., the blue line in Figure 11a and the yellow line in Figure 11c. The reasons can be explained as follows. First, the initial SOH of these three batteries is quite different, indicating 0.92%, 1.01%, and 0.94% for B05, B06, and B07, respectively, as shown in Figure 2c. Second, the discharging protocols for these three batteries are different, as summarized in Table 2. Different discharging cut-off voltages would result in different depths of discharge, which is essential for battery aging. The initial differences are bound to result in inconsistencies in the aging paths of different batteries, and different cycling protocols exacerbate the inconsistencies. Therefore, as shown in Figure 2, it can be observed that the differences between different aging paths of the NASA aging dataset are more evident than the other two aging datasets.

Since the aging path of B06 is quite different from B05 and B07, the fine-tuning-based TL strategy is used to improve the accuracy. Figure 12 compares the estimation results of B06 with or without using the fine-tuning-based TL strategy, and the corresponding results are given in Table 7. After fine-tuning the base model, the re-trained model can perform much better. The estimated SOH curves can follow the actual SOH curve well. The MAE and RMSE after using the TL strategy are 1.39% and 1.86%, respectively, decreasing by 59% and 52%, respectively. In summary, the effectiveness of the proposed LSTM-FC-NN model and fine-tuning-based TF strategy is verified toward the NASA aging dataset.

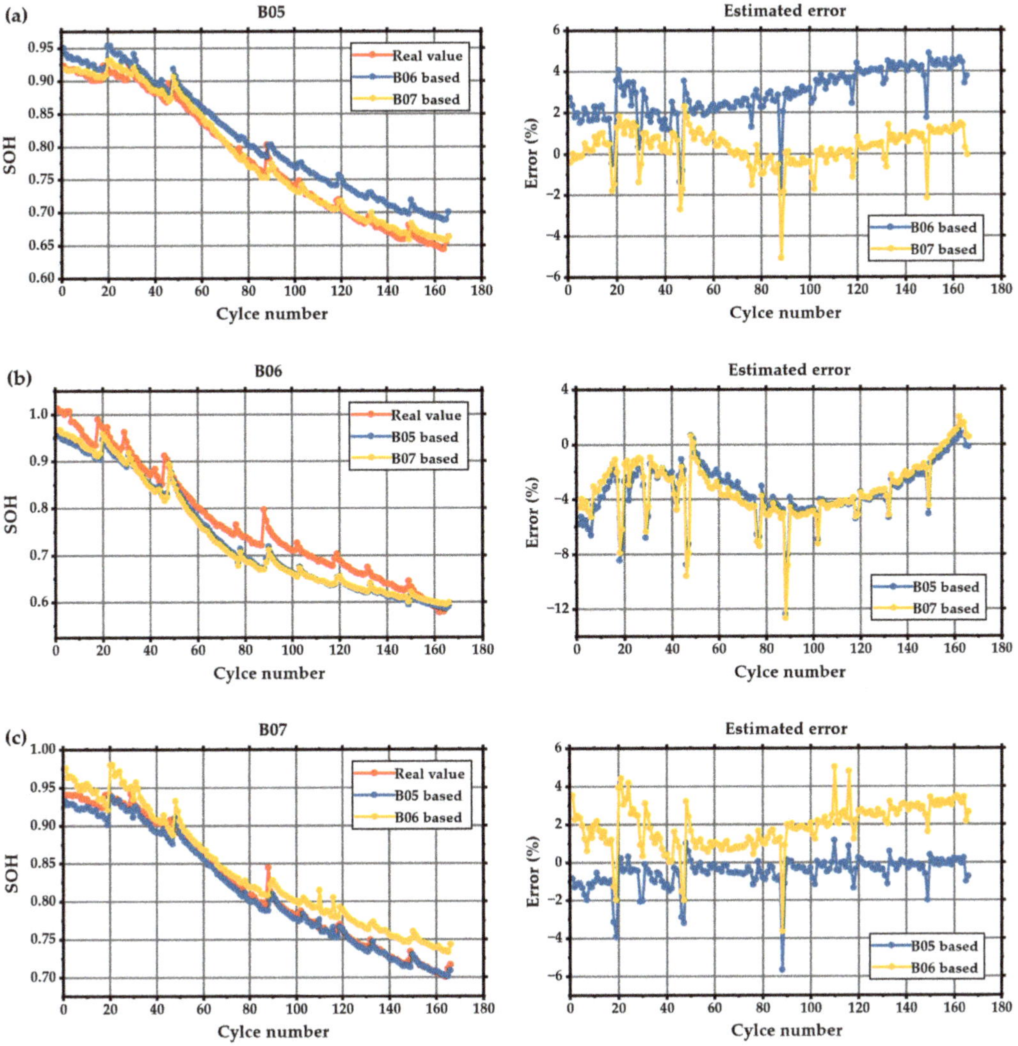

Figure 11. Estimation results of the NASA aging dataset: (**a**) B05; (**b**) B06; (**c**) B07.

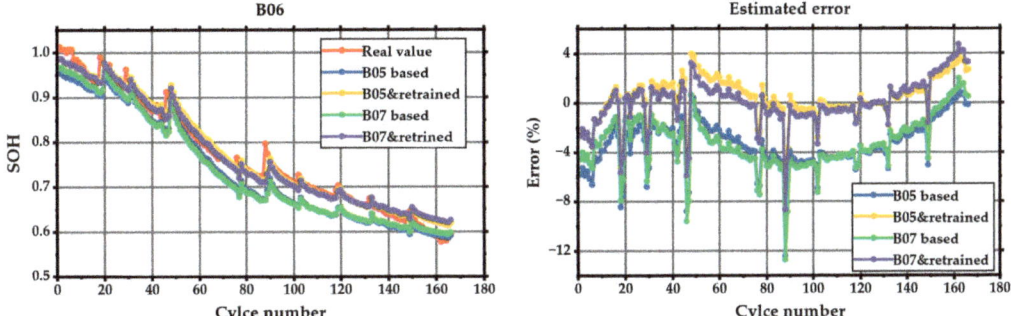

Figure 12. Estimation results of B06 when using the TL strategy.

In summary, three open-source aging datasets are used separately to evaluate the effectiveness of the proposed universal HF extraction strategy and LSTM-FC-NN model for SOH estimation in this section. On the one hand, from the results of the Oxford and CALCE aging datasets, it can be concluded that when the training and testing batteries are the same battery type and have the same aging protocol, the estimations are accurate and reliable, whatever the training dataset. On the other hand, from the results of the NASA aging dataset, it can be concluded that when the testing battery has a significantly different aging path from the training one, the estimations cannot be guaranteed. In this case, the fine-tuning-based TL strategy can be used to improve the adaptivity of the trained model toward the testing battery, which builds the foundation for the following results.

4.2. Results of the Proposed Transfer Learning Strategy between Different Aging Datasets

This section gives the results of the proposed fine-tuning-based TL strategy between three open-source aging datasets in detail. As shown in Figure 8, the evaluation principle is that one of the three aging datasets is randomly selected as the source domain, and the other two aging datasets are then used separately as the target domain. Then, a randomly selected battery from the source domain is used to train the base model. Next, one of the batteries in the target domain is selected randomly, and its cyclic data is used to fine-tune the base model. Finally, the re-trained model is tested using other batteries in the target domain. The following three subsections explain the results in detail.

4.2.1. Transfer Learning between Oxford and CALCE Aging Datasets

Figure 13 gives the results that Cell 1 is used to train the base model, CS38 is randomly selected to re-train the base model using the fine-tuning strategy, and other batteries in the CALCE aging datasets are used to test the re-trained model. Note that the blue lines (labeled as Before TL) give the estimated SOH curves using the base model directly, and the yellow lines (labeled as After TL) give the estimated SOH curves using the fine-tuned base model.

It can be observed that although the overall trend of the estimated SOH curves without using the TL strategy has the same downtrend as the actual SOH curves, there are significant estimation errors between them. That is because the LSTM layer can learn the long-term dependency between the input features and SOH, but the last FC layer determines the final output. Therefore, the re-trained model can adapt to the target domain when the last FC layer is updated. Then, the estimated SOH curves can track the real SOH curves quite well, even with slight fluctuation. Specifically, according to Table 8, the MAE and RMSE of CS35 after using the TL strategy are 0.68% and 0.95%, respectively, which is even better than the results in Table 6.

Table 8. The results of the TL strategy from Oxford to CALCE.

Source/Target	Test	Method	MAE (%)	RMSE (%)	MaxE (%)
Cell 1/CS38	CS35	Before TL	10.61	11.40	19.88
		After TL	0.68	0.95	3.49
	CS36	Before TL	10.96	12.12	20.45
		After TL	0.74	1.08	4.71
	CS37	Before TL	10.68	11.26	19.68
		After TL	0.59	0.72	3.06

Figure 14 illustrates that CS35 is used to train the base model, and Cell 4 is used to fine-tune the base model. The corresponding results are summarized in Table 9. Similar results can be observed in Figure 14 that the estimated SOH curves based on the base model have the same downtrend as the actual SOH curves, but there are significant errors between them. After fine-tuning the base model, the estimated SOH curves can track the actual

SOH curves quite well. The MAE and RMSE are limited to 0.46% and 0.57%, respectively. Overall, the above results between the Oxford and CALCE aging datasets demonstrate the validity of the proposed LSTM-FC-NN model and the fine-tuning-based TL strategy.

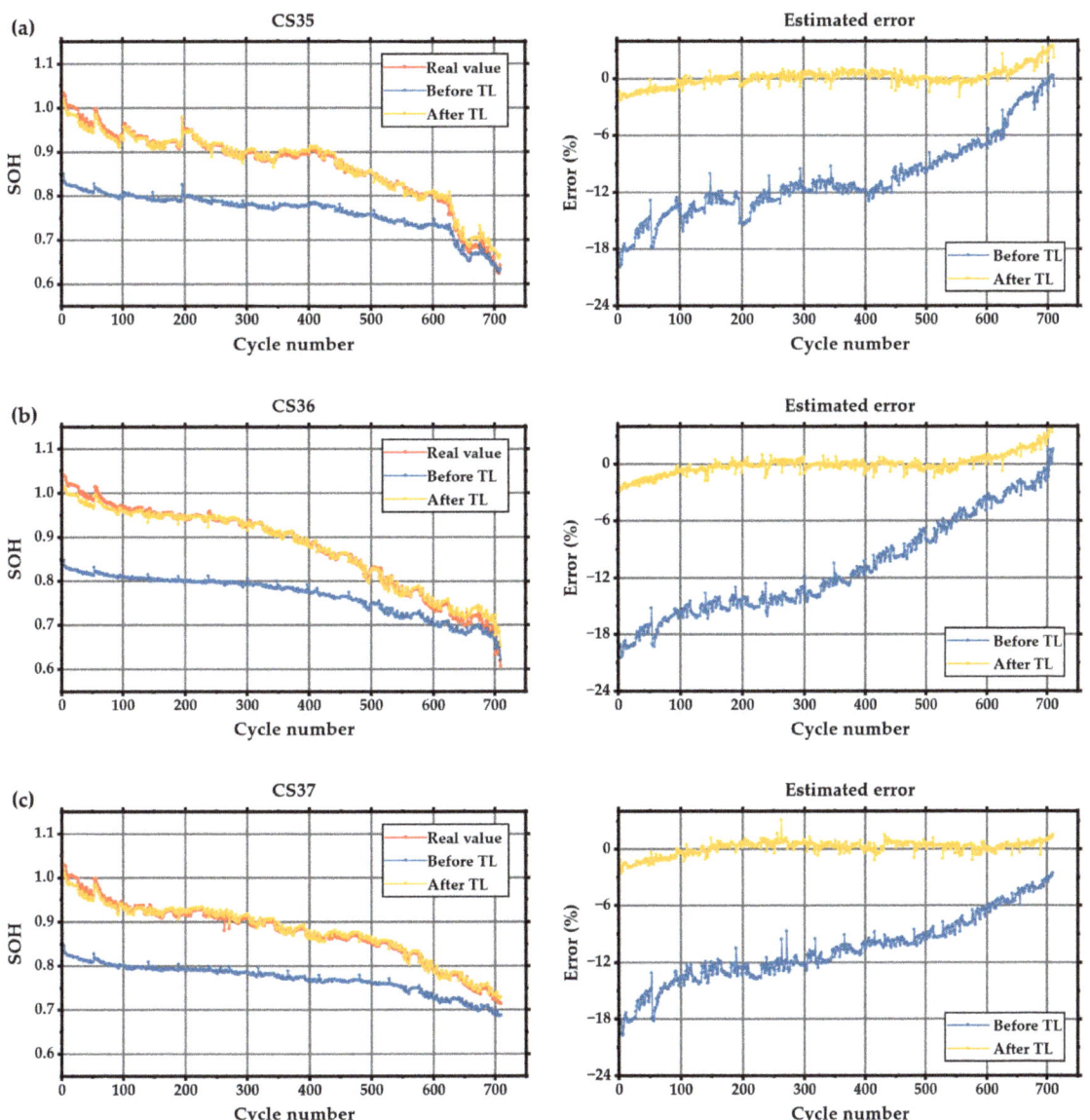

Figure 13. The results of the TL strategy from Oxford to CALCE: (**a**) CS35; (**b**) CS36; (**c**) CS37.

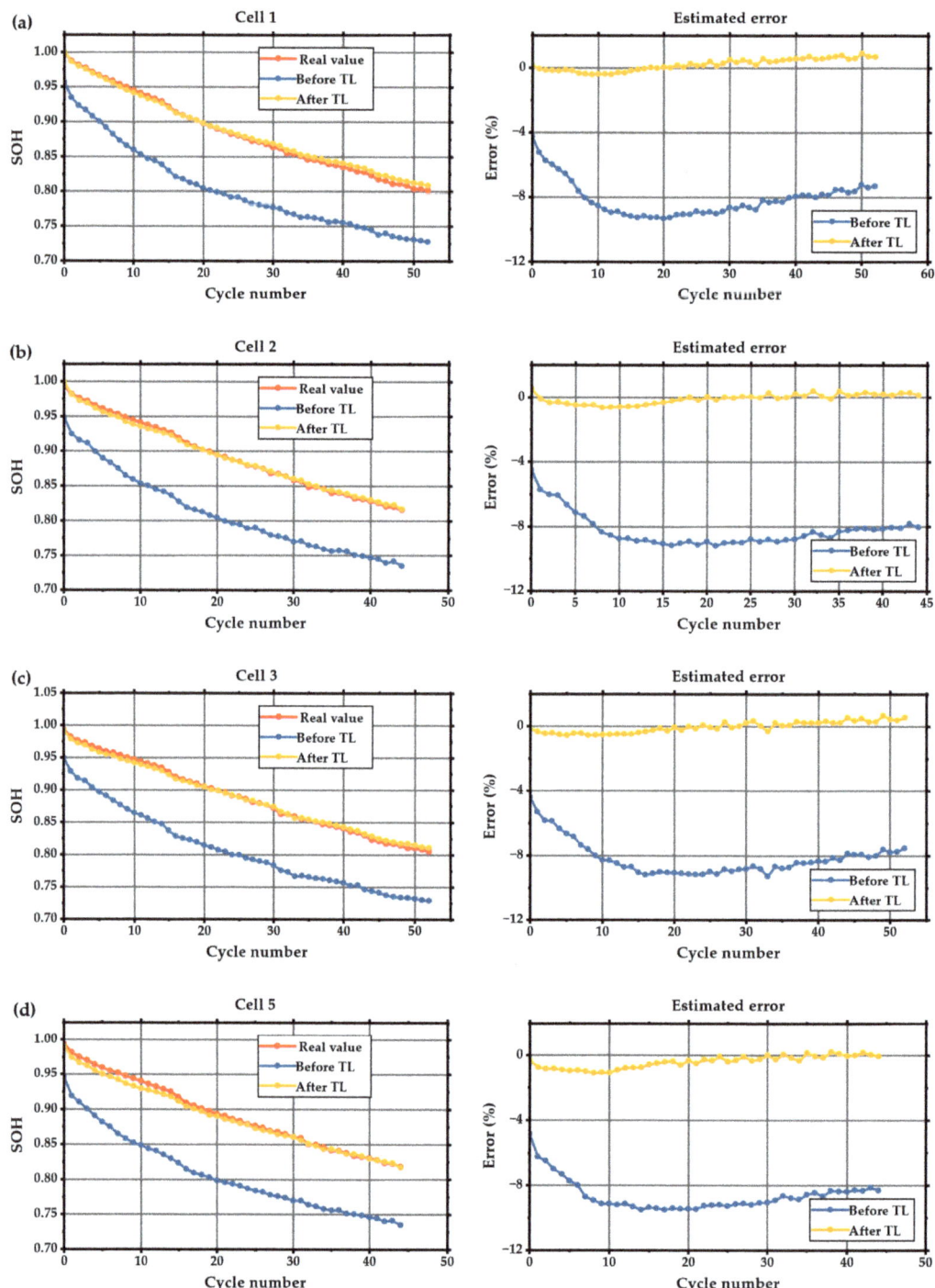

Figure 14. The results of the TL strategy from CALCE to Oxford: (**a**) Cell 1; (**b**) Cell 2; (**c**) Cell 3; (**d**) Cell 5.

Table 9. The results of the TL strategy from CALCE to Oxford.

Source/Target	Test	Method	MAE (%)	RMSE (%)	MaxE (%)
CS35/Cell 4	Cell 1	Before TL	8.11	8.19	9.31
		After TL	0.35	0.42	0.90
	Cell 2	Before TL	8.26	8.32	9.21
		After TL	0.26	0.32	0.64
	Cell 3	Before TL	8.15	8.22	9.27
		After TL	0.32	0.36	0.67
	Cell 5	Before TL	8.63	8.69	9.52
		After TL	0.46	0.57	1.11

4.2.2. Transfer Learning between CALCE and NASA Aging Datasets

The mutual estimation results of the TL strategy between the CALCE and NASA aging datasets are given in Figures 15 and 16, respectively. As an example, Figure 15 shows the estimation results of CS36, CS37, and CS38 with or without using TL. The corresponding results are summarized in Table 10. The overall trend of the estimated curves before using the TL strategy is similar to the real ones, but significant gaps can be found. After updating the last FC layer of the base model, the re-trained model can track the descent SOH curves well. Furthermore, the proposed LSTM-FC-NN model showcases robustness, with the MAE and RMSE constrained to 0.80% and 0.95%, respectively.

Table 10. The results of the TL strategy from NASA to CALCE.

Source/Target	Evaluation	Method	MAE (%)	RMSE (%)	MaxE (%)
B05/CS35	CS36	Before TL	5.66	6.45	13.42
		After TL	0.58	0.85	3.15
	CS37	Before TL	4.26	4.99	12.59
		After TL	0.63	0.78	3.16
	CS38	Before TL	4.20	5.04	12.83
		After TL	0.80	0.95	3.40

Figure 16 gives the estimation results of B05 and B06 with or without using TL, and the corresponding results are summarized in Table 11. As analyzed in Section 4.1.3, the model trained using data of B07 can achieve accurate estimation toward B05 but not B06. Similar results can be found in this TL case. As shown in Figure 16a, the B07-based model can achieve accurate estimation toward B05 after implementing the TL strategy. As for B06, the B07-based model improves the estimation accuracy after implementing the TL strategy, and the MAE and RMSE decreased by 68.80% and 73.43%, respectively.

Table 11. The results of the TL strategy from CALCE to NASA.

Source/Target	Test	Method	MAE (%)	RMSE (%)	MaxE (%)
CS35/B07	B05	Before TL	19.91	20.97	28.29
		After TL	0.70	0.95	5.12
	B06	Before TL	10.77	14.42	27.38
		After TL	3.36	3.83	12.48

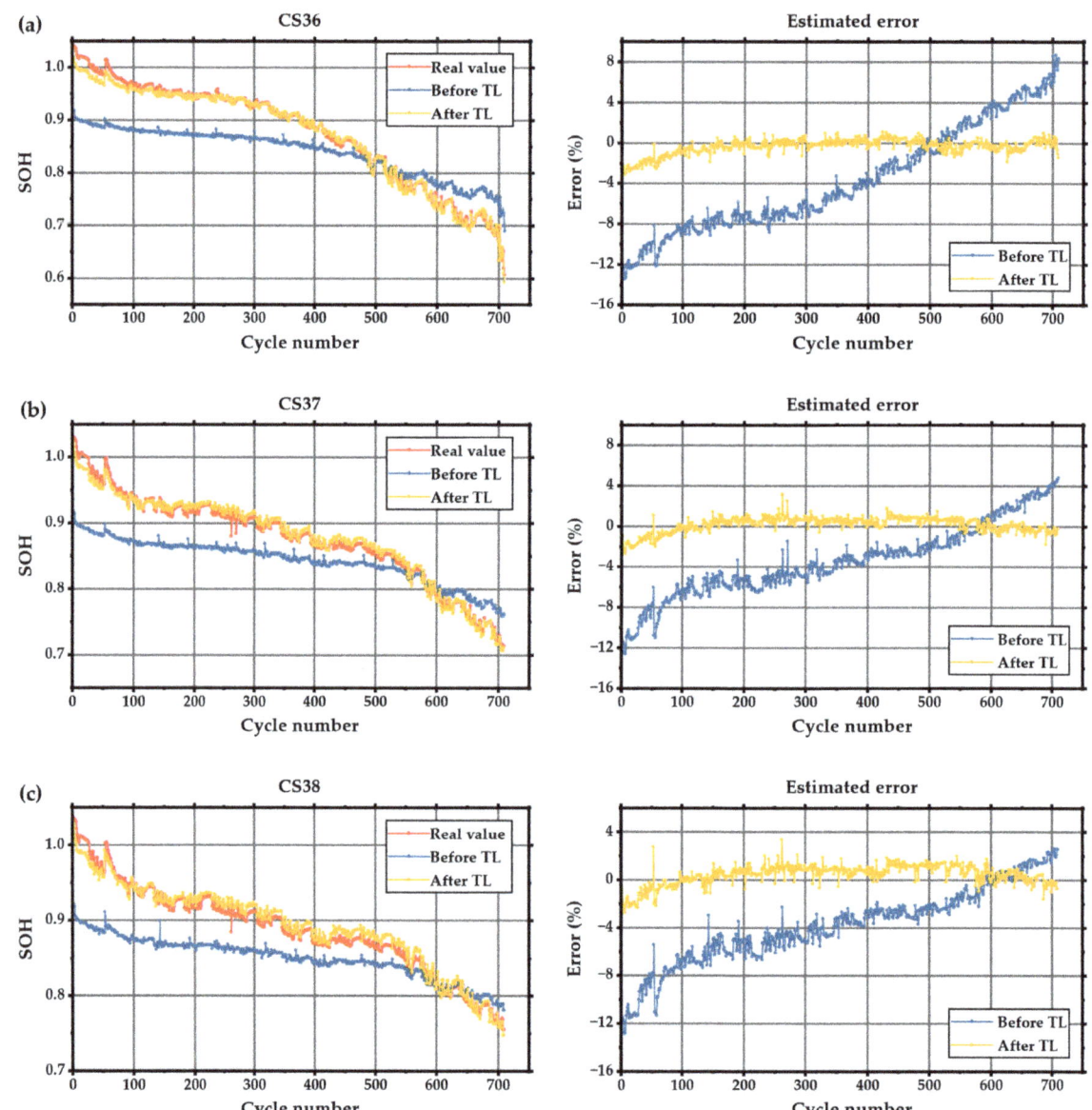

Figure 15. The results of the TL strategy from NASA to CALCE: (a) CS36; (b) CS37; (c) CS38.

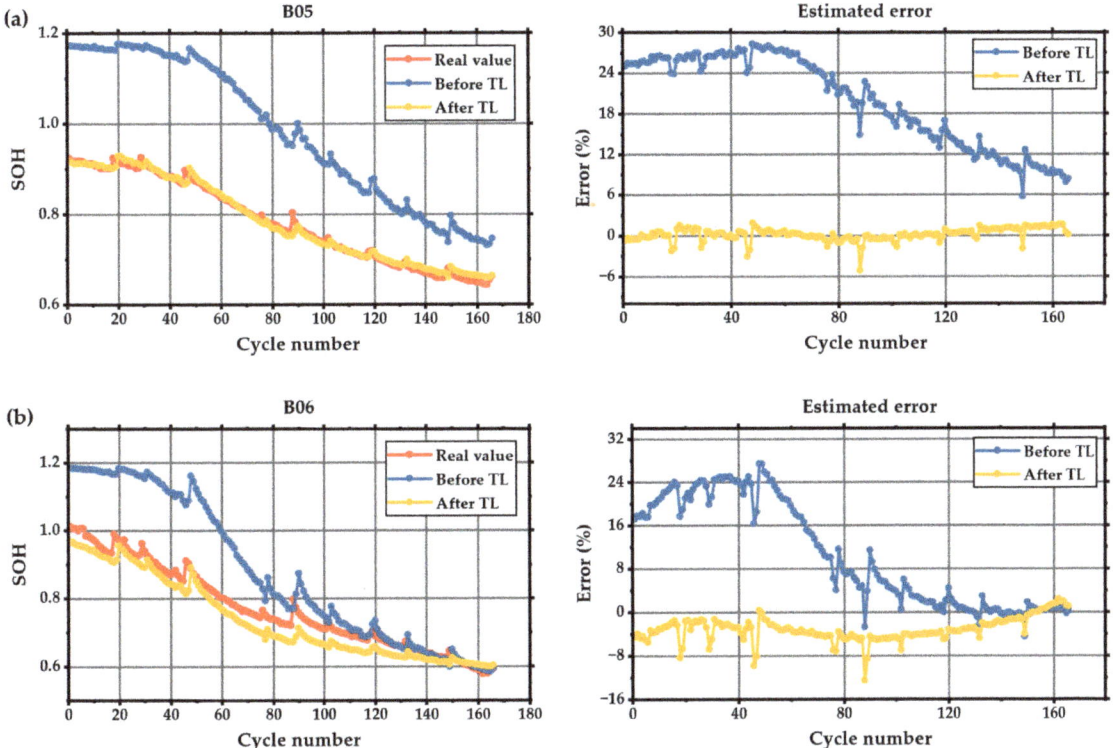

Figure 16. The results of the TL strategy from CALCE to NASA: (**a**) B05; (**b**) B06.

4.2.3. Transfer Learning between NASA and Oxford Aging Datasets

The mutual estimation results of the TL strategy between the NASA and Oxford aging datasets are shown in Figures 17 and 18, respectively. The same estimation results can be found that although the estimated SOH curves before using the TL strategy can follow the overall trend with the real SOH curves, there are significant gaps between them. After fine-tuning the base model, the estimated SOH curves can accurately track the real ones. The results corresponding to Figures 17 and 18 are given in Tables 12 and 13, respectively. For the TL strategy from NASA to Oxford, the MAE, RMSE, and MaxE are limited to 0.4%, 0.47%, and 0.87%, respectively, demonstrating the effectiveness of the proposed deep learning neural network and fine-tuning-based TL strategy.

Table 12. The results of the TL strategy from NASA to Oxford.

Source/Target	Test	Method	MAE (%)	RMSE (%)	MaxE (%)
B05/Cell 2	Cell 1	Before TL	19.22	19.59	26.06
		After TL	0.25	0.29	0.52
	Cell 3	Before TL	19.34	19.70	25.86
		After TL	0.17	0.20	0.38
	Cell 4	Before TL	19.26	19.60	25.41
		After TL	0.32	0.36	0.65
	Cell 5	Before TL	20.04	20.31	25.99
		After TL	0.40	0.47	0.87

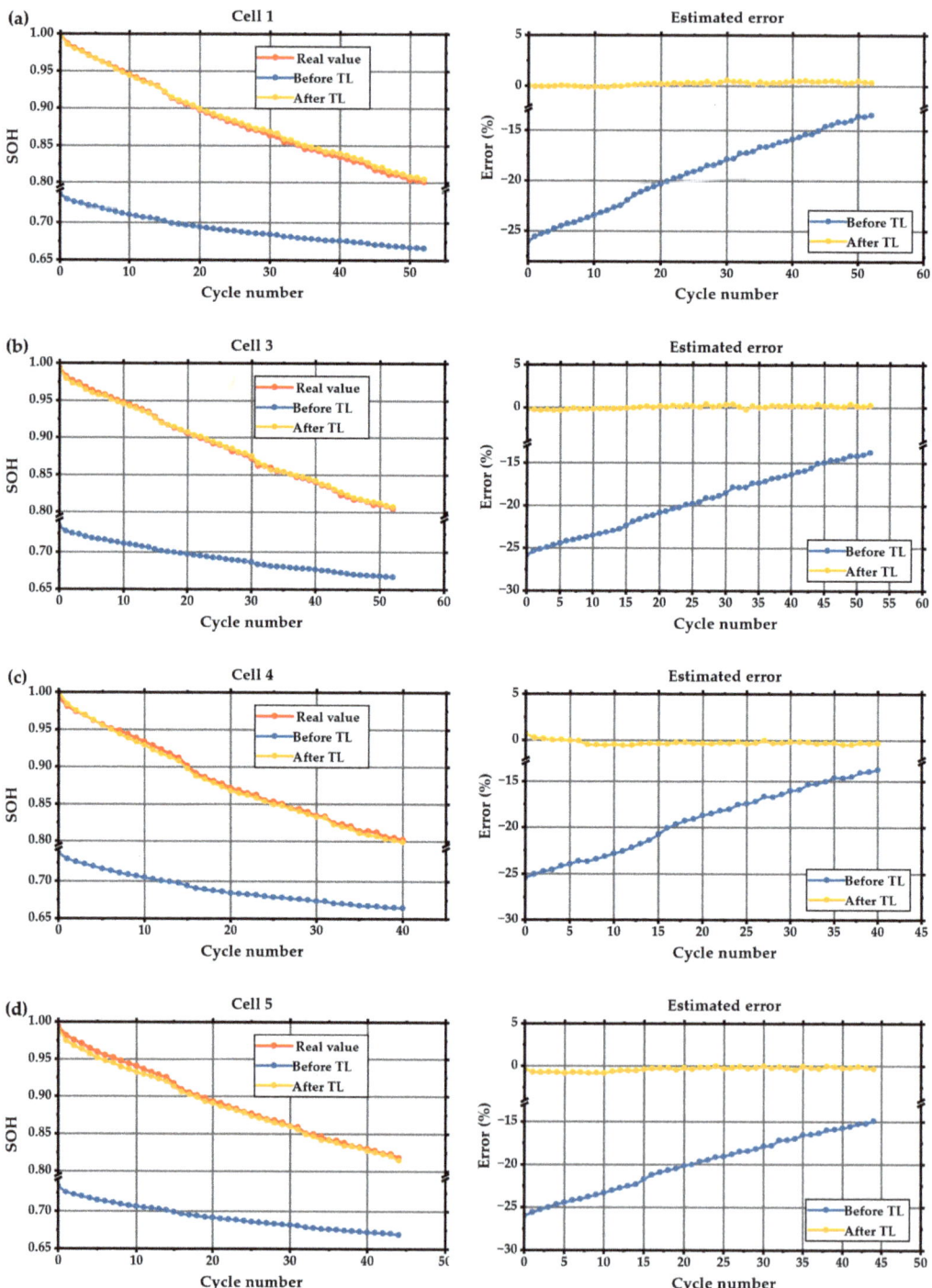

Figure 17. The results of the TL strategy from NASA to Oxford: (**a**) Cell 1; (**b**) Cell 3; (**c**) Cell 4; (**d**) Cell 5.

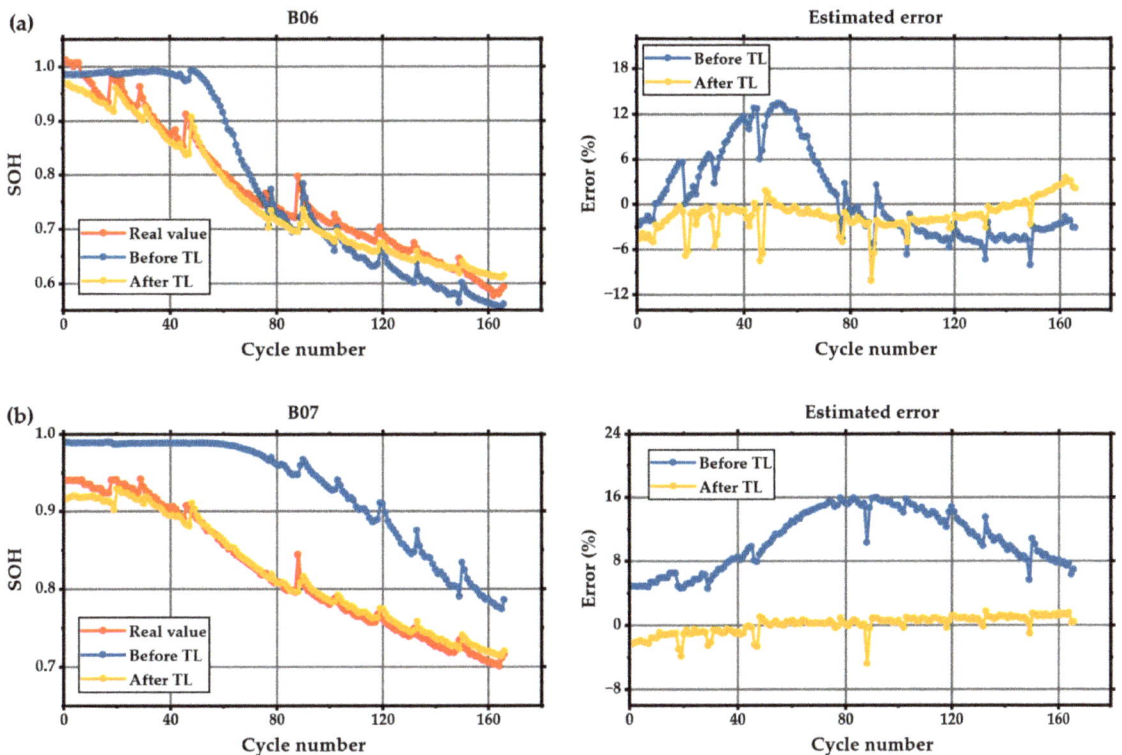

Figure 18. The results of the TL strategy from Oxford to NASA: (**a**) B06; (**b**) B07.

Table 13. The results of the TL strategy from Oxford to NASA.

Source/Target	Test	Method	MAE (%)	RMSE (%)	MaxE (%)
Cell 6/B05	B06	Before TL	4.96	6.03	13.38
		After TL	2.00	2.53	10.19
	B07	Before TL	10.66	11.26	15.94
		After TL	0.87	1.12	4.84

In summary, this section evaluates the effectiveness of the proposed LSTM-FC-NN model and fine-tuning-based TL strategy using three open-source aging datasets. It can be concluded that although the estimated curves based on the base model can follow the overall downtrend of the actual curves, there are significant gaps between them. This result is because the LSTM layer can learn the long-term dependency between the inputs and outputs, and the overall degradation trend of different types of batteries is similar. However, the final outputs are determined by the last layer of the base model. Therefore, after using the fine-tuning strategy to update the last FC layer of the model, the re-trained model can adapt to the target domain quite well and achieve accurate and robust estimation results. Table 14 compares the results with or without using the TL strategy between three aging datasets. The first column represents the source domain, and the first row represents the target domain. The MAEs and RMSEs in Table 14 are the mean values of all evaluation cases from the source domain to the target domain. It can be concluded that with the fine-tuning-based TL strategy, the accuracy of SOH estimation improves significantly, and the MAE and RMSE are limited to 1.09% and 1.41%, respectively. In addition to the model accuracy, the number of learnable parameters and training time are used to analyze the algorithm complexity with or without the TL. According to the hyperparameter setting, the proposed deep learning neural network has 43,100 learnable parameters. If the whole model needs to be re-trained toward the target domain, 43,100 learnable parameters must be involved in the training process. However, only the last FC layer is updated for the fine-tuning-based TL method, meaning only 11 learnable parameters are involved. As for the training time, take training B05 as an example. When training the entire model, it took 14.43 s, and when re-training the model using the fine-tuning method, it took only 7.08 s. Note that this is the average time when training 10 times.

Table 14. The comparison of results with or without the TL strategy.

Unit: (%)		Oxford		CALCE		NASA	
		MAE *	RMSE *	MAE *	RMSE *	MAE *	RMSE *
Oxford	Before TL	-	-	12.78	13.55	7.81	8.65
	After TL			0.93	1.18	1.09	1.41
CALCE	Before TL	8.88	8.97	-	-	15.34	17.70
	After TL	0.42	0.51			0.92	1.24
NASA	Before TL	19.83	20.13	6.55	7.37	-	-
	After TL	0.41	0.48	0.86	1.09		

* The MAE and RMSE are the mean values of all evaluation cases.

4.3. Future Work

This work comprehensively verifies the effectiveness of the proposed LSTM-FC-NN model using a universal HF extraction strategy and the fine-tuning-based TL strategy for SOH estimation using three open-source aging datasets with different material types and working conditions. However, there are some limitations beyond this research. First, this research does not consider the LiFePO$_4$ (LFP) cell, which is a commonly used type of LiB. Second, the batteries used in this research were cycled under CC–CV charge–CC discharge protocols, which are different from real-world working conditions. Third, the results of the NASA aging dataset are worse than the other two because of severe inconsistency and capacity regeneration. In our forthcoming research, we plan to investigate a broader range of battery types, including those cycled under dynamic operational conditions. The accuracy and robustness of the proposed method will be improved according to the specific characteristics of different types of batteries.

5. Conclusions

A universal HF extraction strategy and an LSTM-FC-NN model are proposed in this paper for accurate and robust SOH estimation. In addition, the fine-tuning-based TL strategy is developed considering the difference in data distribution between different types of batteries. Sixteen batteries, featuring two distinct cathode material types, are subjected to cycling under five different operating conditions to assess the efficacy of the proposed methods. The main conclusions are summarized as follows:

(1) Four straightforward HFs, including charged time, charged capacity, peak value of the IC curve, and area under the IC peak, were extracted from partial voltage curves to reflect the battery degradation of different battery types. The Pearson correlation analysis demonstrates that the four extracted HFs are highly related to battery capacity.

(2) A deep learning neural network consisting of one LSTM layer, one dropout layer, and three FC layers was proposed to estimate the SOH accurately. The effectiveness of the proposed LSTM-FC-NN model with the HFs was verified using three open-source aging datasets. The MAE and RMSE are less than 1.21% and 1.83%, respectively.

(3) The fine-tuning method was used to transfer learning between three aging datasets where a random-selected battery from the target domain was utilized to re-train the

base model. The mutual evaluation between the three aging datasets demonstrates the validity of the TL strategy, with the MAE and RMSE limited to 1.09% and 1.41%, respectively.

Author Contributions: Conceptualization, Z.R. and C.D.; Funding acquisition, C.D. and Y.Z.; Methodology, Z.R.; Software, MATLAB 2022b, Z.R.; Supervision, C.D. and Y.Z.; Validation, Z.R.; Visualization, Z.R.; Writing—original draft, Z.R.; Writing—review and editing, Z.R., C.D., and Y.Z. All authors have read and agreed to the published version of the manuscript.

Funding: This research was funded by the National Key R&D Program of China, grant number 2022YFB4003703; Foshan Xianhu Laboratory of the Advanced Energy Science and Technology Guangdong Laboratory, grant number XHD2020-003; and the 111 project, grant number B17034.

Data Availability Statement: The three open-source aging datasets are available on the following websites: https://ora.ox.ac.uk/objects/uuid:03ba4b01-cfed-46d3-9b1a-7d4a7bdf6fac (accessed on 22 October 2023); https://calce.umd.edu/battery-data (accessed on 22 October 2023); and https://www.nasa.gov/intelligent-systems-division/ (accessed on 22 October 2023).

Conflicts of Interest: Author Yifang Zhao was employed by the company SAIC-GM-Wuling Automobile. The remaining authors declare that the research was conducted in the absence of any commercial or financial relationships that could be construed as a potential conflict of interest.

References

1. Dai, H.; Jiang, B.; Hu, X.; Lin, X.; Wei, X.; Pecht, M. Advanced Battery Management Strategies for a Sustainable Energy Future: Multilayer Design Concepts and Research Trends. *Renew. Sustain. Energy Rev.* **2021**, *138*, 110480. [CrossRef]
2. Ge, M.F.; Liu, Y.; Jiang, X.; Liu, J. A Review on State of Health Estimations and Remaining Useful Life Prognostics of Lithium-Ion Batteries. *Meas. J. Int. Meas. Confed.* **2021**, *174*, 109057. [CrossRef]
3. Liu, S.; Nie, Y.; Tang, A.; Li, J.; Yu, Q.; Wang, C. Online Health Prognosis for Lithium-Ion Batteries under Dynamic Discharge Conditions over Wide Temperature Range. *eTransportation* **2023**, *18*, 100296. [CrossRef]
4. Xiong, R.; Li, L.; Tian, J. Towards a Smarter Battery Management System: A Critical Review on Battery State of Health Monitoring Methods. *J. Power Sources* **2018**, *405*, 18–29. [CrossRef]
5. Gao, Y.; Liu, K.; Zhu, C.; Zhang, X.; Zhang, D. Co-Estimation of State-of-Charge and State-of- Health for Lithium-Ion Batteries Using an Enhanced Electrochemical Model. *IEEE Trans. Ind. Electron.* **2022**, *69*, 2684–2696. [CrossRef]
6. Chen, L.; Lü, Z.; Lin, W.; Li, J.; Pan, H. A New State-of-Health Estimation Method for Lithium-Ion Batteries through the Intrinsic Relationship between Ohmic Internal Resistance and Capacity. *Measurement* **2018**, *116*, 586–595. [CrossRef]
7. Du, C.Q.; Shao, J.B.; Wu, D.M.; Ren, Z.; Wu, Z.Y.; Ren, W.Q. Research on Co-Estimation Algorithm of SOC and SOH for Lithium-Ion Batteries in Electric Vehicles. *Electronics* **2022**, *11*, 181. [CrossRef]
8. Schwunk, S.; Armbruster, N.; Straub, S.; Kehl, J.; Vetter, M. Particle Filter for State of Charge and State of Health Estimation for Lithium-Iron Phosphate Batteries. *J. Power Sources* **2013**, *239*, 705–710. [CrossRef]
9. Niri, M.F.; Liu, K.; Apachitei, G.; Román-Ramírez, L.A.; Lain, M.; Widanage, D.; Marco, J. Quantifying Key Factors for Optimised Manufacturing of Li-Ion Battery Anode and Cathode via Artificial Intelligence. *Energy AI* **2022**, *7*, 100129. [CrossRef]
10. Cui, X.; Garg, A.; Trang Thao, N.; Trung, N.T. Machine Learning Approach for Solving Inconsistency Problems of Li-Ion Batteries during the Manufacturing Stage. *Int. J. Energy Res.* **2020**, *44*, 9194–9204. [CrossRef]
11. Mao, J.; Miao, J.; Lu, Y.; Tong, Z. Machine Learning of Materials Design and State Prediction for Lithium Ion Batteries. *Chin. J. Chem. Eng.* **2021**, *37*, 1–11. [CrossRef]
12. Wei, Z.; Yang, X.; Li, Y.; He, H.; Li, W.; Sauer, D.U. Machine Learning-Based Fast Charging of Lithium-Ion Battery by Perceiving and Regulating Internal Microscopic States. *Energy Storage Mater.* **2023**, *56*, 62–75. [CrossRef]
13. Jiang, B.; Berliner, M.D.; Lai, K.; Asinger, P.A.; Zhao, H.; Herring, P.K.; Bazant, M.Z.; Braatz, R.D. Fast Charging Design for Lithium-Ion Batteries via Bayesian Optimization. *Appl. Energy* **2022**, *307*, 118244. [CrossRef]
14. Ng, M.-F.; Zhao, J.; Yan, Q.; Conduit, G.J.; Seh, Z.W. Predicting the State of Charge and Health of Batteries Using Data-Driven Machine Learning. *Nat. Mach. Intell.* **2020**, *2*, 161–170. [CrossRef]
15. Ren, Z.; Du, C. A Review of Machine Learning State-of-Charge and State-of-Health Estimation Algorithms for Lithium-Ion Batteries. *Energy Rep.* **2023**, *9*, 2993–3021. [CrossRef]
16. Lin, M.; Wu, D.; Meng, J.; Wu, J.; Wu, H. A Multi-Feature-Based Multi-Model Fusion Method for State of Health Estimation of Lithium-Ion Batteries. *J. Power Sources* **2022**, *518*, 230774. [CrossRef]
17. Hu, X.; Che, Y.; Lin, X.; Onori, S. Battery Health Prediction Using Fusion-Based Feature Selection and Machine Learning. *IEEE Trans. Transp. Electrif.* **2021**, *7*, 382–398. [CrossRef]
18. Cao, M.; Zhang, T.; Wang, J.; Liu, Y. A Deep Belief Network Approach to Remaining Capacity Estimation for Lithium-Ion Batteries Based on Charging Process Features. *J. Energy Storage* **2022**, *48*, 103825. [CrossRef]

19. Deng, Y.; Ying, H.; Jiaqiang, E.; Zhu, H.; Wei, K.; Chen, J.; Zhang, F.; Liao, G. Feature Parameter Extraction and Intelligent Estimation of the State-of-Health of Lithium-Ion Batteries. *Energy* **2019**, *176*, 91–102. [CrossRef]
20. Yang, D.; Zhang, X.; Pan, R.; Wang, Y.; Chen, Z. A Novel Gaussian Process Regression Model for State-of-Health Estimation of Lithium-Ion Battery Using Charging Curve. *J. Power Sources* **2018**, *384*, 387–395. [CrossRef]
21. Goh, H.H.; Lan, Z.; Zhang, D.; Dai, W.; Kurniawan, T.A.; Goh, K.C. Estimation of the State of Health (SOH) of Batteries Using Discrete Curvature Feature Extraction. *J. Energy Storage* **2022**, *50*, 104646. [CrossRef]
22. Zhang, Y.; Liu, Y.; Wang, J.; Zhang, T. State-of-Health Estimation for Lithium-Ion Batteries by Combining Model-Based Incremental Capacity Analysis with Support Vector Regression. *Energy* **2022**, *239*, 121986. [CrossRef]
23. Mohtat, P.; Lee, S.; Siegel, J.B.; Stefanopoulou, A.G. Comparison of Expansion and Voltage Differential Indicators for Battery Capacity Fade. *J. Power Sources* **2022**, *518*, 230714. [CrossRef]
24. Lyu, Z.; Wang, G.; Tan, C. A Novel Bayesian Multivariate Linear Regression Model for Online State-of-Health Estimation of Lithium-Ion Battery Using Multiple Health Indicators. *Microelectron. Reliab.* **2022**, *131*, 114500. [CrossRef]
25. Yang, D.; Wang, Y.; Pan, R.; Chen, R.; Chen, Z. State-of-Health Estimation for the Lithium-Ion Battery Based on Support Vector Regression. *Appl. Energy* **2018**, *227*, 273–283. [CrossRef]
26. Cao, M.; Zhang, T.; Yu, B.; Liu, Y. A Method for Interval Prediction of Satellite Battery State of Health Based on Sample Entropy. *IEEE Access* **2019**, *7*, 141549–141561. [CrossRef]
27. Hu, X.; Jiang, J.; Cao, D.; Egardt, B. Battery Health Prognosis for Electric Vehicles Using Sample Entropy and Sparse Bayesian Predictive Modeling. *IEEE Trans. Ind. Electron.* **2016**, *63*, 2645–2656. [CrossRef]
28. Lin, M.; Zeng, X.; Wu, J. State of Health Estimation of Lithium-Ion Battery Based on an Adaptive Tunable Hybrid Radial Basis Function Network. *J. Power Sources* **2021**, *504*, 230063. [CrossRef]
29. Doughty, D.H.; Roth, E.P.; Nagasubramanian, G.; Ong, M.D.; Robinson, D.B.; Arslan, I.; Situ, A.-I.; Sullivan, J.P.; Subramanian, A.; Huang, J. Degradation of Commercial Lithium-Ion Cells as a Function of Chemistry and Cycling Conditions. *J. Electrochem. Soc.* **2020**, *167*, 120532. [CrossRef]
30. Christoph, R.B. Diagnosis and Prognosis of Degradation in Lithium-Ion Batteries. Ph.D. Thesis, Department of Engineering Science, University of Oxford, Oxford, UK, 2017.
31. Xing, Y.; Ma, E.W.M.; Tsui, K.L.; Pecht, M. An Ensemble Model for Predicting the Remaining Useful Performance of Lithium-Ion Batteries. *Microelectron. Reliab.* **2013**, *53*, 811–820. [CrossRef]
32. He, W.; Williard, N.; Osterman, M.; Pecht, M. Prognostics of Lithium-Ion Batteries Based on Dempster–Shafer Theory and the Bayesian Monte Carlo Method. *J. Power Sources* **2011**, *196*, 10314–10321. [CrossRef]
33. Lau, S. Intelligent Systems Division. Available online: https://ti.arc.nasa.gov/project/prognostic-data-repository/ (accessed on 21 September 2021).
34. Severson, K.A.; Attia, P.M.; Jin, N.; Perkins, N.; Jiang, B.; Yang, Z.; Chen, M.H.; Aykol, M.; Herring, P.K.; Fraggedakis, D.; et al. Data-Driven Prediction of Battery Cycle Life before Capacity Degradation. *Nat. Energy* **2019**, *4*, 383–391. [CrossRef]
35. Zhu, J.; Wang, Y.; Huang, Y.; Bhushan Gopaluni, R.; Cao, Y.; Heere, M.; Mühlbauer, M.J.; Mereacre, L.; Dai, H.; Liu, X.; et al. Data-Driven Capacity Estimation of Commercial Lithium-Ion Batteries from Voltage Relaxation. *Nat. Commun.* **2022**, *13*, 2261. [CrossRef] [PubMed]
36. Ma, G.; Xu, S.; Jiang, B.; Cheng, C.; Yang, X.; Shen, Y.; Yang, T.; Huang, Y.; Ding, H.; Yuan, Y. Real-Time Personalized Health Status Prediction of Lithium-Ion Batteries Using Deep Transfer Learning. *Energy Environ. Sci.* **2022**, *15*, 4083–4094. [CrossRef]
37. Fan, L.; Wang, P.; Cheng, Z. A Remaining Capacity Estimation Approach of Lithium-Ion Batteries Based on Partial Charging Curve and Health Feature Fusion. *J. Energy Storage* **2021**, *43*, 103115. [CrossRef]
38. Fu, S.; Tao, S.; Fan, H.; He, K.; Liu, X.; Tao, Y.; Zuo, J.; Zhang, X.; Wang, Y.; Sun, Y. Data-Driven Capacity Estimation for Lithium-Ion Batteries with Feature Matching Based Transfer Learning Method. *Appl. Energy* **2024**, *353*, 121991. [CrossRef]
39. Yang, Y.; Zhao, L.; Yu, Q.; Liu, S.; Zhou, G.; Shen, W. State of Charge Estimation for Lithium-Ion Batteries on Cross-Domain Transfer Learning with Feedback Mechanism. *J. Energy Storage* **2023**, *70*, 108037. [CrossRef]
40. Liu, K.; Peng, Q.; Che, Y.; Zheng, Y.; Li, K.; Teodorescu, R.; Widanage, D.; Barai, A. Transfer Learning for Battery Smarter State Estimation and Ageing Prognostics: Recent Progress, Challenges, and Prospects. *Adv. Appl. Energy* **2023**, *9*, 100117. [CrossRef]
41. Huang, K.; Yao, K.; Guo, Y.; Lv, Z. State of Health Estimation of Lithium-Ion Batteries Based on Fine-Tuning or Rebuilding Transfer Learning Strategies Combined with New Features Mining. *Energy* **2023**, *282*, 128739. [CrossRef]
42. Tan, Y.; Tan, Y.; Zhao, G.; Zhao, G. Transfer Learning with Long Short-Term Memory Network for State-of-Health Prediction of Lithium-Ion Batteries. *IEEE Trans. Ind. Electron.* **2020**, *67*, 8723–8731. [CrossRef]
43. Shu, X.; Shen, J.; Li, G.; Zhang, Y.; Chen, Z.; Liu, Y. A Flexible State-of-Health Prediction Scheme for Lithium-Ion Battery Packs with Long Short-Term Memory Network and Transfer Learning. *IEEE Trans. Transp. Electrif.* **2021**, *7*, 2238–2248. [CrossRef]
44. Pan, S.J.; Yang, Q. A Survey on Transfer Learning. *IEEE Trans. Knowl. Data Eng.* **2010**, *22*, 1345–1359. [CrossRef]
45. Li, Y.; Sheng, H.; Cheng, Y.; Stroe, D.I.; Teodorescu, R. State-of-Health Estimation of Lithium-Ion Batteries Based on Semi-Supervised Transfer Component Analysis. *Appl. Energy* **2020**, *277*, 115504. [CrossRef]
46. Ma, Y.; Shan, C.; Gao, J.; Chen, H. Multiple Health Indicators Fusion-Based Health Prognostic for Lithium-Ion Battery Using Transfer Learning and Hybrid Deep Learning Method. *Reliab. Eng. Syst. Saf.* **2023**, *229*, 108818. [CrossRef]
47. Pastor-Fernández, C.; Yu, T.F.; Widanage, W.D.; Marco, J. Critical Review of Non-Invasive Diagnosis Techniques for Quantification of Degradation Modes in Lithium-Ion Batteries. *Renew. Sustain. Energy Rev.* **2019**, *109*, 138–159. [CrossRef]

48. Ren, Z.; Du, C.; Ren, W. State of Health Estimation of Lithium-Ion Batteries Using a Multi-Feature-Extraction Strategy and PSO-NARXNN. *Batteries* **2023**, *9*, 7. [CrossRef]
49. Kolen, J.F.; Kremer, S.C. Gradient Flow in Recurrent Nets: The Difficulty of Learning LongTerm Dependencies. In *A Field Guide to Dynamical Recurrent Networks*; Wiley-IEEE Press: New York, NY, USA, 2010; pp. 237–243. [CrossRef]
50. Li, X.; Wang, Z.; Zhang, L. Co-Estimation of Capacity and State-of-Charge for Lithium-Ion Batteries in Electric Vehicles. *Energy* **2019**, *174*, 33–44. [CrossRef]
51. Understanding LSTM Networks—Colah's Blog. Available online: http://colah.github.io/posts/2015-08-Understanding-LSTMs/ (accessed on 11 October 2021).
52. Zhang, C.L.; Luo, J.H.; Wei, X.S.; Wu, J. In Defense of Fully Connected Layers in Visual Representation Transfer. In *Pacific Rim Conference on Multimedia*; Springer International Publishing: Cham, Switzerland, 2018; Volume 10736, pp. 807–817. [CrossRef]

Disclaimer/Publisher's Note: The statements, opinions and data contained in all publications are solely those of the individual author(s) and contributor(s) and not of MDPI and/or the editor(s). MDPI and/or the editor(s) disclaim responsibility for any injury to people or property resulting from any ideas, methods, instructions or products referred to in the content.

MDPI
St. Alban-Anlage 66
4052 Basel
Switzerland
www.mdpi.com

Batteries Editorial Office
E-mail: batteries@mdpi.com
www.mdpi.com/journal/batteries

Disclaimer/Publisher's Note: The statements, opinions and data contained in all publications are solely those of the individual authors(s) and contributor(s) and not of MDPI and/or the editor(s). MDPI and/or the editor(s) disclaim responsibility for any injury to people or property resulting from any ideas, methods, instructions or products referred to in the content.

www.ingramcontent.com/pod-product-compliance
Lightning Source LLC
LaVergne TN
LVHW070437100526
838202LV00014B/1615